THE PHYSICS

OF *THE HEALING*

BOOKS III & IV

Avicenna

The Physics of *The Healing*

الشفاء : السّماع الطبيعي

Books III & IV

A parallel English-Arabic text
translated, introduced, and annotated by

Jon McGinnis

Brigham Young University Press • Provo, Utah • 2009

Library of Congress Cataloging-in-Publication Data is available.

ISBN: 978–0–8425–2747–7 (alk. paper)

PRINTED IN THE UNITED STATES OF AMERICA.

2 3 4 5 6 7 8 9 17 16 15 14 13 12

First Edition

Contents

Volume 1

Foreword to the Series xiii

Foreword to the Volume xv

Acknowledgments xvii

Translator's Introduction xix
A Note on the Source Texts xxix
Sigla and Conventions xxxii

◆ ◆ ◆

First Book

On the Causes and Principles of Natural Things

Preface 1

Chapter One 3
Explaining the means by which to arrive at the science of natural things from their first principles

Chapter Two 13
Enumerating the principles of natural things by assertion and supposition

Chapter Three 27
How these principles are common

Chapter Four 33
Examination of what Parmenides and Melissus said regarding the principles of being

Chapter Five 37
On defining nature

Chapter Six 45
On nature's relation to matter, form, and motion

Chapter Seven 50
Of certain terms derived from nature and an explanation of their status

Chapter Eight 54
On how the science of physics conducts investigation and what, if anything, it shares in common with the other sciences

Chapter Nine 61
On defining the causes that are of the greatest interest to the natural philosopher in his investigation

Chapter Ten 64
On defining each of the four kinds of causes

Chapter Eleven 71
On the interrelations of causes

Chapter Twelve 74
On the divisions of causal states

Chapter Thirteen 81
Discussion of luck and chance: The difference between them and an explanation of their true state

Chapter Fourteen 91
Some of the arguments of those who were in error concerning chance and luck and the refutation of their views

Chapter Fifteen 103
How causes enter into investigating and seeking the why-question and the answer to it

◆ ◆ ◆

Second Book

On Motion and That Which Follows It

Chapter One 107
On motion

Chapter Two 128
The relation of motion to the categories

Chapter Three 136
Concerning the list of those categories alone in which motion occurs

Chapter Four 152
Establishing the opposition of motion and rest

Chapter Five 157
Beginning the account of place and reviewing the arguments of those who deny and those who affirm it

Chapter Six 163
The various schools of thought about place and a review of their arguments

Chapter Seven 170
Refuting the view of those who say that place is matter or form or any indiscriminate contacting surface or an interval

Chapter Eight 177
The inconsistency of those who defend the void

Chapter Nine 201
The essence of place and its confirmation and the refutation of the arguments of those who deny and are in error about it

Chapter Ten 219
Beginning the discussion about time, the disagreement of people concerning it, and the refutation of those erring about it

Chapter Eleven 229
Identifying and affirming the essence of time

Chapter Twelve 237
Explaining the instant

Chapter Thirteen 248
The solution to the skeptical puzzles raised about time and the completion of the discussion of things temporal, such as being in time and not in time, everlasting, eternity, [and the expressions] suddenly, right away, just before, just after, *and* ancient

Volume 2

◆ ◆ ◆

Third Book

Concerning What Belongs to Natural Things Owing to Their Quantity

Chapter One 260
The manner of investigation peculiar to this book

Chapter Two 262
On succession, contiguity, following immediately, interpenetration, cohesion, continuity, intermediate, limit, being together, and being separate

Chapter Three 273
The state of bodies with respect to their division and a report of the various arguments on which the detractors rely

Chapter Four 281
Establishing the true opinion and refuting the false

Chapter Five 302
Solution to the puzzles of those who prattle on about the atom

Chapter Six 311
On the interrelations of distances, motions, and times with respect to this topic, and an explanation that no first part belongs to them

Chapter Seven 320
The beginning of the discussion about the finitude and infinitude of bodies and people's opinions concerning that

Chapter Eight 325
On the impossibility that either a body or magnitude or number in an ordered series is infinite, and that it is impossible that there be some infinite body that is moved either in its entirety or partially

Chapter Nine 337
An explanation of the way that the infinite does and does not enter into existence, and a refutation of the arguments of those who defend the existence of an actual infinite

Chapter Ten 344
That bodies are finite with respect to influencing and being influenced

Chapter Eleven 359
That nothing precedes motion and time save the being of the Creator (may He be exalted) and that neither of the two has a first [moment] of its being

Chapter Twelve 371
Following up on the claim that there is a point of smallness at which natural bodies are divested of their forms and that, in fact, each one of them has a certain limiting point less than which its form is not preserved; likewise, following up on the claim that no motion is the least, slowest, or shortest

Chapter Thirteen 380
On the directions of bodies

Chapter Fourteen 389
The natural directions of rectilinear motions

◆ ◆ ◆

Fourth Book
On the Accidents of These Natural Things and Their Interrelations, as Well as the Things That Are Necessary Concomitants of Their Interrelations

Chapter One 402
Of the subjects contained in this book

Chapter Two 403
On the numerical unity of motion

Chapter Three 410
On motion that is one in genus and species

Chapter Four 418
Resolving the doubts raised against motion's being one

Chapter Five 424
On motions that are and are not in concert

Chapter Six 432
On the contrariety of motions and their opposites

Chapter Seven 446
Of the opposition of motion and rest

Chapter Eight 450
An explanation of whether one motion can really be continuous with another or whether that is impossible for them, such that there must be a state of rest between them

Chapter Nine 463
On the motion that is naturally prior and a catalogue of the specific differences of motions

Chapter Ten 471
The way in which space and other things are natural to the body

Chapter Eleven 477
On establishing that every body has a single natural space, and [on] the way space belongs to the body's collective kind and to its individual instances as well as to simple and composite [bodies]

Chapter Twelve 485
Establishing that every natural body has a principle of motion with respect to either place or position

Chapter Thirteen 497
Accidental motion

Chapter Fourteen 504
On forced motion and the mobile's spontaneous motion

Chapter Fifteen 513
The states of motive causes and the interrelations between the motive and mobile causes

◆ ◆ ◆

Glossary of Arabic-English Terms 523

Subject Index 565

Sigla and Conventions

Z *al-Ṭabīʿīyāt, al-samāʿ al-ṭabīʿī*, ed. Saʿīd Zāyid (Cairo: The General Egyptian Book Organization, 1983)

Y *al-Ṭabīʿīyāt, al-samāʿ aṭ-ṭabīʿī*, ed. Jaʿfar al-Yāsīn (Beirut: Dār al-Manāhil, 1996)

T *al-Shifāʾ*, 2 vols. (Tehran: n.p., 1886), vol. 1, 2–159

In this work, terms of Arabic derivation found in *Webster's Third New International Dictionary* generally follow the first spelling given and are treated as regular English words. Otherwise, Arabic or Persian words and proper names have been transliterated following the romanization tables established by the American Library Association and the Library of Congress (*ALA-LC Romanization Tables: Transliteration Schemes for Non-Roman Scripts*. Compiled and edited by Randall K. Barry. Washington, D.C.: Library of Congress, 1997; available online at www.loc.gov/catdir/cpso/roman.html).

Passages from *The Physics* are referenced by book, chapter, and paragraph number, e.g. (3.9.7).

THE PHYSICS

OF *THE HEALING*

◆

BOOKS III & IV

THIRD BOOK:
CONCERNING WHAT BELONGS
TO NATURAL THINGS
OWING TO THEIR QUANTITY

Chapter One

The manner of investigation peculiar to this book

(1) Natural things are bodies and the states of bodies, and both are mixed up with quantity in some way. The quantity belonging to bodies is dimensions, while that which belongs to the states of bodies is, for example, time and certain other things that follow upon [bodies], whether essentially or accidentally. Quantity follows upon the states of bodies owing to either the bodies' quantity, which belongs to them or is together with them; or the time, just as it follows upon motion; or the relation to a certain number or its measure that arises from them. The latter is the

المقالة الثالثة

في الأمور التي للطبيعيات من جهة ما لها كم

<الفصل الأول>

في كيفية البحث الذي يختص بهذه المقالة

(١) إنَّ الطبيعيات هي أجسامٌ وأحوال الأجسام، وللكمية مخالطة ما للصنفين. فالكمّيّة التي للأجسام هي الأقطار، وأما التي لأحوال الأجسام فمثل الزمان، ومثل أشياء أخرى تلحقها بالذات أو بالعرض. وأحوال الأجسام يلحقها الكم؛ إمّا من جهة كمّيّة الأجسام التي لها أو معها، وإمّا من جهة الزمان كما يلحق الحركة، وإمّا من جهة القياس إلى عددٍ ما يصدر عنها أو مقداره، وهذا أبعد أنحاء لحوق الكم، وهذا كما يقال قوة

most far-reaching manner of including quantity and is like saying that a power is finite or infinite. The states that are considered to belong to bodies from their quantity are either certain states that, in fact, involve bodies taken individually (as, for instance, the state of the finite and infinite in magnitude and the finite and infinite with respect to division and smallness), or certain states that involve an interrelation among [bodies] (as, for instance, succession, contiguity, following immediately, continuity, and what is analogous). As for the states of bodies under which motion and time fall, [what] is considered concerning the states of their quantity is whether [motion and time] had a beginning and will come to an end or are not like that and, instead, are infinite. As for [the states of bodies] under which powers fall, [what] is considered concerning the states of the quantities in them is how they might correspond with finite and infinite things and how that could be in them.

متناهيـــة وقوة غير متناهية . والأحوال التي تعتبر للأجســـام من كميتها ، إمّا أحوال يصح أنْ تكون بانفراد جسم جســـم مثل حال التناهي واللاتناهي في العظم ، وحال التناهي واللاتناهي في الانقســـام والصغر ، وإمّا أحوالٌ تكون بقياس بعضها إلى بعض ؛ مثل التتالي والتماس والتشافع والاتصال ، وما يجرى مجراها . وأمّا أحوال الأجسام ؛ فالحركة والزمان منهـــا تعتبر من أحوال كميتها ، أنهما هل لهما ابتداءان وهل ينقطعان أو ليس كذلك ؛ بل لا نهايـــة لهما . وأمّا القوى منها فتعتبر من أحوال الكميات فيها أنّها كيف تحاذي أموراً ذوات نهايةٍ أو غير ذوات نهاية ، وكيف يمكن ذلك فيها .

Chapter Two

On succession, contiguity, following immediately, interpenetration,[1] *cohesion, continuity, intermediate, limit, being together, and being separate*

(1) Before we speak about finite bodies and their states with respect to largeness, we should speak about the finite and infinite with respect to smallness and divisibility; and before that we should define *succession, contiguity, interpenetration, following immediately, cohesion,* and *continuity,* as well as defining *intermediate, limit, together in place,* and *being separate.* Now, we say that two things *follow in succession* when there is nothing generically like the two between the first one and the second as, for example, houses in succession, for the one following the first in succession is that one between which it and the first there is nothing generically like them. Sometimes they are alike in species—as for instance, one house after another. [At other times,] they are different in species as, for instance, a line[2] consisting of a man, a horse, a boulder,[3] and a tree. In that case, they are found following in succession, not *qua* varying in species, but inasmuch as they are made up of something common, whether essentially (like corporeality) or accidentally (like whiteness or standing in a

1. "Interpenetration," which is part of the chapter title in both **Y**'s edition and **T**, does not appear in the title in **Z**'s edition or the Latin.

2. Reading *ṣaff* with **Z** and **T** for **Y**'s *ṣanf* (kind); the Latin omits the term altogether.

3. The Arabic *jabal,* which appears in **Y**, **T**, and the Latin (*mons*), most commonly means "mountain" but can also mean any rocky elevation regardless of how small. It can even mean "rock" or "stone"—hence my translation, "boulder." For this use of *jabal,* see Wright *s.v.* Alternatively, **Z** reads *ḥabl* (rope).

<الفصل الثاني>

في التالي والتماس والتشافع
والتداخل والتلاصق والاتصال والوسط
والطرف ومعاً وفرادى

(١) وقبل أنْ نتكلم في أمر تناهي الأجسام وأحوالها في الأعظام، فحقيق أنْ نتكلم في تناهيها ولا تناهيها في الصغر والانقسام. وقبل ذلك، فحقيق بنا أنْ نعرف التالي والتماس والتشافع والتداخل والتلاصق والاتصال، وأنْ نعرف الوسط والطرف، وأنْ نعرف معاً في المكان وفرادى. فنقول: إنَّ المتتالين هما اللذان ليس بين أولهما وثانيهما شيء من جنسهما، مثل البيوت المتتالية، فإنَّ التالي منهما للأول هو الذي ليس بينه وبين الأول شيءٌ من جنسهما. وقد تكون متفقة النوع مثل بيت وبيت، وتكون مختلفة النوع مثل صف من إنسانٍ وفرسٍ وجبلٍ وشجرة، فحينئذ لا توجد متتالية؛ من حيث هي مختلفة النوع، بل من حيث يجمعهما أمرٌ عامٌ ذاتي كالجسمية أو عرضي كالبياض، أو القيام صفاً أو

line, or ascending in size). When there is nothing between them that is said to designate some common feature,[4] then it is said of the one taken to be second that this one following in succession is the co-mate of the other. For example, when these things are taken as individuals ascending [in size], the horse follows the man in succession, and [then] the boulder, and [then] the tree. If they are taken inasmuch as they are animals, then the horse follows the man in succession, while the boulder and tree do not follow in the succession. If they are taken *qua* man, then there is nothing following the individual man in succession; there is just the man.

(2) Something is *contiguous* when, between its limit and the limit of that with which it is said to be contiguous, there is nothing possessing a position. So two things are contiguous whose limits are together, not with respect to place, but with respect to the denotable position that occurs. [That] is because limits are not at all in place; but they do have some position, just as the point has a certain position, where *position* is something such that one can denote its being in some specific location. [In the case of] two contiguous things, this denoting applies to their two limits being together.

(3) When two things coincide with each other such that one trespasses the other's limit to the point that the one completely coincides with the other entity, then that is not an instance of contiguity but, rather, *interpenetration.* Indeed, interpenetration is nothing but the thing itself wholly entering into the other, where that entering is precisely that one of the two coincides with the whole of that which it is said to interpenetrate. If they are exactly equal, then there will be nothing of the one that does not coincide with the other; while, if one of them exceeds [the other], then it will not enter into the whole of it but will enter whatever

4. **Y** reads an *illā* (except), which does not appear in **Z**, **T**, or the Latin. It has been omitted in the translation; but, if retained, the sense would be "when nothing is predicated of [what] is between them except something considered to be general."

الشـــخوص حجماً . فإذا لم يكن بينهما من المقــول عليه الأمر المعتبر عاماً ، قيل للمأخوذ منهما ثانياً إنَّ هذا يتلو صاحبه . مثلاً ، إذا أُخذت هذه الأشياء ، من حيث هي أشخاص منتصبة ، كان الفرس يتلو الإنســان والجبل والشجرة . وإنْ أُخذت من حيث هي حيوانات كان الفرس يتلو الإنسان ، ولم يكن الجبل والشجرة يتلوان ، وإنْ أُخذت من حيث هي ناس لم يكن هناك شيء يتلو شخص الإنسان إلاّ الإنسان .

(٢) وأمّا المماس فهو الشـــيء الذي ليس بين طرفه وبين طرف ما قيل انّه مماسٌ له ، شـــيء ذو وضع . فالمتماسان هما اللذان طرفاهما معاً لا في المكان ، بل في الوضع الواقع عليه الإشـــارة ، فإنَّ الأطراف ليســت في مكانٍ البتة ، ولها وضع مــا ، والنقطة أيضاً لها وضع ما . والوضع أنْ يكون الشـــيء بحيث يمكن أنْ يشـــار إليه أنّه في جهةٍ مخصوصة ، والمتماسان تقع هذه الإشارة على طرفيهما معاً .

(٣) وإذا كان شيئان يتعدى لقاء كل واحد منهما طرف الآخر حتى يلقى ذات الآخر بأسره ، لم يكن ذلك مماسّة ، بل كان مداخلة . فإنّه ليست المداخلة إلاّ أنْ تدخل كلية ذاتٍ في الآخر ، وليس ذلك الدخول إلاّ أنْ يلقى أحدهما كل ما قيل إنّه مداخل فيه . فإنْ ساواه كان لا شـــيء مــن هذا إلاّ وهو ملاقٍ للآخر ، وإن فضــل أحدهما لم يكن داخله كله ، بل

[part] of it that it exactly equals. The true nature of interpenetration, then, is that nothing belongs to this object that does not coincide with the other object, and so nothing is observed that does not coincide with the other. Now, [although] interpenetration necessarily entails that two interpenetrating things be in a single place, [that] is not what is understood by [*interpenetration*]; rather, what is understood is *to encounter completely*. When one thing completely coincides with another and the other does not exceed it, then what[ever] coincides with the other will coincide with the first. Otherwise, part of the first would not be found in it with the encounter; but it was said that the first encountered the whole of it and none of the second exceeded it, which is a contradiction. So two things that encounter [each other] completely are [such that] [(1)] anything encountering one of the two encounters the other; [(2)] one of them does not hinder [something's] being contiguous with the other;[5] and [(3)] joining a thousand [interpenetrating things] does not increase the size (where this is the way a thousand points would be joined). When one thing encounters another, and what is encountered coincides with something that does not coincide with the first thing, then there will be a certain excess in [that second thing] itself beyond [that part] that is coinciding with the first thing, where the second encountered thing provides that excess as something unoccupied as a result of the original encounter.[6] All of these things are evident to the intellect. Likewise, when the thing is occupied by the encounter such that the encounter prevents it from encountering something else, then it occupies either the whole of it or part of it. If it [occupies] the whole of it, then no third thing is contiguous with it, while, if it [occupies] part of it, neither the occupation nor the contiguity occurs completely.

5. **Y** (inadvertently) omits the phrase *lā yaḥjubu wāḥid minhumā ʿan mumāssah al-ākhar*, which appears in **Z**, **T**, and the Latin (*et unum non prohibet contingi aliud*).

6. Unfortunately, the Arabic is at least as awkward as the English, although Avicenna's point is relatively clear. The whole of his point seems to be thus: if, for example, a smaller disk is imposed upon a larger one, the part of the larger disk that is not covered by the smaller one represents the excess.

داخله ما يســاويه منه . فحقيقة المداخلة أنْ يكون لا شــيء من ذات هذا ؛ إلاّ ويلقى ذات الآخر ، فلا يتراءى شــيءٌ لا يلقى الآخر . وأمّا كون المتداخلين في مكانٍ واحد ، فهو أمر يلزم المداخلة وليس هو مفهومها ، بل مفهومها الملاقاة بالأســر . وإذا كان شيء يلقى الآخر بالأسر والآخر لا يفضل عليه ؛ فما يلقى الآخر يلقى الأول ، وإلاّ فسيوجد فيه بالملاقاة شيءٌ خالياً عن الأول ، وقيل إنَّ الأول لاقاه كله ولم يفضل من الثاني عليه ، هذا خُلْف . فالمتلاقيان بالأسر ، أي شيءٍ لاقى أحدهما ، لاقى الآخر ، ولا يحجب واحد منهما عن مماسة الآخر ، ولا يــزداد الحجم باجتماع ألفٍ منها ، وهذا هو ســبيل ألف نقطةٍ لو اجتمعت . وإذا كان شــيء يلاقي شــيئاً ، ويلقى الملاقي شــيءٌ لا يلقى الأول ، فهناك فضل في ذاته عمّا لاقى الأول ، وذلك الفضل يناله الملاقي الثاني فارغاً عن الملاقاة الأولى . وهذه الأشياء كلها بيّنة في العقل . وكذلك إذا كان الشــيء مشغولاً بالملاقاة ، حتى تكون الملاقاة تمنعه عن ملاقاة شــيء آخر ، فإمّا أنْ يكون مشــغولاً كله أو بعضه . فإنْ كان كله لم يماسّه ثالث ، وإنْ كان بعضه فلا يكون - لا الشغل ولا المماسّة - شغلاً بالأسر ، أو مماسّة بالأسر ،

(4) These premises are self-evident, and whatever is alleged to refute them refutes premises more general than they are. For instance, it is said that the whole of something, without being divided, might be known relative to and vis-à-vis one thing, while being unknown relative to and vis-à-vis another thing;[7] and something might be to the right of something and not to the right of something without being divided. Likewise, something, without being divided, might be completely occupied relative to one thing, while completely unoccupied relative to another. Where they[8] primarily go wrong concerning this is that this is a refutation of someone who says[9] that something cannot completely have two opposite features relative to a given thing; but we grant this. What is denied is only its specific relation, for instance, to this premise. In other words, [what we are denying is] that, when [something] completely occupies [another] by being [completely] contiguous with [it], [that occupying thing] will not be contiguous with one side [of that which it occupies] while falling short of another side contiguous with [it] that is proper to it. In that case, if it is unoccupied on one side while occupied on another, then there is some excess in the object beyond the occupation. This premise has been neither refuted nor invalidated. Instead, [the purported refutation] has shown that [premises] generically like it and similar to it are not necessary. This premise, however, neither is required nor is affirmed by first reflection because of its generic sense, but because it is

7. **Y** additionally has *wa-yakunu al-shayʾ yumayyizu shayʾan wa-laysa yumayyizu shayʾan min ghayr inqasām* (and something is distinguished from something and not distinguished from something without being divided), which is absent in **Z**, **T**, and the Latin. It appears to be a result of dittography and a confusion of the texts' *yamīn* (to the right of), in the next phrase, with *yumayyizu* (to distinguish), which are quite similar orthographically.

8. It is not clear whether "they" refers merely to Avicenna's imagined objectors or to some real group. If *they* does refer to some actual group, it might possibly be the Stoics and their doctrine of blending (*krasis*). Although there is no evidence that Stoic sources were available to Avicenna in Arabic translation, he may have known of their views through either Galenic sources or Peripatetic commentaries. Alternatively, the text's "they" might refer to the followers of Ibrāhīm al-Naẓẓām (d. ca. 840), who maintained that all purported "accidents," with the exception of motion, were, in fact, bodies, and that these bodies interpenetrate one another as well as corporeal bodies. For this latter possibility see, for instance, *Maqālāt al-Islāmīn wa-ikhtilāf al-muṣallīn*, ed. Helmut Ritter (Wiesbaden: Franz Steiner, 1963), 309.

9. Both **Z** and **T** additionally have *min jihah ukhrá* ([coming] from another perspective). The phrase does not appear in **Y** or the Latin.

(٤) وهذه مقدمات بيّنة بأنفسها ، وما يورد من النقض لها فهو نقض مقدماتٍ أعمّ منها . وهو ما يقال من أنَّ الشيء قد يكون كله معلوماً بالقياس إلى شيء وعند شيء ، ومجهولاً بالقياس إلى آخر وعند آخر ، من غير انقسام ، وكذلك الشيء يمين شيء وليس يمين شيء من غير انقسام ، مشغولاً بأسره بالقياس إلى شيء ، فارغاً بالقياس إلى آخر من غير انقسام . فأول ما يغلطون في هذا ، أنَّ هذا نقْض قول قائلٍ إنَّه لا يجوز أنْ يكون الشيء بأسره ذا أمرين متقابلين بالقياس إلى شيءٍ ؛ وهذا مسلّم . وإنّما المنكر ما نسبته إلى هذه المقدمة نسبة النوع مثلاً ، وهو أنّه إذا اشتغل بأسره عن أنْ يماس لم يماس في جهةٍ دون جهةٍ مماسّة تخصّه . فإنْ فرغ من جهةٍ واشتغل في جهةٍ ، ففي ذاته فضل عن الاشتغال . وهذه المقدمة لم تتناقض ولم تبطل ؛ بل دلَّ على أنَّ جنسها ليس بواجبٍ ، ولها أشباه ليست بواجبة . وهذه المقدمة لم تجب ولم تثبت في العقل الأول من حيث المعنى الجنسي لها ، بل من حيث هي مخصوصة بالملاقاة ، فإنَّ الملاقاة هذا موجبها ، ولو كان بدل الملاقاة معنى

specific to *encountering;* for this is necessary of encountering. If *encountering* were taken in some other sense, then the whole of something could have one state relative to one perspective and a different state[10] from that former one relative to some other perspective—[that is,] provided that that state in no way requires an occupation while preventing [it], nor does it require an occupation that involves [both] the state of the whole and the state of the part; or [provided that] the occupation of the whole is something relative and not *per se.* Indeed, what is occupied so as to be prevented from being contiguous with something else is not occupied by one thing to the exclusion of some other, for, *qua* occupied, nothing at all is contiguous with it, while, *qua* unoccupied, the whole thing is contiguous with it.[11]

(5) As for what is unknown, its being unknown is not some absolute feature about [the thing]; rather, it is relative to a given thing. Thus, it is not impossible that any given knower know it in any number of ways, unlike the *part* of those [raising the objection], for they limited the possibility of its contiguity to numerable things. In short, that raises no obstacle whatsoever with respect to knowing; and, even if does raise an obstacle to getting at something that is indivisible in every way, it would not be one's knowledge of a thing. Still, we do not need to explain this difference. [That] is because what we said about *encountering completely* (namely, that

10. Reading *bi-ḥāl mukhālifah* with **Z**, **T**, and the Latin (*diversa*).

11. Avicenna's point seems to be the following: if, for example, a smaller disk is imposed on a larger one, then the larger disk inasmuch as it is occupied by the smaller one is not contiguous with it, while inasmuch as it is not occupied by the smaller one it is contiguous with the whole of the disk. An illustration may make the point clearer.

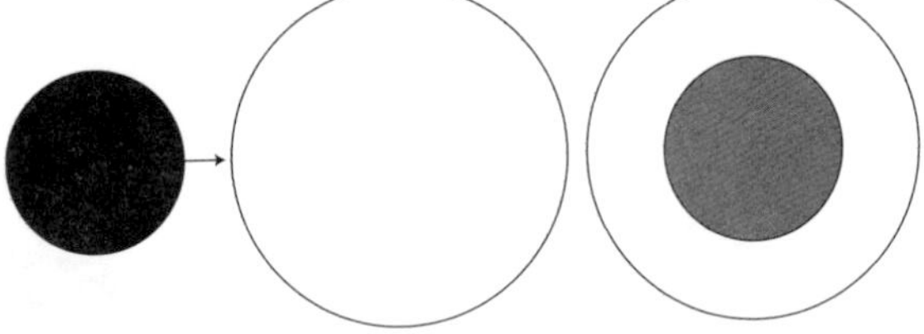

The grey disk represents where the two disks interpenetrate, or encounter each other completely, and here the two are not contiguous; they are only contiguous where the white ring, representing the excess of the white disk, touches the limit of the grey disk.

آخر لكان يجوز أنْ يكون كل الشــيء بالقياس إلى جهةٍ بحالٍ، وبالقياس إلى جهةٍ أخرى بحال مخالفة لتلك الحال؛ إذا كانت تلك الحال لا توجب شغلاً ومنعاً أصلاً، أو كانت لا توجب شــغلاً يتعاطى بحال الكل وبحال البعض، أو كان الشغل للكل أمراً بالقياس ليس أمراً في نفســه. فإنَّ المشــغول الممنوع عن مماسّة شيء آخر، لا يكون مشغولاً عن شيءٍ دون شــيء، فإنّه من حيث هو مشغول لا يماسّه شيءٍ البتة، ومن حيث هو فارغ يماسّه كل شيء.

(٥) وأمّا المجهول، فكونه مجهولاً، ليس أمراً يســتقر فيه البتة، بل هو مضافٌ إلى شــيء، ولذلك لا يمنــع أنْ يعلمه أي عالم كان بأي عددٍ كان مــن العلم، لا كالجزء الذي لهم، فإنهم قصروا إمكان مماسّــته على أشــياء معدودة. وبالجملة لا يوجب ذلك في العلم منعــاً البتة، ولو أوجب منعاً متناولاً لأمرٍ غير متجزّ بوجهٍ من الوجوه، لما عُلم شــيء. وعلى أنّه لا حاجة بنا إلى إبانة هذا الفرق، فإنَّ الذي نقوله في أمر الملاقاة بالأســر من أنّه

when it occupies, the entire thing is occupied, and when it does not occupy, then it does not occupy something) is something obvious in itself and obviously different from the case of knowing. The alleged counter-examples refute something different from what was intended, while rendering permissible something more general than what was intended. So it makes what was intended permissible. To be [more] exact, *to encounter completely* is simply in no way to be occupied as a result of being contiguous. [That is] because, when something prior to the state of contiguity occupies the newly arriving contiguous thing, then what is occupied is prevented from being contiguous, whereas what is occupying is not prevented from being contiguous. So the occupying object gets to be contiguous, while the occupied object does not. Also, the occupied object does not entirely encounter the occupying object, and so they do not completely encounter one another. As for when the encounter occurs completely, there, in fact, is interpenetration, where one of the interpenetrating things will inevitably not occupy the other as a result of touching what touches; and so the judgment rendered concerning encountering completely will be the same. When contiguity is different from interpenetration, and [when] each of the two contiguous things is separated by some proper position (in which case there is one object without another), then *contiguity* is to encounter at the limits of the two objects—that is, between their two limits, there is absolutely no interval. Interpenetration, however, is a complete encounter from which it necessarily follows that the position and place of the two become one. On the most rudimentary reflection, you know that when something is contiguous and then your estimative faculty next imagines it to begin interpenetrating, [that thing] needs to move until it encounters some part of that object into which it is penetrating that it had not encountered and continue to do so until the encounter

إذا شــغل شَغلَ الجميع، وإنْ لم يشــغل لم يشغل شيئاً هو بيِّنٌ بنفسه، وبيِّنٌ خلافه في أمر العلم. وما أوردوه من الأمثلة للمناقضة يناقض غير المطلوب، ويوجب تجويزاً في أمرٍ أعمّ من المطلوب، فيجعل تجويزاً في المطلوب. وبالحري أنْ تكون الملاقاة بالأســر، لا محالة، لا تشــغل البتة عن المماسّة؛ فإنَّ الوارد المماس إذا شغله المتقدّم السابق إلى المماسة امتنع عن مماسّــة المشــغول ولم يمتنع عن مماسّة الشــاغل، فأصاب ذات الشاغل بالمماسّة دون ذات المشغول، وكان ذات المشغول غير ملاقٍ لجميع ذات الشاغل فما كانت بينهما ملاقاة بالأســر. أمّا إذا كانت الملاقاة ملاقاة بالأسر، كانت مداخلة بالحقيقة، والمتداخلان – لا محالة – لا يشــغل أحدهما الآخر عن مسّ ماسّ، فالملاقاة بالأسر كذلك حكمها. وإذا كانت المماسّــة غيــر مداخلةٍ، وكان كل واحد من المتماســين منفرداً بوضعٍ مخصوص؛ فهناك ذاته دون ذات الآخر، فتكون المماسّــة ملاقاة بأطــراف الذاتين، وهو أنْ لا يكون بين طرفيهما بُعْدٌ أصلاً، وتكون المداخلة ملاقاة بالأســر، فيلــزم ذلك أنْ يصير وضعهما ومكانهما واحداً. وأنت إذا تأملت أدنى تأملٍ علمت أنَّ الشيء إذا كان مماسّاً فلو توهمته صــار مُداخلاً، احتاج أنْ يتحرك إلى ملاقاة أمرٍ من ذات المداخل ينفذ فيه لم يكن لاقاه،

is completed throughout, so that it has interpenetrated. Our present discussion does not concern whether interpenetration exists or not, but concerns conceptualizing the meaning of the term and how what is conceptualized about it differs from what is conceptualized about *contiguity,* as well as how to distinguish it (should it exist) from contiguity.

(6) As for *following immediately,* it is a state of a contiguous thing following in succession *qua* in succession. Now, some believed that a condition of that is that it share the species in common.[12] I, however, don't believe that what is understood by the term requires that, unless that is initially agreed upon as a technical term. Besides, we would still need some term corresponding with this more general sense.

(7) What *coheres* is something contiguous that adheres to the thing during locomotion to the point that it is very difficult to separate the two. [Cohesion might occur in one of two ways.] It might be because two surfaces so conform to one another that is impossible to separate one from the other without producing a void, which, as has been explained, cannot exist.[13] That occurs when the limit of neither one of the two corporeal surfaces is more apt suddenly to give way, or one does give way only by distorting the surface's form through becoming convex, concave, or the like, and so complying only by force. Alternatively, it might be because parts of this one sink deeply into parts of that one. Cohesion frequently takes place between two bodies by means of some body that can conform very closely with the two surfaces, owing to its fluidity, and it can sink deeply into each one of them and be such as to dry and become hard so that both bodies adhere; thus the adhesion of the two bodies is assumed to be by means of it as, for example, glue and other similar things.

---

12. The reference may be to Aristotle and his commentators, since, at *Physics* 5.3.227–10, Aristotle says that what is between (*metaxu*) must involve contraries. Since contraries are always of the same species, Avicenna may have just made the logical inference from Aristotle's claim.

13. See 2.8.

حتى إذا استوت الملاقاة صار مُداخلاً. وليس كلامنا الآن في المداخلة على أنّها موجودة أو معدومة، بل على تصور معنى لفظها، وأنَّ المتصوّر منها كيف يخالف المتصوّر من المماسّة، وأنّها لو كانت موجودة كيف كانت تفارق المماسّة.

(٦) وأمّا التشافع فهو حال مماس تالٍ من حيث هو تالٍ، وظنّ بعضهم أنَّ من شرط ذلك أنْ يشارك في النوع؛ وأظنّ أنَّ مفهوم اللفظ لا يقتضي ذلك؛ اللّهم إلاّ أنْ يصطلح على ذلك من رأسٍ، وبعد ذلك، فنحتاج أنْ يكون لهذا المعنى الذي هو أعمّ منه، لفظ بحسبه.

(٧) وأمّا الملتصق فهو المماس اللازم للشيء في الانتقال حتى يصعب الفصل بينهما؛ إمّا لانطباق السطحين حتى لا يمكن أنْ يفارق أحدهما الآخر إلاّ مع وقوع الخلاء - المبيّن استحالة وجوده. وذلك يكون إذا كان ليس طرف أحد سطحي الجسمين أولى بالسَبْق إلى الانفتاح أو يكون إنّما ينفتح بزوال صورة السطح عن كيفيته، باستحداث تقبيب أو تقعير أو غير ذلك، وهو غير مجيبٍ إليه إلاّ بعنف، وإمّا لانغراز أجزاء من هذا في أجزاء في ذلك. وقد يحدث الالتصاق بين جسمين بتوسط جسم من شأنه أنْ ينطبق جيداً على كل واحدٍ من السطحين لسيلانه، وأنْ ينغرز أيضاً في كل واحدٍ منهما كذلك، ثم من شأنه أنْ يجفّ ويصلب، فيلزم كل واحدٍ من الجسمين، ويفرض لذلك التزام الجسمين بوساطته، وهذا كالغراء وما يشبهه.

(8) *Continuous* is an equivocal term that is said in three senses, which we have noted in [certain] places: Two of them are said of the thing relative to another, while one is said of the thing in itself and not relatively. One of the two [relative senses] is said of the magnitude [in the sense] that it is continuous with another when its limit and that of the other are one; and so both what is continuous and that with which it is continuous must actually be continuous, whether absolutely or accidentally. If there is [continuity] absolutely and is itself [something actually existing] in reality, then it has an absolute limit, like the one [belonging] to two lines of an angle. In this case, one is continuous with the other, since one actually existing line is different from the other, having an actual limit, and yet that limit is the same one for the other line as well. What occurs by positing falls under that which is accidentally [continuous]. So it is like what happens when our estimative faculty imagines or, we posit, two parts for a line that is actually one, where we distinguish one [part] from the other by positing. In that way, a limit is distinguished for [the line] that is the same as the limit of the other division. In that case, both are said to be continuous with one another other. Each one of the two, however, exists individually only as long as there is the positing, and so, when the positing ceases, there is no longer *this* and *that* [part]; rather, there is the unified whole that actually has no division in it. Now, if what occurs through positing were to be something [really] existing in the thing itself and not by [merely] positing, then it would be possible for an actually infinite number of parts to exist within the body (as we shall explain),[14] but this is absurd. Again, in short, there is something that is a *this* in the parts of what is continuous only by pointing in the direction of a certain area

14. See 3.3.12.

(٨) وأمّا المتصل فإنّه لفظٌ مشترك؛ يقال على معانٍ ثلاثة ذكرناها في مواضع، إثنان منها يقال للشـــيء بالقياس إلى غيره، وواحد يقال للشيء في نفسه لا بالقياس إلى غيره. فأمّـا أحـد الإثنين فإنَّه يقال للمقدار إنَّه متصل بغيـره إذا كان طرفه وطرف غيره واحداً؛ فيجب أنْ يكون كل واحدٍ من المتصل والمتصل به محصّلاً بالفعل؛ إمّا مطلقاً وإمّا بالعرض. فإنْ كان مطلقاً، وفي الوجود نفسه كان له طرف مطلق كأحد خطّي الزاوية، فإنّه متصل بالآخـر؛ لأنّه خطٌ موجودٌ بالفعل غير الآخر، وله طـرفٌ بالفعل لكنه بعينه طرف الخط الآخـر. وأمّا الذي بالعرض، فمنه ما يكون بالفـرض، فكما يعرض إذا توهمنا أو فرضنا الخـط الواحد بالفعل ذا جزئين وميّزنـا أحدهما عن الآخر بالفرض، فتميّز بذلك له طرفٌ هو بعينه طرف القســم الآخر، فيقال لكل واحدٍ منهما إنَّه متصل بالآخر. وإنّما يكون كل واحـدٍ منهما موجوداً بعينــه ما دام الفرض، فإذا زال الفرض لـم يكن ذاك ولا هذا، بل كان الواحد الكل، ولا قســمة فيه بالفعل، ولو كان ما يقع بالفرض موجوداً في نفس الأمر وإنْ لم يفرض – لم يمتنع وجود أجزاء بالفعل لا نهاية لها بالجســم، على ما سنبيّن، وهذا محال. وبالجملة أيضاً إنّما يكون في أجزاء المتصل شــيء هو هذا باتجاه الإشــارة – بعد

subsequent to the posit; and, likewise there is a *that* only on account of pointing in the direction of some other posited area. This one is *this* and that one is *that* inasmuch as the two acts of pointing are directed toward one [or the other]; for, if the two [acts of pointing] cease, then it is absurd to say that *this* and *that* remain *qua this* and *that*—that is, unless some other distinguishing cause is posited. As for what happens on account of the positing, it ceases when the positing ceases. As will become apparent later,[15] what is continuous has no actual part, and so it comes to have a part that is *this* and a part that is *that* without [that part] having actually existed before. In other words, it is something following upon the pointing. Now, when that pointing ceases, then the effect of that pointing no longer remains. So it is absurd to say afterwards that, even though the pointing has ceased, *this* is inevitably distinguished from *that.* [That] is because it is only by means of the pointing that there is a *this* and *that* during [the pointing]. So it would be like saying that, even though the pointing has ceased, there inevitably is a pointing. The state concerning the parts of something continuous is not like the state of other things whose parts are discontinuous from one another and exist in actuality, for, in the latter case, pointing indicates but does not make, while, in the former case, it makes and then indicates. In some cases, what occurs accidentally[16] specifies some accidental state in some portion [of the thing] but not another, so that when that accident ceases, so does the specification. For example, the whole of a body [might] not be white or hot, and so, on account of the white, [the body] is posited as having some part; [but] when that white ceases, the positing of [the part] also ceases.

---

15. See 3.4.

16. Reading *bi-l-ʿarḍ* with **Z**, **T**, and the Latin (*per accidens*) for **Y**'s *bi-l-farḍ* (by positing).

الفرض – إليه على نحوٍ. وكذلك ذاك، إنّما يكون ذاك لاتجاه الإشـــارة على نحوٍ آخر من الفرض إليه، وهذا هذا، وذاك ذاك من حيث الإشارتان متجهتان إليه، فإنْ بطلتا فمحال أنْ يقـــال إنَّ هذا وذاك باقيان، من حيث هما هذا وذاك؛ اللّهم إلاّ أنْ يُفرض ســـببٌ آخر مميّـــزٌ، وأما ما كان يعرض بالفرض فيبطل بـــزوال الفرض. والمتصل لا جزء له بالفعل، كما يظهـــر مـــن بعد، فيكون حدوث جزء له هو هذا، وجزء له هو ذاك، من غير أنْ كان قبل موجوداً بالفعل، وهو أمرٌ يتبع الإشـــارة، وإذا زالت الإشارة لم يبق معلول الإشارة فمحال أنْ يقال بعد ذلك إنّه وإنْ بطلت الإشارة، فلا بُدّ من تميّز ذاك من هذا، فإنَّ كون هذا وذاك فيها إنّما هو بالإشارة، فيكون كأنه قيل إنْ بطلت الإشارة فلا بُدّ من أنْ تكون إشارة. وليس الحال في أجزاء المتصل كالحال في أجزاء الأشـــياء الأخـــرى المنفصل بعضها عن بعض؛ موجوداً بالفعل، فإنَّ الإشـــارة هناك تدلّ ولا تفعل، وها هنا تفعل فتدل. ومن الذي يكون بالعـــرض اختصاص العرض الحالّ ببعض دون بعض، حتى إذا زال ذلك العرض، زال ذلك التخصيص؛ مثل جسم يبيضّ لا كله، أو يسخن لا كله، فينفرض له بالبياض جزءاً إذا زال ذلك البياض زال افتراضه.

(9) The second [relative] sense of *continuous* is said of that [instance in] which, when one side of the continuous thing is moved in a direction away from the other, the other follows it. So this is something more general than the [sense of] *continuous* that we just noted and [more general] than what coheres. The two extremities can be two in actuality, and there can be something actually contiguous after adhering during the motion. It is also possible[17] that the extremity of what is continuous and that with which it is continuous are one, but it is termed *continuous* in the present sense not inasmuch as its extremity and that of the other are one, but only inasmuch as it follows it during the motion in the aforementioned way.

(10) Something is said to be *continuous* in itself when it is such that you can posit parts for it between which there is the continuity that is in the first sense [mentioned above]—that is, between [the parts] there is a common limiting point that is a limit for *this* and *that,* where this is a limiting point of this continuous thing. Its being said to be divisible into things that are always susceptible to division is its definite description, and that is because this is different from that which constitutes its essence. Because the continuous is truly and really understood in the first sense and one is not aware of whether this latter account [namely, being infinitely divisible] is a concomitant of it or not, except through demonstration, it is one of the necessary accidents of the continuous that needs a middle term in order to prove that it does belong to what is continuous.

17. **Y** (inadvertently) omits the phrase *yakuna talāzum fī l-ḥarakah wa-yajūzu an* after *baʿd an,* which appears in **Z**, **T**, and the Latin; it corresponds with the translation following *after* in the previous line up to this footnote.

(٩) وأمّـا الوجـه الثاني ؛ فيقال متصل للذي إذا نقل مـا قيل إنّه متصل به في جهةٍ تبعده عن الآخر ، تبعه الآخر ، فيكون هذا أمر أعمّ من المتصل الذي قلناه قبل هذا ، ومن الملتصـق . ويجوز أنْ تكـون النهايتان اثنتين بالفعل ، وأنْ يكون هناك تماسّ بالفعل ، بعد أنْ يكـون تلازم في الحركة ، ويجوز أن تكون نهاية المتصل والمتصل به واحدة ، ولكن لا يكون إيقاع اسـم المتصل ها هنا بهذا المعنى عليه ؛ من حيث نهايته ونهاية الآخر واحدة ، بل من حيث يتبعه في الحركة على النحو المذكور .

(١٠) ويقال متصل للشيء في نفسه ؛ إذا كان بحيث يمكن أن تفرض له أجزاء بينها الاتصال الذي بالمعنى الأول ، أي بينها حدّ مشـترك هو طرفٌ لهذا وذاك ، وهذا هو حدّ هذا المتصل . وأمّا الذي يقال إنَّه المنقسم إلى أشياء تقبل القسمة دائماً فهو رسمه ، وذلك لأنَّ هذا غير مقوّم لماهيته ، لأنَّ المتصل يفهم بالمعنى الأول فهماً حقيقياً ، ولا يدُريَ أنَّ هذا المعنـى يلحقه أو لا يلحقه إلاّ ببرهان ، فهو من الأعـراض اللازمة للمتصل المحتاج في إبانة وجودها للمتصل ، إلى حدٍ أوسط .

(11) We use the expression *being separate* of things each one of which has a proper place whose part is not a part of some place that is common to it and another. *Being together in place* is not said in the way that *being together in time* is said inasmuch as a place of either of two [things] would be the very same place as the other's in the way that the time of one is the time of the other; for this is impossible with respect to place, but not so with respect to time. Instead, *being together in place* is said only of things combined together as a single thing that is [in] a place through its totality, but each one of them has a proper place, part of that proper place being part of the common place. The *intermediate* and *between* are that into which the change occurs before changing into something else during the time [of] any type of change.

(12) These things are useful for our purposes, besides being among the states that necessarily belong to natural things insofar as they possess a quantity.

(١١) وأمّا قولنا فرادى، فإنّما يقال لأشياء لكل واحدٍ منها مكان خاص ليس جزؤه جـزءاً من مـكانٍ عام له وللآخر. ويقال معاً في المكان ليـس كما في الزمان، بأنْ يكون مكان كل واحدٍ منهما هو بعينه مكان الآخر، كما زمانه زمان الآخر، فإنَّ هذا مسـتحيل في المكان وغير مسـتحيل في الزمان. بل إنما يقال معاً في المكان لأشياء مجتمعة كشيءٍ واحدٍ يكون بجملتها مكان، ويكون لكل واحدٍ منها مكان خاص، جزء من ذلك المكان الخاص جزء من المكان العام. والوسـط والبيْن هو الذي يقع التغيّر إليه قبل التغيّر إلى غيره في الزمان؛ أي تغيّرٍ كان.

(١٢) فهذه الأشياء نافعة في غرضنا، ومع ذلك فإنّها من الأحوال التي تلزم الطبيعيات من حيث هي ذوات كمّ.

Chapter Three

The state of bodies with respect to their division and a report of the various arguments on which the detractors rely

(1) People have different opinions concerning these perceptible bodies. [(1)] Some believe that they are an aggregation of atoms[1] and that each body contains a finite number of [these atoms], while [(2)] others believe that the body contains an infinite [number] of parts. [(3)] Still others believe that, in every body, either there is a finite number of actually existing parts, or it does not have actual parts at all; and, when it does have parts, each one of its separate parts is also a body lacking actual parts. So, in their opinion, the body is either a body lacking parts, or it is aggregated of bodies lacking parts. The meaning of *lacking parts* is that [the body] presently has no part that one can posit as distinct, but, instead, [the body] is one by way of continuity, which does not mean that it is not such as to be divided. Instead, their view is that it is always susceptible to division; and, whenever it is divided, what results from the division is itself a body that is divisible. Sometimes, however, you cannot divide it because of the absence of something by which to divide [it], or [because] it is outside the power of the one doing the dividing, or owing to [the body's]

1. The Arabic *al-juzʾ alladhī lā yatajazzaʾu* literally means "the part that cannot be partitioned"; however, it was also the standard locution among the *mutakallimūn* for an atom and is so translated here.

<الفصل الثالث>

في حال الأجسام في انقسامها وذكر ما اختلف فيه وما تعلّق به المبطلون من الحجج

(١) فنقول: وقد اختلف الناس في أمر هذه الأجسام المحسوسة، فمنهم مَنْ جعل لها تأليفاً من أجزاء لا تتجزأ البتّة، وجعل كل جسم متضمناً لعدةٍ منها متناهية، ومنهم مَنْ جعل الجسم من أجزاء لا نهاية لها. ومنهم مَنْ جعل كل جسم إمّا متناهي الأجزاء الموجودة فيه بالفعل، وإمّا غير ذي أجزاء بالفعل أصلاً، وإذا كان ذا أجزاء بالفعل؛ كان كل واحدٍ من أجزائه المنفردة جسماً أيضاً لا جزء له بالفعل، فالجسم عنده إمّا أنْ يكون جسماً لا جزء له بالفعل، وإمّا أنْ يكون مؤلفاً من أجسام لا جزء لها. ويعني بقوله لا جزء له إنّه ليس في الحال له جزء مُفترضٌ متميّزٌ، بل هو واحد بالاتصال، وليس يعني أنّه ليس من شأنه قبول الانقسام، بل عنده أنّه يقبل القسمة دائماً، وكلّما قُسّم فالخارج بالقسمة جسم له في نفسه أنْ ينقسم. لكنه ربّما لم يمكن قسمته بسبب عدم ما يقسم به، أو فواته تقدير

hardness or the impossibility of its being broken up, though, in itself, something intermediate can be posited in it. Before the division, then, every body lacks parts entirely, and, instead, it is the existence of division that makes the part, whether that division is by severing the continuity, or by some accident through whose occurrence we distinguish one part from another (whether it be a nonrelational accident, such as white, or a relational accident, such as being opposite and parallel),[2] or by the act of the estimative faculty and positing.

(2) Those who say that bodies terminate at atoms include some who think that those parts are in themselves bodies, others who make them indivisible lines, and still others who make them neither bodies nor lines nor anything that has in itself dimensions or intervals. Proponents of the first of the former two doctrines [that is, (1) in the main division]—namely, the followers of Democritus, Proclus,[3] and Epicurus—differ from the true doctrine in that they say that combining these bodies occurs only by way of contiguity; that nothing continuous comes to be from them at all; that perceptible bodies are not, in fact, continuous (for those primary bodies actually existing in perceptible bodies are distinct from one another); and that [the primary bodies] are not susceptible to division by separation, but only division by an act of the estimative faculty, and, nevertheless, some are smaller and others larger. Proponents of the truth [that is, opinion (3)] concede that a certain large perceptible body

2. **Y** (inadvertently) omits the phrase *aw ʿaraḍ muḍāf ka-l-muḥādhāh wa-l-muwāzāh* corresponding with "such as being opposite and parallel," which occurs in **Z**, **T**, and the Latin.

3. Although it is odd to see Proclus (Abrūqilūs) alongside of Democritus and Epicurus, Proclus was an Atomist of sorts, even if not a corpuscularian. He followed Plato's *Timaeus* in holding that the ultimate building blocks of the physical world were atomic triangles; and, moreover, his commentary on Plato's *Timaeus* had been translated into Arabic. Perhaps Avicenna is extending Proclus's line of thought and reasoning that, if these basic triangles are atomic, then there must likewise be atomic or indivisible lines (a position Avicenna explicitly mentioned) from which these triangles, as it were, are constructed. So Proclus, in Avicenna's mind, may very well correspond with those Atomists who affirm indivisible lines. Alternatively, Jules Janssens has suggested to me that the texts Abrūqilūs is a corruption of Alūqībūs—that is, Leucippus, whom Aristotle mentions as an Atomist in his *Physics*. The suggestion clearly has merit, although it would mean that three scribal errors were made in copying the name and that those errors occurred very early in the transmission process in order to explain the complete absence of this reading of the name in all of the manuscripts.

القاســم ، أو لصلابته أو اســتحالة انكساره ، وهو في نفسه يحتمل أنْ يُفرض فيه وسط . وكل جسم ، فإنّه قبل القسمة لا جزء له البتة ، بل الفاعل للجزء وجود القسمة ، والقسمة إمّــا بتفريق الاتصــال ، وإمّا بعرض يميّز بحلوله جزءاً عن جــزء ، إمّا عرض غير مضاف كالبياض أو عرض مضاف كالمحاذاة والموازاة ، وإمّا بالتوهّم والفرض .

(٢) وأمّا الذين يقولون إنَّ الأجسام تنتهي إلى أجزاء لا تتجزأ ؛ فمنهم مَنْ يجعل تلك الأجزاء أجســاماً في أنفسها ، ومنهم مَنْ يجعلها خطوطاً غير منقسمة ، ومنهم مَنْ يجعلها غير أجسام ولا خطوط ولا أشياء لها في أنفسها أقطار وأبعاد . ويفارق أصحاب المذهب الأول من هذين المذهبين - وهم شيعة ديموكريتس وأبيقورس وابروقلس - المذهب الحق ؛ أنَّ هؤلاء يقولون إنَّ التركيب من هذه الأجسام هو بالتماس فقط ، وأنّه لا يحدث منها متصل البتة ؛ وأنَّ الأجسام المحسوسة ليست بحقيقية الاتصال ؛ فإنَّ تلك الأجسام الأُولى موجودة بالفعل في الأجسام المحسوسة ، متميّزاً بعضها عن بعض ، وأنّها لا تقبل القسمة المفرّقة ، بل القســمة المتوهمة ، وهي مع ذلك بعضها أصغر وبعضها أكبر . وأمّا أصحاب الحق ؛ فإنهم يجوّزون أنْ يكون جســم كبير من المحسوســات لا جزء له بالفعــل ، ويجوّزون أنْ تكون

does not have some part in actuality. They also grant that when parts that are actually separate happen to encounter [one another] at some time, a single thing comes to be from them, but the specific property of each one of the [formerly separate] parts then ceases and so no longer remains in itself. Getting back to where we were, we say that the followers of Democritus differ from other Atomists in that the others do not make their atom a body.

(3) Each of these [groups] has its own specific arguments. So one of the arguments of those defending atoms that are not bodies[4] is that every body is separable into parts, and also, when it is so separated, its parts can be recomposed as they were. Consequently, there is an aggregation in every body before being separated into parts, otherwise, bodies would not differ in how difficult or easy it is to break [them] up. [This difficulty or ease in breaking certain bodies up], they maintained, is not because [the bodies] differ in genus (meaning by *genus* the specific nature), nor owing to the agent's being different nor [due to] an absence of something, nor because of any of the other options they mention. Hence, it is due to the aggregation. Given that there is an aggregation in [a body], there is nothing absurd in our estimative faculty's imagining it to have passed away; and, when it does so entirely, what remains has no composition in it. What has no aggregation in it, however, is not a body, because every body is divisible, whereas what has no aggregation in it is not divisible. This way of arguing has its origins in Democritus, but [it comes] with a slight twist, which we'll come to understand when we present his argument.[5]

4. The present form of Atomism is that of the *mutakallimūn*. The earliest study of the Atomism of *kalām,* which is still quite valuable, is Shlomo Pines, *Beiträge zur islamischen Atomlehre* (Berlin: A. Heine G.m.b.H.,1936), translated by M. Schwarz as *Studies in Islamic Atomism,* (Jerusalem: The Magnes Press, 1997). For a more recent study, which incorporates new material and, in many ways, corrects Pines's earlier study, see Alnoor Dhanani, *The Physical Theory of Kalām, Atoms, Space, and Void in Basrian Muʿtazilī Cosmology* (Leiden: E. J. Brill, 1994). See also Abdelhamid I. Sabra, "*Kalām* Atomism as an Alternative Philosophy to Hellenizing *Falsafa,*" in *Arabic Theology, Arabic Philosophy: From the Many to the One: Essays in Celebration of Richard M. Frank,* ed. James Montgomery (Leuven: Peeters, 2006), 199–272.

5. See par. 12.

الأجزاء – إذا حصلت بالفعل منفصلة – تلتقي مرة أخرى، فيحصل منها شيء واحد فتبطل خاصّة كل واحدٍ منها، فلا يكون ثابتاً بعينه. ونعود إلى ما كنا فيه؛ ونقول: لكن أصحاب ديموكريتس يفارقون الآخرين من أصحاب الجزء، بأَنَّ الآخرين يجعلون جزءهم غير جسم،

(٣) لكل واحدٍ من هؤلاء حججٌ تخصّه. أمّا القائلون بجزء لا يتجزأ، ولا هو جسم؛ فمن حججهم أَنَّ كل جسم فإنّه قابل للتفريق، فإذا تفرّق فأجزاؤه قابلة للتأليف كما كانت. فإذا كان كذلك، فكل جسم ففيه قبل التفريق تأليف، ولولا أَنَّ فيه تأليفاً لكان لا تختلف الأجسام في صعوبة التفكيك وسهولته. قالوا؛ وليس ذلك لأَنَّ جنسها مخالف – ويعنون بالجنس الطبيعة النوعية – قالوا، ولا لاختلاف الفاعل ولا لعدم شيءٍ ولا لأقسام يذكرونها؛ فإذن هو للتأليف. وإذا كان فيه تأليف فتوهمناه زائلاً لم يكن محالاً، وإذا زال بكليته بقي ما لا تأليف فيه، وما لا تأليف فيه فليس بجسم؛ لأَنَّ كل جسم ينقسم، وما لا تأليف فيه لا ينقسم وهذا الاحتجاج مبدؤه لديموكريتس إلاّ أنّه حُرّف عنه بشيء يسير، نفهم ذلك إذا أوردنا حجته.

(4) They further say that if the parts of the body were not finite, then they would be infinite; but then a body would be divisible into half, and again into half, and so on infinitely.[6] When something in motion intends to cross a given distance, it would need to cross half, but, before that, half of half of it; and, in a finite time it would need to cross an infinite number of halves. So it could not cross the distance at all. Also, the fleet-footed Achilles could never catch up with the plodding tortoise, and the ant would never completely cross a sandal over which it travels (the first example is from the Ancients, the second is from the Moderns).[7] Motion, however, exists. So the body's divisions are finite.

(5) They further claimed that, if the body could be divided infinitely, it would follow from that necessarily that the mustard seed is divisible into parts that would completely cover the face of the Earth.

(6) [Similarly,] they said that, if the body were infinitely divisible, the parts of a mustard seed would equal the parts of an enormous mountain, which is absurd.

(7) Again, they said that the point must be either a substance subsisting in itself or not. On the one hand, if it subsists in itself, then it is, in fact, the atom. Moreover, that which encounters it will be another point, and so the points following in succession will make up a body (or [they will make up] a line, which makes up a surface, which, in its turn, makes up that body). If it is an accident, on the other hand, then it inheres in a substrate, and everything that inheres in a substrate does so in what is similarly equal to [the substrate]. Thus, the point would inhere in an indivisible substance.

---

6. The arguments of this paragraph rely heavily on Zeno's paradoxes; see Aristotle, *Physics* 6.9 for the classical understanding of Zeno's paradoxes, and G. E. L. Owen, "Zeno and the Mathematicians," in *Logic, Science and Dialectic: Collected Papers in Greek Philosophy* (Ithaca, NY: Cornell University Press, 1986), 45–61, for a contemporary discussion.

7. The reference to the "Ancients" is clearly to Zeno, while the "Moderns" refers to the *mutakallimūn,* and specifically Abū Hudhayl (d. 841), who is credited with formulating the paradox in terms of the ant and the sandal.

(٤) وقالوا أيضاً ، إنَّه لو لم تكن أجزاء الجسم متناهية لكانت غير متناهية ، فكان للجسم أقسام وأنصاف في أقسام ، وأنصافٌ من غير نهاية . فكان المتحرّك إذا أراد أنْ يقطع مسافةً احتاج أنْ يقطع نصفها ، وقبل ذلك نصف نصفها ، واحتاج في زمانٍ متناهٍ أنْ يقطع أنصافاً بلا نهاية ، فكان يجب أنْ لا يقطع المسافة أبداً ، ويجب أن لا يلحق ‹أخيل› السريع العدو ، السُلحفاة البطيئة العدو . وكانت الذرة لا تفرغ من قطع نعْلٍ يسير عليها . فالمثل الأول للقدماء والثاني للمحدثين ؛ لكن الحركة موجودة ، فأقسام الجسم متناهية .

(٥) وقالوا أيضاً إنّه لو جاز أنْ ينقسم الجسم إلى غير نهاية ، لوجب من ذلك أنْ تكون الخردلة تنقسم أقساماً تبلغ إلى أنْ تغشى أديم الأرض كله .

(٦) قالوال ولو كان الجسم ينقسم إلى غير نهاية ؛ لكانت الخردلة في أقسامها متساوية لأقسام الجبل العظيم ، وهذا محال .

(٧) وقالوا أيضاً ؛ إنَّ النقطة لا تخلو إمّا أنْ تكون جوهراً قائماً بنفسه أو لا تكون ، فإنْ كانت قائمة بنفسها ، فقد حصل الجزء الذي لا يتجزأ ، ويكون الذي يلقاها أيضاً نقطة أخرى ، فتتوالى النقط فاعلةً لجسمٍ أو لخطٍ فاعلٍ لسطحٍ فاعلٍ لذلك الجسم . وإنْ كانت عرضاً ، فهي تحلّ محلاً ؛ وكل حالٍّ في محلٍ فهو يحلّ فيما يساويه ويكون مثله ، فتكون النقطة تحلّ جوهراً لا يتجزأ .

(8) They additionally said that, if the body could be divided into an infinite number of parts, it could be composed of an infinite number of parts, or, along with something else, it could make up an infinite composition.[8]

(9) They could also say that, when we assume that one line is made to coincide with another so that the point [on one line] parallels the point [on the other]—or encounters, or interpenetrates, or whatever term you want to use to indicate what is understood by the idea—and then the line moves, the point that was contiguous will no longer be such, and the contiguity will cease all at once such that, in a single instant, [the point] will no longer be contiguous—that is, at that instant, it will encounter some point that follows the first point in succession. So the points on the line follow in succession, and from them the line is composed, since the [same] account will apply to the loss of contiguity with the second point, just as it did to the loss of contiguity with the first point, and so on.

(10) Also among their arguments is the existence of an indivisible angle—namely, the one that Euclid deemed the smallest acute angle.[9]

(11) Likewise they asked: What do they say about a sphere's rolling over a smooth surface? Isn't it contiguous with one point after another such that the line that the sphere maps out is composed of points?[10]

---

8. The argument is enthymematic but seems to be that, if a body is infinitely divisible, then the body is composed of an infinite number of parts (or it is part of something that has an infinite number of parts), in which case there is an infinity of parts; but an infinite is impossible (a common premise in many *kalām* arguments). Thus, since what gives rise to something impossible is itself impossible, the assumption that the body is infinitely divisible is impossible.

9. This is the so-called horn angle, whose vertex is the point of tangency between a line and a circle. In his *Elements* 3.16, Euclid provided a proof that the horn angle is smaller than any rectilinear angle. For Avicenna's treatment of angles and, specifically, his criticism of Euclid's proof, see Irina Luther, "The Conception of the Angle in the Works of Ibn Sīnā and al-Shīrāzī," in *Interpreting Avicenna: Science and Philosophy in Medieval Islamic,* ed. Jon McGinnis (Leiden: E. J. Brill, 2004), 112–125.

10. Reading *nuqaṭ* with **Z**, **T**, and the Latin (*punctis*) for **Y**'s *nuqṭa* (meaning point, sing.).

(٨) وقالوا أيضاً: إنْ جاز أنْ ينقسم الجسم إلى أجزاء غير متناهية، جاز أنْ يتركب من أجزاء غير متناهية، أو أنْ يُركّب مع غيره تركيباً بلا نهاية.

(٩) ولهم أنْ يقولوا أيضاً: إنّا إذا فرضنا خطاً منطبقاً على خط، حتى تكون النقطة محاذية للنقطة أو ملاقية أو مداخلة، أو أي إسم شئتم أنْ تدلّوا به على المعنى المفهوم، ثم تحرّك الخط، فقد صارت النقطة المماسّة غير المماسّة، وزوال المماسّة دفعة، فتكون في آنٍ واحدٍ صارت غير مماسّة؛ وهي في ذلك الآن ملاقية لنقطةٍ تالية للنقطة الأولى. فتكون النُقط متتالية في الخط، والخط مؤلفاً منها، إذ الكلام على زوال مماسّة النقطة الثانية، كما هو في زوال مماسّة النقطة الأُولى كذلك، وهلمّ جرا.

(١٠) ومن حججهم وجود زاويةٍ غير منقسمةٍ، وهي التي جعلها أقليدِس أصغر الحادّات.

(١١) وقالوا أيضاً: ما يقولون في حركة الكرة على سطح أملس، أليس تكون بمماسةٍ واقعةٍ بنقطةٍ بعد نقطة، فيؤلّف الخط، الذي تمسحه الكرة، من نُقط؟

(12) Those who make this ultimate part a body—namely, the followers of Democritus—said that the body must either be wholly divisible, such that nothing remains of it that is not divided, or it is not wholly divisible. If it is in its nature to be divided, then that [such a division] should occur is not impossible. When something that is not impossible is assumed to exist, no absurdity arises from it. (Indeed, some nonabsurd falsity might result, but a nonabsurd falsity does not entail an absurdity.) So let us assume that every possible division in the body has actually resulted. In that case, either nothing exists, or points exist, or indivisible bodies exist. It is absurd, however, that they terminate at nothing or at points. [That] is because, if [the body] breaks down into nothing, then it is an aggregation of nothing, which is absurd; whereas, if it breaks down into points, then it is an aggregation of points, which is also absurd. (The consensus among the learned is that, no matter how many points are joined together, they do not exceed the size of a single point; [points] encounter one another completely, not with some of them hindering others from encountering one another; they are not moved so as to form a composition and so become something occupying a place; and nothing continuous comes to be from them). So it remains that [the body] breaks down into bodies whose nature is not to be separated and divided [further], except through the act of the estimative faculty and positing.

(١٢) وأمّــا الذين جعلوا الجزء المنتهى إليه جســماً - وهو شــيعة ديموكريتس - فقالوا إنَّ الجســم لا يخلو إمّا أنْ ينقســم كله حتى لا يكون منه ما لا ينقسم، أو لا ينقسم كله. فإنْ كان في طباعه أنْ ينقسم، فغير ممتنع أنْ يقع، وغير الممتنع - إذا فرض موجوداً لــم يعرض منه محال، بل ربّما عرض منه كذب غيــر محال، والكذب غير المحال لا يلزمه المحال. فلنفرض أنَّ كل قســمة ممكنة في الجسم فقد خرجت بالفعل، فحينئذ لا يخلو إمّا أنْ يحصل لا شيء؛ أو تحصل نقطٌ، أو تحصل أجسامٌ لا تنقسم. لكن من المحال أنْ تنتهي إلى لا شــيء أو إلى النقط، فإنّه إنْ كانت انتقاضه إلى لا شيء فتأليفه من لا شيء وهذا محال. وإنْ كان إنتقاضه إلى النقط فتأليفه من النقط، وهذا أيضاً محال. قد أجمع العلماء على أنَّ النقط كمّ اجتمعت، لا تزيد على حجم نقطة واحدة، وأنّها إنّما تتلاقى بالأســر، ولا يحجب بعضها بعضاً من الملاقاة، ولا تتحرّك إلى التأليف فتصير شــاغلة مكاناً، ولا يحــدث منها متصل، فيبقى أنْ يكون إنتقاضه إلى أجســامٍ ليــس في طبيعتها أن تنفصل وتنقسم، اللّهم إلاّ بالوهم والفرض.

(13) Those who said that the body has an infinite number of existing parts[11] [that is, group (2) noted at the beginning] were driven to this account by the impossibility of composing bodies out of indivisible parts and indivisible bodies. They said: Since bodies in themselves also possess divisions (even if they are not actually separated), if it is stipulated and assumed that they are divided into parts, then each one of [those parts] is a portion and part of the body (even if it is not at all separated). It remains, they continued, that the parts of the body are infinite, and, because of that, the body is infinitely divisible. [This is so] since the divisions (whether [resulting from] positing or actual separation) occur only at parts that exist adjacent to one another in the body, in which case the parts of the body are commensurate with the capacity for the divisions. So, if the capacity for divisions is infinite, [the body] possesses an infinite number of parts.

(14) Since the Atomists harassed those [advocating that the body has an infinite number of parts] and forced their hand with the problem of the sandal and the ant, as well as the tortoise and Achilles, and, in general, that motion would proceed over an infinite number of halfway points such that the final end would never be reached, [those advocating that the body has an infinite number of parts] took refuge with Epicurus and so advocated the leap.[12] In other words, the body might cross a given distance in order to arrive at an intended end point from a point of departure without encountering or passing directly over the intermediate [space].

11. This position seems to be that of Ibrāhīm al-Naẓẓām. See Sabra, "*Kalām* Atomism as an Alternative Philosophy to Hellenizing *Falsafa*," esp. 226 and 262–263.

12. The author of the theory of the leap is, in fact, the *mutakallim* Ibrāhīm al-Naẓẓām (d. ca. 840).

(١٣) وأمّــا الذيــن قالوا بوجودِ أجزاء غير متناهيةٍ للجســم؛ فقد دفعهم إلى هذا القول امتناع تركيب الأجسام من الأجزاء غير المتجزئة ومن الأجسام غير المتجزئة. قالوا، فإنَّ الأجســـام أيضاً في نفسها ذوات أقسام وإنْ لم تفصل بالفعل، فهي إذا جزئت بالتعيين والفرض، كان كل جزء منها بعضاً وجزءاً من الجسم وإنْ لم ينفصل البتة. قالوا؛ فبقي أنْ تكون أجزاء الجسم بلا نهاية، وبسبب ذلك ينقسم الجسم إنقساماً لا يتناهى، إذ الانقسام الفرضي أو التفريقي إنّما يرد على أجزاء موجودة في الجسم متجاورةٍ، فتكون أجزاء الجسم بحســب احتمال الانقســام، فإنْ احتمل انقساماً غير متناهٍ؛ كان ذا أجزاء غير متناهية.

(١٤) ولمّــا ضيّق أصحاب الجزء علــى هؤلاء وألجأوهم إلى مســألة النعل والذرّة والسُلحفاة و<أخيل>، وبالجملة إلى أنْ تكون الحركة تأتي على أنصافٍ لا تتناهى فلا تبلغ الغاية البتة، التجأوا إلى ما التجأ إليه <ابيقورس> فقالوا بالطفرة؛ وهي أنَّ الجسم قد يقطع مسافة حتى يحصل في حدّ منها مقصود عن حدّ متروك، ولم يلاقِ ولم يحاذِ ما في الوسط.

(15) The first of Epicurus's foreign imitators produced an example of that from the rotations of two circles:[13] one near the edge of a spinning millstone, and the other near the center. They noted that, if the motion of the part at the edge were equal to the motion of the part that is near the center, the two together would cross an identical distance. Now, it is absurd that what is in the middle would be at rest, because the [whole millstone] is continuous, with each part adhering to one another. So, clearly, what is in the middle is moved, but its leaps are few; and equally clear is [the fact] that the part at the edge is moved, but it leaps more frequently in order to cover a larger interval than the interval of that which is in the middle.

(16) Since the first of the noted foreigners clung to this account but considered the leap repulsive and, further, did not allow that one continuous motion is faster than another motion without the intermediacy of a rest, they were forced to make that which is near the center rest more frequently than that which is at the edge and were forced into the possibility that what is in the middle rests. They were also forced to conclude that the millstone fragments while [it is] moving, with some of its parts breaking away from others such that one of them does not need to be moved together with the other but, instead, one of them rests while the other is moved. Thus, one [group] would not forsake the repulsive leap, while the other [would not forsake] the repulsive fragmentation.

---

13. The example is that of al-Naẓẓām.

(١٥) وأورد، أوّل من تشبه بأبيقورس من الخارجين لذلك، مثالاً من دوران الدائرة القريبة من طرف الرحى والدوّامة، والأخرى القريبة من المركز، وذكروا أنّه لو كان الجزء الذي عند الطرف يتحرّك مع حركة الجزء الذي عند الوسط بالسواء؛ لقطعا معاً مسافةً واحدةً، ومحال أنْ يسكن الذي في الوسط لأنّه متصلٌ ملتزمٌ بعضه لبعض. فبيّنٌ أنَّ الذي في الوسط يتحرك وتقل طفراته، مع أنَّ الذي عند الطرف يتحرك ويطفر أكثر حتى يحصل في بُعْدٍ أكثر من بُعْد الذي في الوسط.

(١٦) ولمّا استشنع الأولون من الخارجين المذكورين الطفرة ولزومهم هذا الكلام، ولم يجوّزوا أنْ تكون حركة متصلة أسرع من حركةٍ بلا توسط سكون؛ اضطروا إلى أنْ جعلوا الذي يلي الوسط يسكن سكونات أكثر من سكونات الذي على الطرف، واضطروا إلى تمكين المتوسط من السكون، وإلى أنْ حكموا بأنَّ الرحى تنفك أجزاؤها عند الحركة، بعضها عن بعض تفكّكاً لا يلزم أحدهما أنْ يتحرك مع الآخر، بل يسكن أحدهما ويتحرك الآخر، فلم يزل أحدهما في شناعة الطفرة، والآخر في شناعة التفكّك!

Chapter Four

Establishing the true opinion and refuting the false

(1) Since we have indicated the different schools of thought concerning this question of ours, let us begin by indicating the soundness of the true doctrine, then take up and resolve the doubts produced by its detractors. The view that maintains that an actually infinite number of parts are in the body is obviously false, because it is impossible to traverse an infinite number of things in a finite period of time and because the assertion of the leap is clearly false in itself. [It is also false] in that any given multiple consists only of its units, and, when one unit does not actually exist, then neither does a multiple. So, when a unit part does not exist, there would not be an infinite number of parts. Now, the unit part, insofar as it is a unit, is indivisible, and so, when units like it are added to it, then the addition must be by either contiguity, interpenetration, or continuity. If it is by continuity, what is continuous comes to be from delimited magnitudes, and so the opinion is falsified. If [the addition] is by way of interpenetration, then no determinate quantity comes to be from it whatsoever, even if there really were an infinite number of additions. If it is by encountering one another [that is to say, by contiguity], then each one of two parts requires a position proper to

<الفصل الرابع>

في إثبات الرأي الحق وإبطال الباطل

(١) وإذ قد دلّلنا على اختلاف المذاهب في مسألتنا هذه، فلنبدأ بالدلالة على صحة المذهب الحق، ثم لنحمل على الشكوك التي أوردها مخالفوه فنحلّها حلاً. فنقول: أمّا المذهب القائل إنَّ الجسم فيه أجزاء بالفعل غير متناهية، فيظهر بطلانه من جهة استحالة قطع أشياء بلا نهاية في زمانٍ متناهٍ، ولأنَّ إثبات الطفرة بيّن البطلان في نفسه وبأنَّ كل كثير فإنّما هو من آحادٍ، وإذا لم يكن واحدٌ موجوداً بالفعل لم يكن كثيراً، فإذا لم يكن جزءٌ واحدٌ موجوداً بالفعل لم تكن أجزاء بلا نهاية. والجزء الواحد لا ينقسم من حيث هو واحد، فإذا أُضيف إليه آحاد أمثاله؛ لم يخل إمّا أنْ تكون الإضافة على سبيل المماسّة أو على سبيل المداخلة، أو على سبيل الاتصال. فإنْ كان على سبيل الاتصال؛ حدث المتصل من مقادير منها محدودة، فبطل الرأي، وإنْ كان على سبيل المداخلة لم يحدث منها قدر وإنْ بلغت أضعافاً لا نهاية لها في الوجود، وإنْ كان على سبيل الملاقاة؛

it and must have some determinate corporeal quantity in itself (as we will explain later),[1] in which case it is a body. Now, when one body is joined with a finite number of bodies like it, the composition of that undoubtedly results in a certain body [*x*] that will have a certain proportion to the body [*y*], [made up of] infinite parts, where the proportion is of one delimited [body] to another with respect to its size. So, when that proportion is increased with respect to the parts, then *x*, which is composed of finite parts, will ultimately reach the level of *y*. So *x* will be a body consisting of a finite number of parts that is equal to [*y*], but then *y*, likewise, consists of a finite number of parts.

(2) We'll set aside discussing the view of those who claim that the division terminates at bodies [whose] continuity[2] cannot be divided by separation, for they do not deny that the bodies at which the division terminates have a certain capacity such that parts are posited of them. They deny only that that occurs in actuality; and we may or may not allow that, for it depends upon another kind of investigation whose proper place is the investigation of the elements.[3]

(3) We need to make clear what is wrong about the position of those who aggregate bodies from [indivisible] nonbodies.[4] We say that, when these parts are combined and a body comes to be from them, their combination must be by means of either mere succession, contiguity, interpenetration, or continuity. [That follows] because there either is or is

---

1. The reference appears to be to *Ilāhiyāt* 2.3, where Avicenna discusses the relation between corporeality and being localized.

2. Paul Littinck, in "Ibn Sīnā on Atomism, Translation of Ibn Sīnā's *Kitāb al-Shifāʾ al-Ṭabīʿiyyāt 1: al-samāʿ al-ṭabīʿī*, Third Treatise, Chapter 3–5," (in *Al-Shajarah* 4 [1999]: 1–51, esp. 30, fn. 52) suggests that the texts *li-l-ittiṣāl* (continuity) be emended to *wa-l-infiṣāl* (and severing). In that case, the text would read, "the division terminates at bodies that cannot be divided by separation *and severing*," which clearly makes more immediate sense. Still, all the manuscripts seem to agree in reading *li-l-ittiṣāl*, which is also the *lectio difficilior*, and so it should probably be retained.

3. See 3.12.1–9.

4. *Kalām* atoms, while being conceptually indivisible and so having no real or even conceptual parts, were envisioned a being cuboidal. Thus, while they had no parts, they did have directions (*jihah*)—namely, up, down, front, back, left, and right. Since these atoms have no parts, despite being cuboidal, a single atom cannot be said to be composed or aggregated of lines and planes. In fact, according to *kalām* Atomists, the smallest possible line consists of two atoms; the smallest possible plane consists of four atoms; and the smallest possible body consists of eight atoms.

فــكل واحدٍ من الجزئين يقتضــي وضعاً مخصوصاً ، ويجب أنْ يكون له في نفســه قدر جسماني ، - على ما يُوضّح من بعد - فيكون جسماً ؛ والجسم إذا قُرن بأجسام أمثاله متناهية العدّة ، كان من تركيب ذلك جســمٌ لا محالة ، له نســبة إلى الجسم غير المتناهي الأجزاء نســبة محدودٍ إلى محدودٍ في عِظمه . فإذا زيد في الأجزاء على تلك النســبة ، بلغ المؤلّف من الأجزاء المتناهية مبلغه ؛ فكان جسماً مساوياً له من أجزاء متناهية العدد ، فكذلك الجسم الأول هو من أجزاء متناهية العدد .

(٢) وأمّا مذهب القائلين بأنَّ القســمة تنتهي إلى أجسام لا تنقسم بالتفريق للاتصال ، فإنّا نؤخّر الكلام في النظر في أمر هذه الأجســام ؛ فإنّهم ليســوا يمنعون كون الأجزاء التي إليهــا تنتهي القســمة ذات احتمال لأنْ يُفرض لها أجزاء ، إنّمــا يمنعون وقوع ذلك بالفعل . وعسانا نجوّز ذلك أو لا نجوّزه ، فيتعلق بنوعٍ آخر من النظر ، إنّما الموضع الأخص به : النظر في الأُسْطُقسّات .

(٣) وأمّا مذهب المؤلفين للاجســام من غير الاجســام ، فيجــب أنْ نوضّح بطلانه فنقــول : إنَّ هذه الأجزاء إذا اجتمعت وكان منها جســم ، فإمّا تجتمع على ســبيل تتالٍ فقط ، أو على ســبيل تماسّ ، أو على ســبيل تداخل ، أو على سبيل اتصال ، إذ الأشياء

not an interval between the aggregated things. If there is no interval between them, either they encounter one another completely or not. If [they encounter one another] completely, they interpenetrate, as we explained,[5] whereas if they do not [encounter one another] completely, then either there is something unique to each of them with which it encounters the other, or that thing [with which one encounters the other] is common to both. If it is unique, then there is contiguity, while if it is common, there is continuity. Thus, when these parts are combined, their combination must be in one of these ways. If they are combined according to mere succession, then no perceptibly continuous bodies result. Our discussion, however, is about [just such bodies]. When they are combined according to either continuity or contiguity, then each one of them is divisible into what is occupied and what is unoccupied, what is being touched and what is not, according to what we explained in the preceding chapters.[6] If they do not interpenetrate, then when one of them [x] meets another [y], and then a third one [z] encounters one of the two [for example, y], then, as a result of [y's] intermediate position, [z] necessarily is hindered from encountering x. In that case, each has conferred on the contact [with y] something of itself that the other has not.[7] This is self-evident. So what is in the intermediate position [that is, y] is divisible. If they encounter one another completely, then they interpenetrate, and so their combination produces no quantitative increase; for, whenever they are combined, they will be like the unit, which has no length, breadth, or depth. Since these atoms do not combine such that

5. See 3.2.3.

6. See specifically 3.2.5.

7. Again, the position that Avicenna is trying to refute is a conception of wholly indivisible atoms, whether actually or even conceptually. Avicenna's example shows that, if such atoms were aggregated in the way illustrated in the diagram below, then x would contact some *part* of y different from the *part* that z contacts; for if they both contacted y at one and the same place, they would, in fact, be interpenetrating y, but it was posited that they were not interpenetrating.

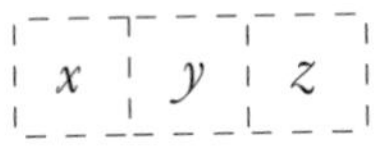

المجتمعة إمّا أنْ يكون بينها بُعْدٌ أو لا يكون. فإنْ لم يكن بينها بُعْدٌ، فإمّا أنْ يكون تلاقيها بالأسـر أو لا بالأسر، فإنْ كان بالأسر كانت مداخلة - على ما أوضحنا - وإنْ كان لا بالأسـر، فإمّا أنْ يختص كل منها بشـيءٍ به يلقى الآخر، أو يكون ذلك الشيء مشتركاً. فإنْ اختص فهو مماسّـة، وإنْ كان مشتركاً فهو اتصال. وكذلك هذه الأجزاء إذا اجتمعت لـم يخلُ اجتماعها من أحد هذه الوجوه؛ فـإنْ اجتمعت على التالي فقط لم تحدث منها الأجسام المتصلة في الحس - وكلامنا فيها - وإنْ اجتمعت على اتصالٍ أو تماسّ، فكل واحد منها ينقسـم إلى مشغولٍ وفارغ وممسوسٍ وخالٍ، على نحو ما شرحنا في الفصول السـالفة. ويجب، إن لم يتداخل، أنْ يكون إذا لقي واحدٌ منها واحداً فجاء ثالث ملاقٍ لأحدهما، أنْ يكون محجوباً عن ملاقيه الآخر بتوسـط هذا الملاقي؛ فيكون كلٌ قد نال بالملاقاة من ذاته ما لم ينله الآخر - وهذا بيّنٌ بنفسـه - فيكون المتوسـط منقسماً. وإنْ كانت الملاقاة بالأسـر كانت مداخلة، فلا يـزداد باجتماعها قدر؛ فتكون كلما اجتمعت كالواحـد الذي لا طول له ولا عرض ولا عمـق. فإذ كانت هذه الأجزاء التي لا تتجزأ لا

their aggregation forms a body, the body is not reducible to them. Thus, the division of bodies does not terminate at parts that cannot be divided by any type of division, and the same holds for all other magnitudes (I mean surfaces and lines).

(4) Also, what sane person would allow us to say that a sheet of atoms that the Sun illuminates on one side (or any other state that happens to it on one side) must be such that the other side is in that state as well? Or [who would allow us] to say that the sheet in itself does not have two sides, and, rather [that] the light falls on one side of the sheet and the side that does not face the Sun is that very same side? So [for example] when one sees *this* side, one has already seen *that* side, since this one and that one are the same when there is no *this* and *that*. In that case, anyone standing on one side of the sheet sees the sheet illuminated from the other side.

(5) In fact, the existence of atoms would necessarily entail that there be no circles, right triangles, or many other [geometrical] figures. [This follows in the first case,] since the circle requires that the outside circumference be larger than any inside circumference that is contiguous with it; but what is contiguous is equal to that with which it is contiguous, not

تجتمع اجتماعاً يتألف به منها جسم فالجسم إذن غير منتقص إليها ، فإذن ليس تنتهي قسمة الأجسام إلى أجزاء لا يمكن أنْ تقسم نوعاً من القسمة ، وكذلك سائر المقادير ، أعني السطوح والخطوط.

(٤) وأي عاقل يرخّص في أنْ نقول إنَّ صفيحة من أجزاء لا تتجزأ أضاءت عليها الشمس ، أو عرض لها حالٌ من جهة ، فيجب أنْ تكون الجهة الأخرى بتلك الحال؟ أو يقول إنَّ الصفيحة ليس لها في نفسها وجهان؛ بل الضوء على كل ما هو وجهٌ للصفيحة ، والوجه الذي لا يلي الشمس هو ذلك الوجه بعينه ، فإنّه إذا أبصر هذا الوجه فقد أبصر ذلك؛ إذ هذا وذلك واحدٌ ، وليس ها هنا هذا وذاك ، فيكون الواقف من جهةٍ من الصفيحة يرى الصفيحة مضيئة من الجهة الأخرى؟

(٥) وقد يجب ، من وجود الأجزاء التي لا تتجزأ ، أنْ لا تكون دائرة ولا مثلث قائم الزاوية ، ولا كثير من الأشكال ، إذ الدائرة توجب أنْ يكون الطوق الخارج أكبر من طوقٍ

larger.[8] [In the second case,] when two sides of a right triangle are each ten units, then the hypotenuse is the square root of two hundred, which [according to the present view] would either be an absurdity that does not exist, or it is true, but parts would be broken up, which [according to the present view] they are not.[9]

---

8. Avicenna's point becomes obvious as soon as one images composing a circle (or any arc) solely from tightly fitting squares (which does seem to be the way that *kalām* Atomists envisioned their atoms)—that is, joining the squares such that neither do the bottom corners overlap nor is there any gap between the upper corners (see diagrams). In the case of the overlap, the indivisible atom would be divided, while in the case of the gaps there would be increments smaller than the smallest measure, both of which the Atomists deny.

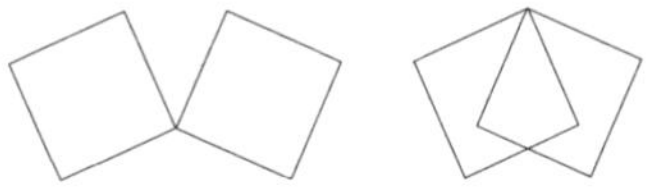

9. Avicenna's argument is that, if atoms were indivisible, the Pythagorean theorem, $A^2 + B^2 = C^2$, could not even be approximately true. According to the *mutakallimūn*'s own description of their atoms, they are cuboidal. So, construct a three-by-three piece of atomic space thus:

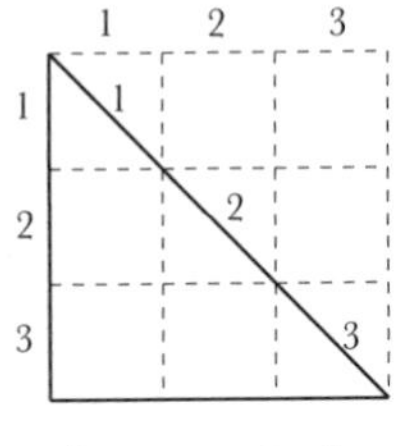

According to the Pythagorean theorem, the hypotenuse should equal $\sqrt{18}$; but according to the figure, the equation becomes $3^2 + 3^2 = 3^2$, or $9 + 9 = 9$, which is clearly false. It also does no good to complain that the hypotenuse of an atom is larger than the side, for whether the side or the hypotenuse of an individual atom is set at unit length 1—which by the Atomists' own assumption would be the smallest possible magnitude—there will be a magnitude smaller than 1, and so there will be a magnitude smaller than the smallest magnitude, which is absurd. This result is obvious if the hypotenuse of the atom is set at unit length 1, since the side of the atom is smaller. Similarly, if the side is set at unit length 1, the hypotenuse's magnitude can be calculated using the Pythagorean theorem as $\sqrt{2}$, which is 1.4142…, and so once more there is a magnitude, 0.4142…, which is less than the indivisible and purportedly smallest possible unit 1.

داخلٍ يماسّــه، والمماسّ مساوٍ للمماس، والمساوي لا يكون أكبر. والمثلث القائم الزاوية؛ إذا كان كل واحدٍ من ضلعيه عشرة عشرة كان وتر القائمة جذر مِئتين، وهو إمّا محالٌ لا يوجد، وإمّا صحاح وكسّر، وأجزاء لا تنكسر.

(6) They say, however, that vision errs with respect to the circle and right triangle, and these figures are, strictly speaking, made up of successively indented layers. They nonetheless do not deny the existence of a square, for example, having the following description. Let one construct a straight line from four atoms, as well as three other lines like it. Now, take one of the lines, AD, and let us superimpose it on another line EH such that there is no space whatsoever between them, and in like manner IL is after EH, and MP is after IL until a surface A[DM]P is produced, according to their doctrine. Now, it is commonly accepted [according to their doctrine] that there is no space left between these atoms in the surface to accommodate another atom, so four atoms—namely, the first, [A] (from line AD),[10] the second, [F] (from line EH), the third, [K] (from line IL), and the fourth, [P] (from line MP)—are the diagonal.[11]

10. **Y** has inadvertently omitted a number of lines found in **Z**, **T**, and the Latin, which read *La-nuṭabbiq bihi ʿalá khaṭṭ* jd, *wa-ḥattá lā yakuna baynhumā saʿah shayʾ wa-kadhālika* hz *baʿd* jd *wa-*ḥṭ *baʿd* hz *ḥattá yaḥdutha saṭḥ* aṭ *ʿalá madhabihum. Fa-maʿlūm annahu laysa yasaʿu bayn hādha al-ajzāʾ fī al-saṭḥ juzʾ ākhar al-battatah, fa-l-ajzāʾ al-arbaʿah allatī hiya al-awal min khaṭṭ* ab…, corresponding with the translated lines "and let us superimpose it on another line EH such that there is no space whatsoever between them, and in like manner IL is after EH, and MP is after IL until a surface A[DM]P is produced, according to their doctrine. Now, it is commonly accepted [according to their doctrine] that there is no space left between these atoms in the surface to accommodate another atom, so four atoms—namely, the first, A (from line AD)…."

11. Reading *huwa quṭr,* which **Y** secludes.

(٦) لكنهـم يقولون إنَّ البصـر يُخطيء في أمر الدائرة والمثلث، وإنّما هي أشـكال مضرّسـة، ومع ذلك فإنّهـم لا يدفعون وجود المربع القائم الزوايـا مثلاً على هذه الصفة. ليركّب من أربعة أجزاء لا تتجزأ؛ خط على الاسـتقامة وليركّب لمثله منه خطوط ثلاثة غيره، وليؤخذ منها خط مثل خط (آ ب) ولنطبق به على خطّ (ج د)، وحتى لا يكون بينهما سـعة شـيء وكذلك (ه ز) بعد (ج د) و(ح ط) بعد (ه ز) حتى يحدث سطح (ا ط) على مذهبهم. فمعلوم أنه ليس يسـع بين هذا الاجزاء في السطح جزء آخر البتة،

There are only two possibilities. One possibility is that these atoms must be touching one another along [a line] projected between atoms A and P, in which case there is a straight line composed from them—namely, the diagonal—but it will be equal to the two equal sides. This, however, is far from acceptable, for it is known from observations that the diagonal in a case like this is longer than the side. The other possibility is that these atoms must be separated from one another. In this case, there is either an empty space between them or not. If there is an empty space between them, then the lines were not superimposed on one another with no empty space between them; but that is what was done. This is a contradiction. If there is no empty space between them, then there must be something between them, whether an atom or more or less than [an atom]. If it is less than an atom, then the atom has been divided. If it is an entire atom or two atoms, it will always follow that the length of the diagonal either does not fall short of the two sides together, or it falls short of the combined length of the two sides by a single imperceptible atom.[12] The diagonal, however, always falls short of the combined length of the two sides, and [the length by which it falls short] is perceptible and a significant magnitude.

12. Avicenna's point here is that the Atomists cannot explain the well-known property of the triangle that the combined length of any two sides of a triangle always exceeds that of the third side. So, if more than a single atom were inserted between each of the squares of the diagonal, the hypotenuse would exceed the combined length of the two sides. If a single extra atom were inserted between each of the squares of the diagonal, the length of the two sides would exceed that of the hypotenuse by only a single atom; but, of course, a right or obtuse triangle puts the lie to this suggestion.

فالاجــزاء الأربعــة التي هي الأول من خطّ (آ ب) والثانــي من خط (ج د) والثالث من خــط (هــ ز) والرابع من خط (ح ط) ، لا يخلو إمّا أنْ تكون هذه الأجزاء يماسّ بعضها بعضاً على ســمتٍ بين جزء (آ) وجزء (ط) ، فيكون خطاً مســتقيماً مؤلفاً منها وهو القطر ، ويكون مســاوياً للضلعين المتســاويين ، وهذا بعيد عن الجواز . فمعلوم بالمشاهدات أنَّ القطــر فــي مثل هذا أطول من الضلع . وإمّا أنْ تكون هذه الأجزاء متباينة ؛ فحينئذ إمّا أنْ يكون بينها فُرج أو لا يكون ، فإنْ كان بينها فُرج فلم تطبق الخطوط تطبيقاً لا فُرج فيه ، وقد فعل ذلك ، هذا خُلْف . وإنْ لم يكن بينها فرجة ، فيكون فيما بينها شيء لا محالة إمّا جزء أو فوقه أو بعض جزء . فإنْ كان بعض جزء فقد قُسّــم الجزء ، وإنْ كان جزء بتمامه أو جــزءآن لزم دائماً أنْ يكون طول القطر إمّا أنْ لا ينقص عن الضلعين معاً ، أو ينقص عن ضعف الضلعين بجزءٍ واحدٍ غير محسوس ، ونقصان القطر عن ضعف الضلعين دائماً فهو أمرٌ محسوس وقدر كبير .

(7) They claim, however, that this line is not straight, but indented according to a form like such: ~o~o~o~o.[13] I mean that there is one atom and another atom diverging from it in some direction, then another atom along the projected path of the first, followed by another diverging atom (as if each diverging atom is placed in the common [space] separating two atoms arranged along a single projected path). It will soon become obvious that [this view] is faulty and futile. That is because the two squares that are arranged along the single projected path are either contiguous or not. On the one hand, if they are contiguous, then all the atoms arranged along the single projected path encounter one another such that a continuous straight line is formed from them. So, from the two arrangements, there would be two straight lines, one of

13. This figure is the one that appears in **Z** and **T** and is here reproduced on the Arabic side; however, there are at least two other different depictions of the situation that Avicenna is trying to represent in the manuscripts: these (represented below) are found in the Bodleian manuscript, which appears in **Y**'s text, and in the margins of **T**. None of the three seems to be particularly helpful.

Bodleian

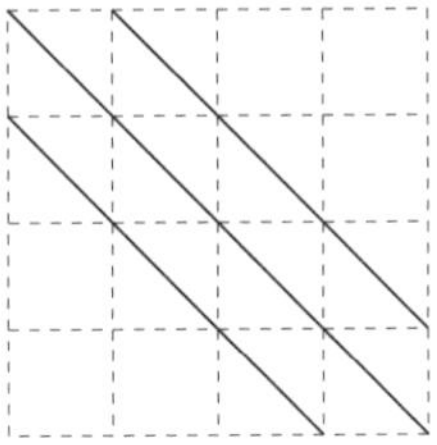

Tehran ms marginalia

(٧) وأمّا ما يقولون إنَّ هذا الخط لا يكون مستقيماً بل مضرّساً على صورةٍ هكذا : ~o~o~o~o . أعني أنْ يكون. جزء وجزء آخر منحرف عنه إلى جهةٍ، ثم جزء آخر في سـمت الأول، ثم جزء ينحرف عنه، وكأنَّ كل منحرف عنه موضوع في الفصل المشترك بين المرتبتين في سمتٍ واحدٍ - فإنّه يظهر فساده وبطلانه عن قريب. وذلك لأنَّه إمّا أنْ يكون المرتّبان في سـمتٍ واحدٍ متماسّين أو غير متماسين، فإنْ كانا يتماسان فكل أجـزاء مرتبة في سـمتٍ واحدٍ متلاقية، بحيث يتصل منها خطٌ مسـتقيم، فيكون من الترتيبين خطان مسـتقيمان موضوعٌ أحدهمـا بجنب الآخر؛ فلا تضريس. وإنْ كانا غير

which is placed next to another. In that case, however, there is no indentation.[14] If, on the other hand, they are not contiguous, then there must be an atom between them, in which case [the line] is not indented, but continuous along a straight line.[15] It would be indented only if part of the atom occurs in that which is in between the two and part of it outside. The atom, however, does not have parts; and, instead, only an entire [atom]

14. The present argument and the ones that follow are based upon different ways by which one might construct the Atomists' indented diagonal. In the present case, each atom must be contiguous with the next, where the series of "projecting" and "diverging" atoms might be presented thus:

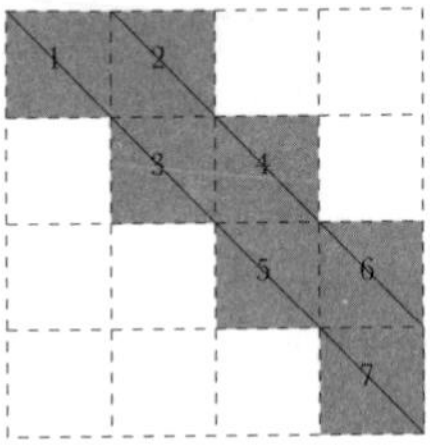

So atom 2 represents the atom diverging from the first atom (1); 3 is the atom along the projection of the first, and so on; however, in this case there is not a single indented line, but two straight lines along the two projections "1, 3, 5, 7" and "2, 4, 6." Thus, the intended indented line is not formed.

15. In this case, the atoms in the series are not contiguous and so might be presented thus:

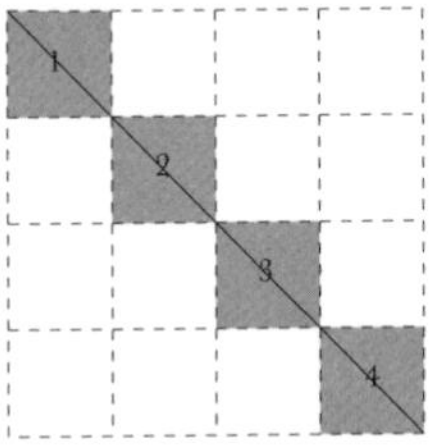

While there is only a single line, it will not be indented in such a way as to avoid the initial objection (presented in paragraph 6) that the Atomist cannot even approximate the Pythagorean theorem.

متماســـين فلا محالة أنْ يكون بينهما جزء ، فلا يكون مضرّساً بل يتصل على الاستقامة . إنّمــا كان يكون مضرّســاً لو كان بعض الجزء واقعاً فيما بينهمــا وبعضه خارجاً والجزء لا يتبعّض ، بل إنّما يكون كله بينهما أو لا شـــيء منه بينهما . وإذا كان كله بينهما فليس هناك

would be between the two or nothing at all.[16] When an entire [atom] is between the two, there will not be anything there apart [from what is in between the two], but neither will there be any indentation; and, instead, it will fall along a straight line in just the way that they agree that atoms are capable of being ordered into a straight line. If they think that in relation to what is between the parts arranged along a single projected path, there are two atoms not in the order of the projected path, but along the width so as to be indented, then each atom would have something in relation to what is between the two atoms and something apart from it; but then they would have made each atom divisible.

(8) What they say about the straight line works, which, again, is that [the line's] first atom is laid out at the point A, with the line being made to coincide with [a line on] the surface [ADMP] so that it meets point P. Now, it is known that this is possible; and so, then, it is clearly possible to construct a straight line between the two atoms, from which it necessarily follows that some direction belongs to the atom different from those that are possible for [the Atomists].[17] Now, when it is, in fact,

16. In this case, we are to imagine that, instead of being separated by a complete atom, the diagonal is formed by slightly offsetting the atoms, thus:

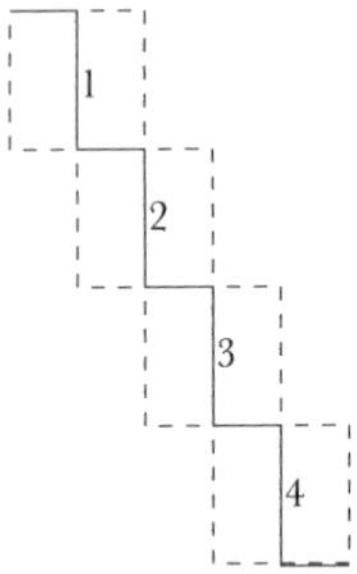

Such a scenario only works if the atoms are divisible or have parts. So, for example, the first atom (1) can be divided into that part that does *not* overlap 2 and that part that *does* overlap 2.

17. The atoms of the Muslim Atomists were cubical. Moreover, they believed that there were only six directions—up, down, front, back, left and right—corresponding with the six surfaces of these atoms. Avicenna's point here, which he argues for in depth at 3.13, is that there are more than just six directions. Thus, as in the present case, there also would be the directions defined by the two extremities of a diagonal running through one of the Atomists' cubic atoms. See Alnoor Dhanani, *The Physical Theory of Kalām*, 113–117, who argues that the atoms of Muslim Atomists are cubical in shape and so have six sides or directions (*jihah*).

زوال وتضريس البتة، بل يكون هناك وضْعٌ على الاستقامة، كما هو مقبولٌ عندهم أنَّ من شأن الأجزاء أنْ ينتظم منها خط مستقيم. فإنْ جعلوا فيما بين المُرتبين على سمتٍ واحدٍ جزءين لا في نظام السمت بل عرضاً حتى يتضرّس، فقد جعلوا من أمر كل جزء شيئاً فيما بين الجزئين؛ وشيئاً زائلاً حتى يكون تضريس، فجعلوا كل جزءٍ منقسماً.

(٨) وما يقولون في خطٍ مستقيم يُعمل ثم يُركّب جزؤه الأول على نقطة (آ) ويطبق على السطح حتى يلقى الخط نقطة (ط)، ومعلوم أنَّ هذا ممكن. فبيّن إذن أنّه يمكن أنْ ينْظمّ بين الجزئين خط مستقيم، فيلزم من ذلك وجود جهةٍ للجزء غير الجهات التي لهما. وإذا صحّ أنّه يمكن أنْ يُفرض بين الجزئين خط مستقيم، على أي وضعٍ كان الجزءان،

possible to posit a straight line between two atoms, whatever the placement of the atoms, we can place two atoms on top of the two atoms A and P, with nothing between the two [atoms placed on top themselves], and construct a line between them, superimposing it on the diagonal. So how are the points that follow the first one that is placed on the point A going to be? Will [the second point of line that they form] fall on the second point of the diagonal, which is the second point [F] of the line EH?[18] Or will it fall at the common space separating [the two] so that it is contiguous with both of them, such that they parallel some common space of separation and do not encounter one another? [In that case,] either that empty space is smaller than the space that [an atom] occupies (in which case there is something smaller than the size of that which cannot be divided), or it is [equal to] the space of it (in which case the contradiction[19] that we previously mentioned arises).[20] If [the second point] falls on itself [that is, on point F], the straight line corresponds with the [original] diagonal [AFKP], and what corresponds with the straight line is straight so as to equal it.

(9) It is simply amazing what they are forced into in this situation—namely, that it is possible for an atom to occur on top of the common division between two atoms and that one and the same [atom] can move a little so that it coincides with only one of them. On the one hand, if that with which it coincides when it is contiguous with the first and the second is the very same thing as that with which it coincides when it is contiguous with only the second, then it is still touching [the first] when it

18. The Arabic text's *ḥ* needs to be corrected to a *j* (rendered as *E* in the translation) in order to conform to Avicenna's earlier lettering in par. 6.

19. The "contradiction" in paragraph 6 is that spaces will exist between the atoms when it was posited that there are no spaces between them. That contradiction does not seem to apply in the present case; the problem here seems to be the next issue mentioned in par. 6, that the length of the diagonal will exceed the combined length of the two sides, in which case there could not be a triangle.

20. **Y** has (inadvertently) omitted the line *fa-yakūnu shayʾ aṣghar min ḥajm mā lā yatajazzaʾu au yakunu yasaʿuhu,* which occurs in **Z** and **T** but is omitted in the Latin. It corresponds with "(in which case there is something smaller than the size of that which cannot be divided), or it is [equal to] the space of it in the translation."

فيمكننا أنْ نضع على جزئي (آ) و(ط) جزئين ولا يكون بينهما شيء ، ونُنظم بينهما خطاً ونطبقه على خط القطر ، فكيف يكون حال النقط التي تتلو النقطة الأولى الموضوعة على نقطة (آ) ؟ أتقع على النقطة الثانية من القطر الذي هو النقطة الثانية من خط (ج د) ، أو تقع في الفصل المشترك حتى يماسّ كلاهما محاذياً لفصلٍ مشتركٍ بينهما لا يتلاقيان عليه؟ أو تكون تلك الفرجة أصغر من أنْ تسعه فيكون شيء أصغر من حجم ما لا يتجزأء أو يكون يسعه فيكون ما قدمنا من الخُلْف. وإنْ وقع عليه نفسه فقد انطبق الخط المستقيم على القطر ، وما طابق المستقيم فهو مستقيم مساوٍ له.

(٩) ومن العجيب ما يضطرون إليه في هذا الموضع ؛ من إمكان وقوع جزء على فصلٍ مشتركٍ بين جزئين ، وهو بعينه يمكنه أنْ يتحرك قليلاً حتى يلقى أحدهما وحده ، فإنْ كان الذي يلقاه وهو يماسّ الأول والثاني هو الذي كان يلقاه بعينه وهو يماسّ الثاني وحده ، فيكون عند التنحي مماسّاً أيضاً ؛ هذا خُلْف. وإنْ كان يلقى منه غير الذي لقيَ أولاً ، فيكون قد

moved away from [it], which is a contradiction.[21] If, on the other hand, it coincides with something different from that with which it originally coincided, then it would have been divided at the [different] places where it coincided. Also that which they tried to avoid necessarily follows for them—namely, the directions exceed six, which they erroneously thought must be six but which is not at all necessary.[22] That is only recognized with respect to the division of animals' [anatomical] directions, but then it was thought that that recognized division is some necessary pronouncement for everything. The truth is that, between any two atoms sharing a common border, there is some other direction and potentially infinitely many so. This is also like the [erroneous] belief of some of them that the body has a determinate length, breadth, and depth, all of which are actual and each one of which has two limits; and so the directions are not more than six. (We ourselves shall discuss this soon;[23] and, in fact, one should know that, while the account that there are six directions is widespread and commonly accepted, it is neither true nor demonstrated.) They contradict themselves in saying that squares are the configuration of these four parts from which a large square is composed [as, for example, the squares making up the large square ADMP in figure 1], and the squares along the diagonal [AFKP] do not encounter one another at their limits that are the line nor at anything between them. Now, they are wrong about that to an unusual degree, and that is because [the squares] encounter each other at points—that is, the limit of the limit[24] is where they truly meet—not at lines, whereas between these lines there are other halves of squares with which they coincide, since the squares are divisible and so fill the empty space, which is not the case for atoms.

---

21. The situation that Avicenna is picturing is something like the following diagram:

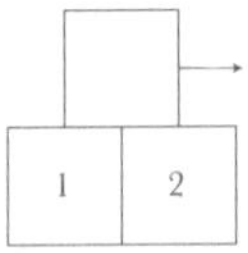

22. See 3.13.3.

23. See 3.13.3–6.

24. In other words, the line is the limit of the square, whereas a point is the limit of the line; and it is at the point defining the corner of the square where the squares of the diagonal meet.

انقســم بمواضع اللقاء ، ويلزمهم ما يهربون منه من أمر زيادات الجهات على الســت ؛ التي يظنون أنّها واجب أنْ تكون ستّاً ، وليس ذلك بواجب البتة . إنّما ذلك أمر قد تعورف في قســمة جهات الحيوان ، فظنّ أنَّ ذلك المتعارف - بحسب القسمة - حكمٌ واجبٌ في كل شيء ، بل الحق أنَّ بين كل جهتين متجاورتين جهة أخرى ، وذلك إلى غير النهاية بالقوة . وهذا أيضاً مثل ما يظنّ بعضهم أنَّ للجسم طولاً وعرضاً معيناً وعمقاً معيناً ، وانَّ كل ذلك بالفعل ، فيكون لكل واحدٍ منهما طرفان ؛ فتكون الجهات ستّاً ليس غيرها . ونحن سنقول في هذا عن قريب ، بل يجب أنْ يُعلم أنَّ القول بكون الجهات ســتّاً أمرٌ مشــهور متعارف ؛ ليس بحقٍ ولا عليه برهان . ولكنهم يناقضون ما قلناه بمربعاتٍ تكون على هيئة هذه الأجزاء الأربعــة ، يؤلّف منها مربعٌ كبير ، فتكون المربعات التي على القطر لا متلاقية بأطرافها التي هي الخطوط ، ولا بينها شــيء . وقد ضلّوا في ذلك ضلالاً بعيداً - وذلك لأنّها متلاقية بالنقط ، وطرف الطرف طرفٌ يصلح به اللقاء ، وغير متلاقية بالخطوط ، وبين تلك الخطوط أنصاف مربعات أخرى تملؤها ، إذ المربعات تنقســم فتســدّ الفُــرج ، ولا كذلك الأجزاء .

(10) One thing that is known with certainty, about which there is no doubt or difference of opinion, is that there is a certain projected path between any two things having some placement such that, when we make a straight line between them, it either fills that projected path

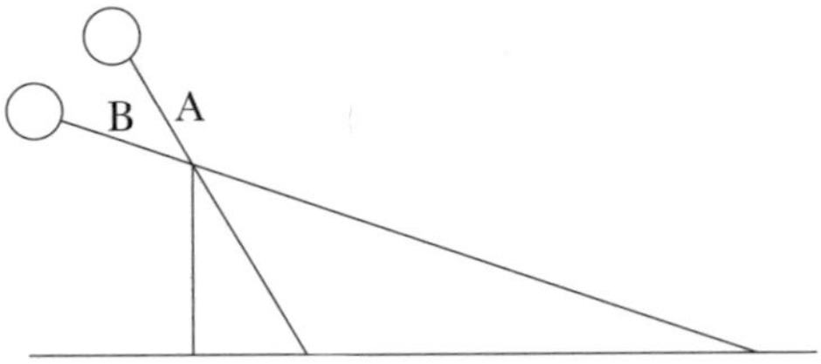

or occurs along it. Now, assume that there is a body, like the Sun, that undergoes a number of motions, and also assign some body opposite it, like the Earth, and erect there a vertical upright (all of which can, in fact, exist in the intellect). Next, assume that the Sun illuminates the Earth and the upright provides a shadow corresponding with the Sun's projected path. In that case, when the Sun moves away one atom, the projected path between the Sun and the upper limit of the upright either ceases [pointing toward the exact same] limit of the shadow [projected onto the Earth] or continues [pointing toward it]. If, on the one hand, it continues [to point toward the very same point on the Earth], then there undoubtedly continues to be a projected path, and the projected path has the status of a straight line. So that other one, issuing in a straight line from the Sun to the upper limit of the upright on the Earth, will also be a straight line, like the line designated A from the two lines A and B. There will then be two distinct straight lines coming together at a single point, after which they unite into one straight line such that that line is straight together with each one of them. So the common part—that is, that

(١٠) وممّا يُعلم يقيناً لا شك فيه ولا اختلاف؛ أنَّ بين كل شيئين ذوي وضْع سمْتاً؛ حتى أنّا إذا عملنا خطاً مستقيماً بينهما فإنّه يملأ ذلك السمت أو يقع في ذلك السمت. فإذا كان جسم كالشمس يتحرّك حركات كثيرة، وقد جُعل جسمٌ بإزائه كالأرض أيضاً، ونُصب هنالك شيء نصباً قائماً - وجميع ذلك صحيح جواز الوجود في العقل - ثم كانت الشمس مضيئة للأرض، وكان المنصوب يستر على قدْر سمْت الشمس، فإذا زالت الشمس جزءاً، فلا يخلو إمّا أنْ يزول السمْت الذي بين الشمس وبين طرف المنتصب عن طرف الظّل، أو يبقى. فإنْ بقي بقي لا محالة سمتاً، والسمّت على حكم خطٍ مستقيم، فيكون ذلك الآخر، المخرج من الشمس على الاستقامة إلى طرف المنتصب على الأرض، أيضاً خطّاً مستقيماً كالخط الذي عليه علامة (ب) من خطي (أ ب) فيكون خطّان مستقيمان متباينان يجتمعان عند نقطةٍ ويتحدان بعد ذلك خطاً مستقيماً، حتى يكون

which is between the two limits of the upright and the point on the Earth—is together with each one of the two projected paths connecting the Sun and the upper limit of the rod as a single straight line. This is known to be impossible. Additionally, they have made the Sun stand opposite a single atom (that is, the one at the upper limit of the rod) from two directions, one of which goes beyond what their [theory] can project. If, on the other hand, the projected path does not remain the same and, instead, ceases [pointing toward the exact same point on the Earth], it moves away either by an atom, or more than an atom, or less. If [the projected path] ceases [pointing toward the exact same point on Earth] by either an atom or more, then the amount of motion of the Sun in the Heavens will equal the amount of motion of the extreme limit of the projected path or the extreme limit of the projected path will cover a larger distance, both of which are manifestly absurd, whereas if it is less than an atom, the atom has been divided.

(11) The same holds when we submit a straight line—like the hypotenuse along a right angle, one of whose sides is shorter—to punishment. So [for example] we yank on the extreme limit of that line—one of its extreme limits being against a wall and the other on the ground (where the height of the wall is less than the space between the [right] angle and the ground)—so that [the line] is stretched out,[25] letting [the part that is stretched out] be cut off.[26] In that case, when we yank on the extreme limit that is on the ground so that this line is stretched out one atom, and then[27] the other extremity is yanked on so as to be stretched

25. The verb *jarrara* is the intensive form of *jarra*. To capture its intensive nature, I have slightly overtranslated it as "to yank so as to stretch out."

26. The text's *jad*ᶜ, which most of the MSS have, literally means "what is cut off [by an act of mutilation] of the anterior parts of the nose, to its further, or utmost part" (see Lane *s.v.* J-D-ᶜ). One of the MSS read *jidh*ᶜ (tree trunk), which is followed by the Latin (*lignum*) and which Paul Lettinck also follows ("Ibn Sīnā on Atomism," 37). While "tree trunk" might appear to make more immediate sense, I believe that *jad*ᶜ, which is clearly the *lectio difficilior,* is the intended sense, as I hope will become clear below.

27. Reading *fa* with **Z**, **T**, the Latin (*si igitur*), and two of the earlier MSS consulted by **Y**, for **Y**'s preferred *wa* (and).

ذلك الخط مستقيماً مع كل واحدٍ منهما. فيكون الجزء المشترك - وهو الذي بين طرفي المنتصب ونقطةٍ على الأرض، هو مع كل واحدٍ من السمتين المتصلين بين الشمس وبين طرف المقياس - خطاً واحداً مستقيماً، وهذا معلوم الاستحالة. ومع ذلك فقد جعلوا جزءاً واحداً، وهو طرف المقياس، توازيه الشمس من جهتين، إحداهما خارجة عن السمْت الذي لهم. فإنْ لم يثبت السمْت بل زال، فإمّا أنْ يزول جزء أو أكثر من جزءٍ أو أقل. فإنْ زال جزء أو أكثر فتكون حركة الشمس في السماء مساوية لحركة طرف السمْت ومسافتاهما متساويتان، أو يكون طرف السمْت بقطعٍ أكثر، وجميع هذا ظاهر الإحالة. وإنْ كان أقل من جزء فقد انقسم الجزء.

(١١) وكذلك إذا أوقعنا خطّاً مستقيماً كالوتر على زاوية قائمة، أحد ضلعيها أقصر، فجررنا طرف ذلك الخط - وليكن جِذعاً - أحد طرفيه على حائط والآخر على الأرض، وارتفاع الحائط أقصر من البُعْد بين الزاوية والأرض. فإذا جررنا هذا الخط من الطرف الذي على الأرض جزءاً؛ فكان ينجّر الطرف الآخر جزءاً؛ وجب من ذلك أنْ

out one atom, it should follow that what is cut off the two sides would be the same. That, however, is not the case; and, instead, less is cut off from the smaller side.[28] Whatever follows as a result of fragmentation or loss of continuity [among the atoms] is nothing but a difference in the iron and adamant used and the wood used; but the two intervals will be equal in all [cases]. Even then, should we start again [by] placing a straight line, corresponding with the length of the stretched body, at the extreme limit to which the stretched body was brought, it would come to precisely where the stretched body comes.

(12) The same thing holds for the atoms at the edge of the millstone, and what we related about the group—[again the one] who were forced to affirm fragmentation and the fragmented thing—does necessarily follow on their [view, namely, the view of the Atomists].[29] [That outcome follows] whether the fragmentation of [the body] occurs inasmuch as its atoms are separated from each other by having gaps [between them] (in which case the measurement of the millstone increases during the motion), or inasmuch as the atoms separate from one another in such a way that there is a mutual replacement of the places, so that the whole retains its size (in which case they cease being distinct and the atoms do not retain the positions that they had vis-à-vis one another in the millstone).

28. In other words, assuming an initial right triangle ABC, where AC is the hypotenuse and AB is less than BC, we are asked to calculate how much longer a new hypotenuse AC′ (i.e., AC + x) would be for (BC + 1) and then again for (AB + 1). Since, on the atomic theory, AC is being stretched only one atom along the ground or one atom along the wall, then *x*—the difference between AC and AC′—should also be only one atom, according to the theory. In fact, however, *x* is smaller when one atom is added to AB than when one atom is added to BC. While the formal proof for this claim is complex, a concrete example will give one a feel for the argument. So, for example, start with a triangle AB = 3, BC = 4, AC = 5, and then calculate AC′ (i.e., (AC + x) when AB is increased by one atom to equal 4, while BC remains the same. Then calculate AC′ when BC is increased by one atom to equal 5, while AB remains the same. When AB is lengthened by one atom, the difference between AC and AC′, is 0.6568542494492, while it is 0.830951894845 when BC is lengthened by one atom. So, as Avicenna notes, what is cut off from the hypotenuse when it is extended toward the shorter side of the triangle is less than what is cut off from it when it is extended toward the longer side. A variation of this argument appears in al-Ījī (see Sabra, "*Kalām* Atomism as an Alternative Philosophy to Hellenizing *Falsafa*," 270–71, "Argument 5.4(3)."

29. For the millstone example, see 3.3.15–16.

يكون ما يُقطع من الجانبين سـواء ، وليس كذلك ، بل يكون القطع من الجانب الأقصر أقل . وليـس ذلك ممّا يتبع تفكّكاً أو يفـرّق الاتصال البتة ، وإلاّ لا ختلف في المعمول من الحديد والألماس والمعمول من الخشب ، بل يستوي البُعْدان في الجميع . وإلاّ لو استأنفنا وضع خطٍ مسـتقيم على الطرف الذي نزل إليه المجرور بقدر طول المجرور ، كان لا يقع إلا حيث يقع عليه المجرور .

(١٢) وكذلك الأجزاء التي في طرف الطاحونة ويلزمهم ما حكيناه عن قوم أحوجوا هـؤلاء إلى القـول بالتفكّك والمتفكّك ؛ إمّا أنْ يكون تفكّكه بـأنْ تتزايل أجزاؤه بفُرجٍ تقع لها ، فيجب أن تزداد مساحة الرحى عند الحركة ، أو بأنْ تتزايل الأجزاء على نحو تبادل الأمكنـة ، حتى تبقى الجملة على حجمها ، فتكون العلامات تزول ولا تبقى الأجزاء التي في الرحى على الأوضاع التي كانت لبعضها عند بعض .

(13) This is a response to those who maintain [that] fragmentation [occurs] and [that] slow motions are interspersed with rests. What do they say about the galloping horse? Do we have any doubt that its motions are greater than its rests? If the number of rests were greater, the slowness and slowing down would be clearly obvious; and, even if they deny this about the horse, they cannot with respect to the shot arrow. In addition, the ratio of the galloping horse and flying arrow to the procession of the Sun is not one that can be squeezed out by increasing the rests [during] the galloping and flight. That is because, if the motions were equal to the number of rests in the galloping and flight, while the Sun has no [rests], but only motions, the galloping horse and flying arrow would [still] be moving at half the speed of the Sun, but that is not the case at all. The fact is that this is not even comparable to that [actual ratio]. The galloping horse is known [to move slower than half the speed of the Sun] from immediate observation. As for the flight of the arrow, it has been tested among a group [of archers] who stand at certain distances from given targets, where each one shoots his arrow exactly when the arrow of his companion hits nearby, and so, in that, the dissimilarity [between the motions of a flying arrow and the Sun] was learned. Were one to consider this ratio closely, one would find that it is less than the ratio of one part of one thousand parts of [the Sun's speed]. From that, it necessarily follows that the horse and the arrow would undergo thousands of rests [for] each single motion. One would simply not be able to see their motion, which would not be apparent because the resting would overwhelm it! Even if some [motion] did appear, it would be minuscule. Reality is different from this, for motion does appear without any apparent rest.

(١٣) هـذا جواب مَنْ يقول بالتفكّك وتخلّل السكونات الحركات البطيئة. وماذا يقولون في فرسٍ شديد العدو؛ هل نشك في أنّ حركاته أكثر من سكناته؟ ولو كانت السكنات أكثر لكان البُطْء أظهر والفتور أوضح. وانْ أنكروا هذا في الفرس، لم يمكنهم أنْ ينكروه في السهم المرمي. ومع ذلك فإنَّ نسبة حركة عدو الفرس وارتماء السهم إلى سير الشمس، ليست نسبة تقتضيها زيادة حركات العدو والارتماء على السكنات. وذلك لأنَّه لو كانت الحركات في الركض والارتماء مساوية للسكنات، وكانت الشمس ليس لها إلاّ الحركات فقط؛ لكان ركض الفرس أو ارتماء السهم نصف سير الشمس، وليس الأمر كذلك، بل لا قياس لهذا إلى ذلك. أمّا ركض الفرس فمعلومٌ بالمشاهدة، وأمّا ارتماء السهم فقد جُرّب في قوم يقفون على أبعاد مرامي سهامٍ، وكل واحدٍ يرسل سهمه مع سقوط سهم صاحبه بالقرب منه، فعُرف التفاوت في ذلك. ولو استقصى المستقصي في هذه النسبة، لوجدها أقل من نسبة جزءٍ من أُلوف أجزاء منه. فيجب من ذلك أنْ يكون الفرس أو السهم يسكن أُلوف سكناتٍ ويتحرك واحدة، وكان يجب أنْ لا تُرى حركته ولا تظهر لغلبة السكون عليها، وإنْ ظهر منها شيءٌ كان قليلاً يسيراً. والوجود بخلاف هذا؛ فإنَّ الحركة هي الظاهرة والسكون لا ظهور له البتة.

(14) One thing that will clear this up is what we know about what is heavy[30]—namely, whenever there is an increase in heaviness, the motion downward is faster. So assume some heavy object is undergoing a downward motion with certain pauses intermingled with it. If we then perpetually increase the magnitude of the body so as to require that the heaviness increase, we shall, in that way, at some moment reach a motion with which no rest is intermingled. In that case, when we double that body, [that larger body] would necessarily move faster without the occurrence of any interspersed rest to account for the slowness [of the initial body's motion]. The same holds if we were to assume that a single atom were undergoing motion without pause, and then [we] joined something heavy to it. Here is something that is amazing. Grant that something is moved through calm air or a void (which they acknowledge). In other words, [it is moved through] something in which there is no opposition. Also, grant that the principle of [the mobile's] motion is a certain inherent inclination or tendency toward some direction. [It is just amazing, then,] that that tendency and inclination should remain the same throughout the distance [covered], while the motion does not; and, in fact, a rest occurs during slow motion, as if the inclination grows tired of working and so chooses to take a rest and then, feeling refreshed, gets back at it. Just how does some obstructing and neutralizing cause come to be in the calm air or a void? Also, in what way can it be said that the inclination and tendency in it are used up and renewed?

30. For a brief discussion of the role of heaviness and speed, see 2.8.11.

(١٤) ومِمّـا يوضّح هذا مـا نعلمه من أنَّ الثقيل، كلما ازداد ثقلاً، كانت حركته إلى أسفل أسرع. فإذا كان ثقيل ما يتحرك إلى أسفل تخالطه وقفات، فإنْ زدنا مقدار الجسم دائماً نطلب زيادة الثقل، بلغنا بذلك وقتاً مّا إلى حركةٍ لا يخالطها سكون. فإذا ضممنا إليها ضعف ذلك الجسم لزم أنْ يتحرك أسرع، من غير تخلّل سكونٍ يكون سبباً للإبطاء. وكذلك لو فرضنا جزءاً واحداً يتحرك الحركة التي لا وقوف لها؛ ثم اتصل به ثِقْل. ومن العجائب أنَّه إذا تحرّك المتحرّك في هواء راكد أو في خلاء مقرون به، وهو ممّا لا مقاوم فيه، ويكون مبـدأ حركته ميلٌ فيه، أو اعتمادٌ إلى جهـة، أنْ يبقى ذلك الاعتماد وذلك الميل في تلك المسافة بعينها ولا تبقى الحركة، بل يحدث سكونٌ يقع فيه البُطء، كأنه يعرض كَسَل متعب، فيميل بالاختيار إلى السـكون ثم 〈يؤوب〉 إليه النشاط. وكيف يحدث سبب يمنع ويبطِل فـي هواءٍ راكدٍ أو خلاء وكيف يمكن أنْ يقال أنَّ الميل والاعتماد يبطلان فيه ويتجددان؟

(15) One of the repugnant consequences of the atom is the following: We know with absolute certainty that, when a moving object is moved from right to left and another is moved from left to right along two parallel, straight lines, the two keep getting closer to one another until they meet opposite each other and then depart from one another. So let

us posit one set of four atoms and another set of four, and construct from each set of the four a line and place one of the lines next to the other (just as we did for the square built of atoms). Now, we posit an atom on the right extreme of one of them and an atom on the left extreme of the other. We set the two atoms in motion until the one that was on the right arrives at its other extreme and the one that was on the left arrives at its other extreme, also imagining that they move at the same speed. In that case, the two will [at some point] be opposite one another, and then they will depart from one another. Now, the two must be opposite one another either along the [first] half or after the half. If the opposition occurs precisely when the latter is at the second atom from the extreme from which it is moved and the former is at the second atom from the extreme from which it is moved, then the two are not yet opposite, because the second atom from both of [the points of departure] is the third from the other, according to what was laid down.

(١٥) ومن الشناعات التي تلزم الجزء أنّا نعلم يقيناً لا نشك فيه أنّه إذا تحرّك متحرّك من اليمين إلى اليسار، ومتحرّك آخر من اليسار إلى اليمين على خطّين متوازيين مستقيمين؛ أنهما لا يزالان يتقاربان حتى يلتقيا متحاذيين، ثم يتفارقان. فإذا فرضنا أربعة أجزاء لا تتجزأ، وأربعة أخرى، وركّبنا من كل أربعةٍ خطّاً، وكان أحد الخطّين موضوعاً بجنب الآخر - كما فعلنا في المربع الذي أنشأنا من أجزاء لا تتجزأ - وفرضنا على طرف أحدهما الطرف الذي على اليمين جزءاً، وعلى طرف الآخر، الطرف الذي يلي اليسار جزءاً، وحركنا الجزئين حتى صار الجزء الذي على أحد الخطّين وعلى طرفه الأيمن نافذاً إلى طرفه الآخر، والجزء الذي على طرف الخط الآخر وعلى طرفه الأيسر نافذاً إلى طرفه الآخر، وتوهمنا أنَّ حركتيهما متساويتان، فتحاذيا وتفارقا؛ فلا يخلو إمّا أنْ يكون تحاذيهما على النصف أو بعد النصف. فإنْ كان التحاذي إنّما يقع إذا كان هذا على الثاني من الطرف الذي تحرّك عنه، وذلك على الثاني من الطرف الذي تحرّك عنه، فبعد لم يتحاذيا، لأنَّ محاذي الثاني من كل واحدٍ منهما هو الثالث من الآخر وما يوضع عليه. فإنْ

If they are opposite one another by each one being at the third atom, then the two departed from one another at the very moment of the opposition. If they are opposite when one is at the second atom on its line and the other is at the third atom of its line, then they were not moving at the same speed.[31]

(16) One of the things forced on them, which is obvious to anyone with an ounce of intelligence, is the well-known fact that, when two things are opposite one another [and] moving toward each other until they meet, and there is absolutely no external obstacle that prevents their meeting, then the two can simultaneously move until they encounter one another. When they encounter one another, they can obstruct one another; but before[32] that, there is no obstacle between them. This is something that is self-evident. So when our estimative faculty imagines three atoms forming a row and two atoms on top of the two extreme [atoms], and each one [of the atoms on top] moves without obstruction until it meets the other, they both will meet each other after not having done so.[33] In that case, their meeting each other occurs either when

31. The argument assumes atomic motion, that is, for example, when a mobile moves along AD in the diagram, at the beginning of the motion it completely occupies A, then in the next instant it completely occupies B, and so on for C and then D. In other words, the mobile cannot continuously move along the atomic line, since, then, when some mobile, *x*, which is itself an indivisible atom, starting at A, is halfway over E and F, both of which are also purportedly indivisible, an atom would be conceptually divided (for *x* would be *half* over E and *half* over F), which is a contradiction. Given the assumption of atomic motion, Avicenna has us construct two atomic lines AD and EH. He then has some mobile, *x*, starting at A move to the right until it reaches D, and another mobile, *y*, moving at the same speed as *x*, start at H and move to the left until it reaches E. Now, maintains Avicenna, *x* and *y* must pass by each other, that is, be opposite one another, at some point during their motion, and he wants to know when. At the first instant after moving, *x* is at B while *y* is at G, and so they are not passing by each other; however, at the next instant, *x* is at C, while *y* is at F, which means they have already passed by each other, but there will be no instant when they passed by each other.

32. Reading *qabl* with **Z**, **T**, and the Latin (*ante*) for **Y**'s (typographical error) *qīla* (it was said).

33. The situation Avicenna is describing is represented in the following diagram:

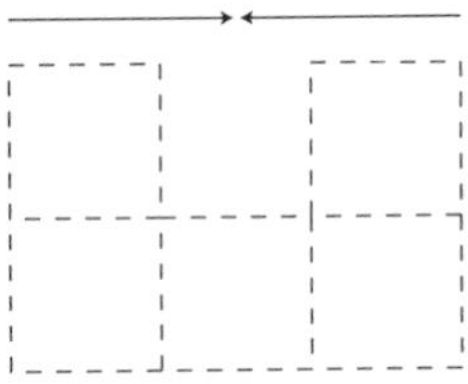

تحاذيا - بأنْ يكون كل واحدٍ منهما على الثالث - فهما في حال التحاذي متفارقان وإنْ تحاذيا ، وأحدهما على الثاني من خطّه والآخر على الثالث من خطّه ، فليست حركتاهما على السواء .

(١٦) وممّا يلزمهم لزوماً يظهر لكل ذي عقل ما نعلمه أنّه إذا تقابل شيئان لكل واحدٍ منهمـا أنْ يتحرك إلى الآخر حتى يلقاه ، ولا مانـع له البتة على لقاء الثاني خارجاً ، فلهما أنْ يتحـركا معاً حتى يلتقيا . فإذا التقيا أمكن أنْ يتمانعا ، وقبل ذلك لا تمانع بينهما ، وهذا شـيءٌ بيّنٌ بنفســه . فإذا توهمنا ثلاثة أجزاء على صف ، وعلـى الطرفين جزءان ، لكل واحدٍ منهما أنْ يتحرك حتى يلقى الآخر ولا مانع ، فإنَّ لهما جميعاً أنْ يتحركا إلى أنْ يلتقيا

each one of them is completely situated at the middle atom, having completely moved into it, in which case the two are interpenetrating one another, or each one of them covers part [of the way] until they encounter one another, and so, consequently, the middle atom, the two extreme atoms, and the two moving atoms are divided. Now, lo and behold, what do they say against this? That it is impossible for both to move together because of the impossibility of being divided! It is as if, when one of them is moved and the other rests, [the other's] motion just peters out.[34] So, if the other wanted to move and the one is aware of that, then either [(1)] the one stops [because of the other's intention], or [(2)] some cause comes from the other that forcibly stops the one, or [(3)] the one stops as a result of encountering the other. [Against (3)], it is no more fitting that one as opposed to the other should get there before the encounter. Also [against (1)] no rational individual is ignorant of the fact that, should both of them want to move at the same time, the intention of one of them to move does not in itself impede its counterpart from moving—[that is,] until they meet one another. It is just absurd, then, to say that this is an impediment because the other had a mind to move. Next [in response to (2)] how is that a cause that hinders [the one's] power to push away the impediment such that both remain at a standstill while neither yields? The two are not touching each other, nor are they stuck to what is underneath them, nor is there some influence to impede the other in either of them, nor is there any external impediment. In short, when the two are hindered and still continue to be so, one must provide some nonvarying state so that the two hinder each other [from moving]; but that state [on the Atomists' view]

34. Reading *nafidat* with **Z** and **T** for **Y**'s *nafadhat* (to pierce).

بعدما ليسا بملتقيين. فلا يخلو التقاؤهما إمّا أنْ يكون وكل واحدٍ منهما مستقر على كمال الوســط وقد انتقل إليه بكماله فيكونان متداخلين، أو كل واحد قطع شيئاً إلى أنْ التقيا. فإنْ كان كذلك، فقد انقسم الجزء الوسط والجزءان الطرفان والجزءان المتحركان. والعجب من قولهم على هذا إنَّ هذا يستحيل فيه أنْ يتحركا معاً لا ستحالة الانقسام، كأنَّ أحدهما إذا تحرّك والآخر ساكن نفدت حركته، فإنْ كان الآخر يريد أنْ يتحرك يشعر بذلك فيقف، أو يقفه سببٌ واردٌ من الآخر إليه يقسره، أو يقفه بملاقاة الآخرِ له، وليس سبق الآخر إلى الملاقاة أولى من ســبق هذا. وليس يخفى على العاقل أنَّه إذا أُريد تحريكهما معاً، لم يكن قصد أحدهما ليتحرّك في نفسه حابساً لصاحبه عن أنْ يتحرك إلى أنْ يلقاه. فمن المحال أنْ يقال إنَّ هذا تحبيسٌ بسبب أنَّ الآخر يهمّ أنْ يتحرك. فكيف يكون ذلك سبباً معاوقاً لقوة الدفع من الحبْس حتى يقفا ولا يطيعان وليســا بمتماســين ولا ملتصقين بما تحتهما، ولا في أحدهما تأثير حابسٍ من الآخر، ولا من خارجٍ حابس؟ وبالجملة؛ يجب أنْ يحدث عند

is nothing but running into each other. Who is satisfied by the claim that the impossibility of division holds the two in check and makes them such that they don't yield to what pushes and sets [them] into motion? Now, if one of them were to have something that pushes it, while the other does not, it would be pushed and respond to some cause; but when it so happens that there is something present pushing the other as well, and neither this one nor that one responds—right, let someone be satisfied with [that]! As for the sane person, he will deem the obvious impossibility of this impeding a reason for abandoning [the view that atoms] are absolutely indivisible, not that the [purported] absolute indivisibility [of the atom] is a reason for this impeding.

(17) Now, when you expand upon and consider closely what we have presented briefly, you will become absolutely certain of the falsity of this school of thought. Also, when this school of thought and its contrary are false, the truth must lie in its contradictory opposite—namely, that the single body does not have an actual part but that it is potentially divisible infinitely.

احتباسـهما - بعد الاستمرار - حالٌ غير التباين حتى يتمانعا، وتلك الحال ليست غير المصادفة. ومَنْ قنع بأنْ يقول إنَّ امتناع القسـمة تحبسهما وتجعلهما غير مطاوعين للتحريك والدفع ولو كان لأحدهما دافع ولم يكن للآخر دافع لاندفع وأجاب بسـبب. لكنه لما اتفق حضـور دافع للآخر صار لا هذا يجيب ولا ذاك: فليقنـع. وأمّا العاقل فإنّه يجعل ظهور استحالة هذا الاحتباس سبباً لبطلان منع الانقسام؛ لا منع الانقسام سبباً لهذا الاحتباس.

(١٧) وأنت إذا بسـطت ما أوجزنا القول فيه وتأملته، أيقنت بطلان هذا المذهب أصـلاً، وإذا بطل هذا المذهب ومضاده معاً، وجَبَ أنْ يكون الحق مقابله بالنقيض؛ وهو أنّه ليس للجسم الواحد جزء بالفعل وأنّه ينقسم إلى غير النهاية بالقوة.

Chapter Five

Solution to the puzzles of those who prattle on about the atom

(1) Let us now begin to solve [the Atomists'] puzzles and complete what is appropriate to this discussion with respect to the potentially infinite divisibility of what is moved, motion, and time. That is what follows.

(2) Since they believed their claim that there is an aggregation in anything that is separable[1] to be true, they built upon it. This, however, is not something [that we] concede, if by *aggregation* one means that there are two actually distinct parts in [the body], between which there is contiguity, and if *to separate* [means] to set one of them at some distance from the other and undo the contiguity. This, then, is not conceded. If it were conceded, there would be no need to resort to *separating* in order to complete their proof; and, instead, it would turn out valid, once aggregation is established as such. [That] is because, in determinate parts, there must be no aggregation, since it impossible that an infinite number of parts actually exist, while it is necessary that there be an actual unit wherever there are many [parts]. Now, if by *aggregation* they mean the

1. See 3.3.3.

<الفصل الخامس>

في حل شكوك المبطلين في الجزء

(١) فلنشرع الآن في حلّ شكوكهم، وفي تتميم ما يليق بهذا الكلام من مناسبة المتحركات والحركات والأزمنة في هذا الانقسام غير المتناهي بالقوة؛ وما يتبع ذلك:

(٢) أمّا قولهم إنّ كل قابل للتفريق ففيه تأليف، فهو الذي لما ظنّوه حقاً بنوا عليه، وليس هذا بمسلّم. فإنْ عنى بالتأليف أنْ يكون فيه جزءان متميزان بالفعل وبينهما مماسّة؛ وأنّ التفريق تبعيد أحدهما عن الآخر وإبطال المماسّة، فهذا غير مسلّم. ولو سُلّم لكان لا يحتاج إلى أنْ يلتجئوا إلى التفريق حتى تتمّ حجتهم، بل كانت تكون صحيحة مع ثبات التأليف تأليفاً، إذ كان يجب أنْ تكون أجزاء حاصلة لا تأليف فيها؛ لاستحالة وجود ما لا يتناهى من الأجزاء بالفعل، ووجوب واحدٍ بالفعل حيث يكون كثيراً. وإنْ عنوا

preparedness to become many in what is a single thing in which there is no multiplicity, then this is conceded. You cannot remove this from the body; otherwise, the body would cease, since there is no way to undo the unity of the actual unit, except by making it altogether cease to be or making it many. When it does not cease to be, but becomes many [for example, it is divided into two], there remain two units whose state is like that of [the initial unit], whereas the entirety of the unity is eliminated completely only by making [the unity] cease to be. Some have reckoned that the existence of bodies that differ in the speed and difficulty with which they can be separated necessarily establishes aggregation. They said: Now, that is not because the bodies differ in genus—that is, in their specificity [as bodies]—nor because of a difference in the agent, nor [due to] the occurrence or nonoccurrence of something—as if they believe that these are the only options and believe that bodies do not differ generically. Let us concede all of that to them. Why, then, must it be solely due to aggregation? Instead, why aren't these two things—namely, the difficulty and speed with which [bodies] are susceptible [to separation]—simply accidents that just happen to bodies by which [the bodies] differ while still agreeing in the specified ways? [In this respect, they] would be just like black, white, and the rest of the accidents. So [for example] you see that, when bodies differ by being black and white, that requires that they differ by some accident other than being black and white—namely, aggregation—since [differing by being black and white] is not due to the genus, the agent, or the occurrence and nonoccurrence of something.

بالتأليف الاستعداد لأنْ تحدث كثرةٌ فيما هو واحدٌ لا كثرة فيه - فهذا مسلّم. وهذا لا تجوز إزالته عن الجسم أو يبطل الجسم؛ إذ لا سبيل إلى إبطال وحدة الواحد بالفعل إلاّ بإعدامه أصلاً أو تكثيره. فإذا لم يعدم بل كثر، بقي واحدان حالهما حاله، وجملة الوحدة لا ترفع عنه البتة إلاّ بإبطاله. وحسب بعضهم أنَّ وجود الأجسام، مختلفة في سرعة قبول التفريق وعُسْر قبوله، يوجب إثبات التأليف؛ قال: وذلك ليس لاختلاف جنس الأجسام أيّ نوعيتها، ولا لاختلاف الفاعل ولا لحدوث شيء ولا لعدم شيء، كأنَّ عندهم ليست الأقسام إلاّ هذه، وعندهم أنَّ الأجسام لا تختلف نوعيتها. فلنسلّم كل ذلك لهم، فلِمَ يجب أنْ يكون للتأليف لا غير؟ . بل لِمَ لا يكون هذان المعنيان - وهو عُسْر القبول وسرعة القبول - عرَضين يعرضان للأجسام، تختلف بهما بعد الاتفاق المذكور؛ كالسواد والبياض وغير ذلك من الأعراض؟ فترى أنَّ الأجسام إذا اختلفت بالسواد والبياض احتاج ذلك إلى أنْ يكون اختلافهما بعرضٍ غير السواد والبياض هو التأليف، إذ ليس للجنس والفاعل والحدوث وعدم الشيء.

(3) There would be something to their argument based upon halving[2] only if we were to say that the body has a certain part that is not divisible into halves, thirds, fourths, or the like, for it would it would have an infinite number of parts. We ourselves, however, in no way require that the body have parts, unless it is divided into parts; but a body cannot have already been divided into an infinite number of halves. So what they said does not follow. Most of what they say here is, "You see that . . ." when nothing is pointed to and no part marked off from another. There is no separate *this* and *that.* They are simply oblivious to the fact that that becomes *this* and *that* only by pointing and, when there is no [pointing], neither is there *this* or *that;* but when there is no *this* and *that,* how can there be a separate *this* and *that*? And [all of this] despite the fact that the distance covered is covered in a time like itself—that is, whose limits are finite, being infinitely divisible into halves through an act of the estimative faculty and positing, while having no actually existing division.[3]

(4) As for the account of the mustard seed and the mountain[4], neither one of them has divisions as long as it is not divided. So, when they are divided together, the number of their divisions is, in fact, equal (and that does go on infinitely), but the [size of] the divisions that belong to the mustard seed are smaller. The repugnancy would result only if [the divisions] were to go on infinitely in them in terms of equal magnitudes. An example of this is to let the mountain and also the mustard seed be doubled infinitely through an act of the estimative faculty or the power of God. So, just because the doubling is equal it does not result from that that the multiplied mountain and the multiplied mustard seed have

2. See 3.3.4.
3. Cf. Aristotle, *Physics* 6.2.233a2ff.
4. See 3.3.6.

(٣) وأمّـا الحجّة المبنية على الأنصاف؛ فإنّما كان يكون من ذلك شـيء لو قلنا إنَّ للجسم جزءاً ما لم يُجزّأ نصفاً أو ثلثاً أو ربْعاً، أو غير ذلك، فكان يكون له أجزاء بلا نهاية. ونحن لا نوجب للجسم جزءاً البتة إلاّ أنْ يجزّأ، ولا يمكن أنْ يكون جسم قد جُزِّئ بأنصافٍ لا نهاية لها، فلا يلزم ما قالوا. وأكثر ما يقولون ها هنا ترى أنك - إذا لم يُشر ولم يعيّن إلى جزء جزء - لا يكون ذلك مفرداً؛ وهذا مفرد، ولا يدرون أنَّ ذلك إنّما صار ذلك وهذا بالإشـارة. فإذا لَمْ تكن لمْ يكن لا ذلك ولا هذا، وإذا لم يكن ذلك ولا هذا، كيف يكـون ذلك مفرداً وهذا مفرداً؟ وعلى أنَّ المسـافة المقطوعة تُقطع بزمانٍ مثلها، متناهي الأطراف، منقسـم بلا نهاية في الأنصاف توهماً وفرضاً، ولا قسـم لـه وجوداً وفعلاً.

(٤) وأمّا حديث الخردلة والجبل، فإنّه لا أقسـام لأحدهما ما لم يُقسم، فإذا قسّما معاً حصلت عدّة أقسـامهما متسـاوية في العدد، وكل واحد من الأقسـام التي للخردلة أصغر، ويذهب ذلك إلى غير النهاية. وإنّما تكون الشناعة لو كان الذهاب إلى غير النهاية فيهما بمقادير متساوية. ومثال هذا أنَّ تُضعّف الجبل في التوهّم، وفي قدرة الله، إلى غير النهاية والخردلة أيضاً، فلا يكون من ذلك أضعاف الجبل مسـاوية، في المقدار، لأضعاف

the same magnitude; rather, they will differ in determinate quantity, even though they are equal in number. What precludes things from being equal in number while not equal in magnitude—[that is,] as individuals but not collectively? The fact is that things capable of going on infinitely can be greater than others—as, for example, multiples of tens in relation to multiples of hundreds.

(5) As for [the possibility that] the divisions of the mustard seed would cover the face of the Earth,[5] let us grant them the existence of the atom and additionally judge that the mustard seed is divided into its indivisibly small atoms such that the number of them existing in the mustard seed does cover the entire Earth, should they be spread over it one by one. In this case, we do not know whether it is true or false. Perhaps there does exist in the mustard seed a significantly large quantity of the atoms so as to cover the earth's surface. Who knows the atom's measure so as to know which is the first body composed of them that includes the requisite number to cover the Earth? The fact is that, when it is conceded that the parts of the mustard seed might cover the Earth, they can offer nothing but an incredulous look. A dogmatic assertion that this is impossible, however, is not a very reliable [argument]. So, [if covering the face of the Earth with the parts of a mustard seed] is not clearly impossible with a finite number of divisions, how could its [purported] impossibility show the impossibility of infinite divisibility? Nonetheless, we are not claiming that what is possible concerning that might pass into actuality.[6]

---

5. See 3.3.5.

6. **Z** and **T** additionally have *bal nusallimu annahu yajūzu an yantahiya ilá aṣghar yaᶜjizu ᶜan tafriqah li-basṭah ᶜalá al-arḍ au ghayrihi, wa-lā yaᶜjizu ᶜan qismatihi bil-farḍ wa-l-tawahhum wa-bi-wujūh ukhrá lā tuᵓaddiy ilá tafrīqihi wa-taqṭīᶜihi;* "rather we are [merely] granting the possibility that [division] terminates at some smallest atom [and that there is enough] to be able to cover the Earth or the like, whether by posit, an act of the estimative faculty, or any other means that lead to its being separated and cut up." The lines are not found in three of the MSS consulted by **Z**, nor are they found in the Latin. **Y** does not note the omission.

الخردلة لأجل أنَّ التضعيف متســـاو ، بل يكونان مختلفين في القدر ، وإنْ تساويا من وجهٍ في العدد . وما الذي يمنع أنْ تكون أشـــياء متســـاوية في العدد ، وليســـت متساوية في المقدار ، أفراداً ولا جملة؟ بل يجوز أنْ يكون في الاحتمال أشـــياء تذهب إلى غير النهاية أكثر من أشياء ، كتضعيف العشرات مع تضعيف المِئين.

(٥) وأمّا تغشية أديم الأرض من أقسام الخردلة ؛ فلنسلّم لهم وجود الجزء ، ومع ذلك لنحْكم بأنَّ الخردلة تنقسم أجزاؤها التي لا تتجزأ في صغرها ؛ بحيث يكون عدد الموجود منها في الخردلة يغشـــى الأرض كلها لو بســـطت عليها واحدة واحدة ، فما كان يدرينا أنَّ هذا حقٌ أو باطل؟ فعســـى أنْ يكون في الخردلة من الأجزاء التي لا تتجزأ ما تبلغ كثرته أنْ تغشـــى به صفحة الأرض . ومَنْ عرف تقدير الجزء الذي لا يتجزأ ، حتى يعرف بذلك الجسم الذي هو أول جسم مركب منها يشتمل على العدد المحتاج إليه في تغشية الأرض؟ بل لا يكون في أيديهم - إذاً سُلِمَ أنَّ أجزاء الخردلة تغشّيّ الأرض - شيءٌ غير التعجّب ! وأمّـــا جزم القول بأنَّ هذا ممتنعٌ ، فأمرٌ غير موثوقٍ به . فالذي لا يكون بيّن الاســـتحالة ، مع فرض تناهي الانقســـام ، فكيف تبين باستحالته استحالة لا تناهي الانقسام؟ على أنّا لسنا نقول إنَّ الممكن من ذلك قد يخرج إلى الفعل .

(6) As for their argument taken from substance and accident,[7] let them know that we do not grant them that it belongs to the true nature of the accident that it itself is equal to the substrate itself, spreading over and coinciding with it. The fact is that the accident is nothing more than an attribute of the thing, which does not make the thing itself subsist in that it is a part of it, as we have said in other places.[8] So perhaps it is not such as to indicate an entity that spreads throughout the very thing of which it is an accident, such as all relations, motion, and a mode of being[9] that they maintain. Indeed, that [type of accident] is not like white that spreads throughout its substrate. So, if by *accident* one means what they mean—namely, that it is an entity belonging to an entity throughout which it is spread—then a point is neither an accident nor a substance, since not every existing thing must either coincide with some entity so as to pervade it or, otherwise, not exist in a subject. [That] is because the two, [that is, coinciding so as to pervade a subject or not existing in it] are not contradictory propositions, nor are they the obvious concomitants of contradictory propositions. If one means by *accident* something belonging to a thing by which [the thing] comes to have a certain attribute but [which does] not [play] some part in its subsistence, then a point is an accident, because it is a certain extremity belonging to what is limited by it, while not being part of its existence. It is an accident of its substance in that it is an attribute in this sense, because it is an extremity of [the substance] and nothing else.

---

7. See 3.3.7.

8. See for instance *Kitāb al-madkhal* 1.14, where Avicenna argues that the accident is not equivalent with the substance, and *Kitāb al-maqūlāt* 1.4 where he discusses the difference between the ways accidents stand to subjects and forms stand to matter—namely, that accidents subsist through their subjects, whereas matter subsists through the form.

9. The Arabic *kawn,* translated "mode of being" is the standard term for accident among the *mutakallimūn;* see Richard Frank, *Beings and Their Attributes: The Teaching of the Basrian School of the Mu'tazila in the Classical Period* (Albany, NY: State University of New York Press, 1978), ch. 5.

(٦) وأمّا الحجّة المأخوذة من الجوهر والعرض، فليعلموا أنّه لا نُسلّم لهم أنَّ العرض من حقيقته أنْ يكون له ذات مساوية لذات المحل؛ فاشية فيه مطابقة له، بل ليس العرض أكثر من وصفٍ يكون للشيء، ليس يقوّم ذاته بأنّه جزء منه، على ما قلنا في مواضع أخرى. فربما لم يكن بحيث يُشار أنَّ ذاته فاشية في ذات الشيء الذي هو له عرض؛ كالإضافات كلّها وكالحركة وكالكون الذي يقولون، فإنَّ ذلك ليس كالبياض المنتشر في محله. فإنْ عنى بالعرض ما يقولون من أنّه ذاتٌ مساويةٌ لذات ما هو فاشية فيه؛ فليست النقطة بعرضٍ ولا بجوهر. إذ ليس يجب أنْ يكون كل موجود إمّا مطابقاً لذاتٍ سارياً فيها، وإمّا موجوداً لا في موضوع؛ لأنّه ليس أحدهما نقيض الآخر، ولا بيّن اللزوم للنقيض. وإنْ عنى بالعرض معنى للشيء يصير به الشيء ذا صفةٍ وليس جزءاً في قوامه؛ فالنقطة عرضٌ لأنّها نهاية ما موجودة لما هو بها متناهٍ، وليست جزءاً من وجوده. وكونها عرضاً لجوهرها هو أنّها صفة بهذه الصفة لأنّها نهاية له، وليس غير هذا.

(7) The account comparing divisibility to composition—whether the composition of the body in itself or together with something else[10]—is unsound. [That] is because division produces the parts, whereas composition requires that parts determinately exist. It is impossible, however, that an infinite number of determinate parts[11] exist such that [the body] is composed from them.

(8) As for the account of contiguity and its passing away,[12] a foundation has already been advanced in the section on time, which, if you recall it, provides an immediate response to [the objection].[13] In brief, noncontiguity does not occur all at once at an instant.[14]

(9) Concerning the report of the celebrated [horn] angle,[15] it is not indivisible, but is, in fact, divisible. There is a potential infinity of angles smaller than it. [Euclid's] demonstration was based solely on the fact that no acute angle [constructed] from two rectilinear lines is smaller than that one; but when one claims that nothing having such a description is smaller than that, it does not prove that nothing at all is smaller than it. Now, anyone who has acquired some knowledge of the fundamentals of geometry knows that that angle is infinitely divisible by arcs.[16]

---

10. See 3.3.8.

11. Following **Z** who secludes the *hā* (their).

12. See 3.3.9.

13. See 2.12.4, where Avicenna specifically treats the issue of the cessation of contiguity.

14. Following **Z**, **T**, and the Latin (*subito*) for **Y**'s *lā mumāssah* [*lā*] *taḥṣulu illā fī ān* (non-contiguity does not occur except at an instant), which is just the opposite of Avicenna's claim at 2.12.4.

15. See 3.3.10.

16. The horn angle is represented by the angle formed from the circle and the tangent line AB, while the curves in the following diagram illustrate Avicenna's point.

(٧) وأمّا حديث تشــبيه الانقسام بالتركيب - سواء كان تركيب الجسم في نفسه أو تركيبه مع غيره - فليس بصحيح؛ لأنَّ الانقسام يُحدِث الأجزاء، والتركيب يحتاج إلى أجزاء حادثة حاصلة، ويســتحيل أنْ توجد أجــزاء حاصلة بلا نهاية حتى يركب منها.

(٨) وأمّا حديث المماسَّــة وزوالها فقد مضى أصلٌ في باب الزمان، إذا تذكرته كان الجواب مقتضباً منه؛ وبالجملة أنْ لا مماسّة لا تحصل دفعة في آن.

(٩) وأمّا حديث الزاوية المذكورة، فإنّها ليســت غير منقســمة، بل هي منقسمة، وهناك زوايا أصغر منها بالقوّة بلا نهاية. إنّما قام البرهان على أنّه لا تكون زاوية من خطّين مســتقيمين حادّة أصغر من تلك، وليس إذا قيل إنّه ليس شــيء بصفة كذا أصغر من كذا دلَّ على أنّه ليس شــيء البتة أصغر منه، وكل مَنْ حصّل علماً بأُصول الهندسة؛ علم أنَّ تلك الزاية تُقسم بالقسي قسمة إلى لا نهاية.

(10) As for what was related about the surface and the sphere,[17] one does not know whether a sphere could exist on a surface having this description in reality or only in the activity of the estimative faculty, in the way that mathematical objects do. Nor is it known (if it exists in reality) whether [the sphere] is, in fact, rolling over [the surface] or not, for perhaps it is impossible for it to roll over it. Aside from all this, it is not necessary, in any case, that the sphere touch the surface and the line at only a point; rather, it may be like that [only] when it is stationary and at rest. In that case, when it is moved, it would touch [the sphere] at the line during the time of the motion, and there simply would be no actual moment at which it touches at the point, save in the estimative faculty. [That] is because the estimative faculty images that [contact at a point] only when it imagines the instant, whereas the instant has no existence in actuality. In summary, this problem does not turn out to be truly admissible, because what is admitted is that the sphere does not meet the surface at a single instant save at a point, from which it does not necessarily follow that the motion involves a locomotion from one point to another point immediately adjacent to it and from one instant to another, immediately adjacent one. Indeed, if this is admitted, there is no need to mention the sphere and the surface; rather, there would be points that encounter one another from which the line is composed, as well as immediately adjacent instants from which time is composed. So what [they take] for granted is that the sphere encounters the surface at some given instant; the difference among motions and times that are not composed of indivisible phases and instants is like the difference with

17. See 3.3.11.

(١٠) وأمّا ما أورد من حديث السطح والكرة، فإنّه لا يدري هل يمكن أنْ توجد كرة على سطح بهذه الصفة في الوجود، أو هو في التوهم فقط على نحو ما تكون عليه التعليمات؛ ولا يدري أنّه إنْ كان في الوجود، فهل يصحّ تدحرجه عليه أو لا يصح، فربما استحال تدحرجه عليه. وبعد هذا كله، فليس يلزم أنْ تكون الكرة مماسّة للسطح والخط في أي حالٍ كان بالنقطة لا غير؛ بل تكون في حال الثبات والسكون كذلك. فإذا تحرّكت ماست بالخط في زمان الحركة، ولم يكن البتة وقت بالفعل يماس فيه بالنقطة إلاّ في الوهم؛ إذ ذلك لا يتوهم إلاّ مع توهم الآن، والآن لا وجود له بالفعل. وبالجملة فإنَّ هذه المسألة لا تتحقّق مسلّمة، لأنَّ المسلّم هو أنَّ الكرة لا تلقى السطح في آنٍ واحدٍ إلاّ بنقطة، وليس يلزم من هذا أنْ تكون الحركة تنتقل من نقطةٍ إلى نقطةٍ مجاورة لها، ومن آنٍ إلى آنٍ مجاورٍ له. فإنّه إنْ سُلّم هذا، لم يحتج إلى ذكر الكرة والسطح، بل صحّ أنَّ هناك نقطاً متلاقية، ومنها تأليف الخط، وآنات متجاورة منها تأليف الزمان. فإذا كان المسلّم هو أنَّ الكرة تلاقي السطح في آن، وكان الخلاف في أنَّ الحركات والأزمنة غير مركبة من أمور غير

respect to spatial magnitude; and only if instants are, in fact, immediately adjacent to one another does it necessarily follow that points are immediately adjacent to one another. Using [all] that, however, to prove that points follow in succession is as good as begging the question! The fact is that this explanation is completed only by what I was saying—namely, that at this state it touches a point, and at the second state it is touching a point, [and so on], where the states are immediately adjacent so that the points are as well. If we do not say this, the argumentation is incomplete. This will become clear to you once you learn that there is no first part among the parts of motion, rest, and distance.[18]

(11) Democritus erred in his argumentation[19] in granting himself one premise—namely, that the whole body is divisible—since this has two senses. One is that [the body] is simultaneously divided in its entirety, while the other is that [the body] is not divisible in a given way unless it leads to parts that are themselves divisible without end. The first is simply not granted, nor is its true opposite that the division of the body terminates at what is indivisible; rather, its opposite is [that the whole of it is not actually divided simultaneously].[20] As for the whole of its not being actually divided simultaneously, this does not preclude its being divided one division after another infinitely. Moreover, neither is it the case that, when each one of [the divisions] taken separately is separately possible, the whole is something whose occurrence is possible. For instance, any numerical doubling can apply to number, but not every numerical doubling can apply simultaneously. The truth, in fact, is that any division

18. See 3.6.3–6.

19. See 3.3.12.

20. The text is apparently incomplete here and seems to require the bracketed material if it is to make sense. The omission may be because there was a repetition in the original text of the immediately following clause, "The whole of it is not actually divided simultaneously," which, as noted, would in fact be the true opposite of "The body is simultaneously divided in its entirety," and so completes the thought. In that case, then, it is quite possible that the first instance was omitted quite early by homeoteleuton.

متجزئة ومن آنات، كالخلاف في المسافة، وكان إنّما يلزم تجاور النقط - لو صحّ تجاور الآنات - كان استعمال ذلك في إثبات تتالي النقط كالمصادرة على المطلوب الأول. فإنّه لا يتم هذا البيان إلاّ بما أقول إنّه في هذه الحال ملاقٍ بنقطة، وفي الحال الثانية ملاقٍ بنقطة، والحالات متجاورة، فالنقط متجاورة، فإنْ لم نقل هذا لم يتم الاحتجاج. وأنت ستتبين هذا إذا علمت أنّه ليس في أجزاء الحركة والسكون والمسافة ما هو أول جزء حركة أو جزء سكون أو جزء مسافة.

(١١) وأمّا احتجاج ديموكريتس فقد ضلَّ فيه، في تسليم مقدمة واحدة لنفسه؛ وهو أنَّ الجسم ينقسم كله، لأنَّ هذا يدلّ على معنيين، أحدهما أنّه ينقسم بكليته معاً، والآخر أنّه لا ينقسم قسمة إلاّ أدت إلى أجزاء هي أيضاً تقبل القسمة ولا تقف. فأمّا الأول فليس ذلك بمسلّم، ولا نقيضه الصادق هو أنَّ الجسم ينتهي في القسمة إلى ما لا ينقسم؛ بل نقيضه. وأمّا أنْ لا ينقسم كله بالفعل معاً، وهذا لا يمنع أنْ يكون ينقسم إنقساماً بعد إنقسام بلا نهاية، وليس أيضاً إذا كان كل واحدٍ من الانفصالات انفصالاً ممكناً، فالكل ممكن الوقوع، كما أنَّه كل تضعيف عددي جائز على العدد، وليس كل تضعيفٍ عددي جائز

you want can occur in the body when each one of the successive divisions is of the potentially infinite sort. We simply do not grant that they all occur [simultaneously], because the first thing that that would require is that the things bringing about the divisions are themselves actually infinite, and this is impossible. In summary, this is one of the errors that occurs as a result of equating the two expressions *the whole* and *each one.* We shall show to a greater extent that the existence of these atoms comes to naught when we take up the more specialized discussions.

(12) As for the proof of those who affirm an infinite number of parts,[21] you are able to resolve it from what you understand.

---

21. See 3.3.13–16.

أنْ يقع معاً . بل الحق أنَّ كلّ قسمةٍ أردتها ، وكل واحد واحد من أصناف قسمة ، هي بلا نهاية بالقوّة ، يجوز أنْ تقع في الجســـم ، ولا نســلّم أنَّ الكل يقع البتة ؛ لأنّه يحتاج أول شيء إلى أنْ يكون الذين يوقعون القســـمة لا نهاية لهم بالفعل ، وهذا مســتحيل . وبالجملة ، فإنَّ هذا من جملة الخطأ الواقع لتشـــابه لفظي الكل وكل واحد ، وسنبلغ في إبطال وجود هذه الأجسام غير المتجزئة إذا شرعنا في الكلام - الذي هو أشدّ تخصصاً من هذا الكلام .

(١٢) وأمّـــا حجة مثبتي أجزاء بلا نهاية ؛ فأنت تقدر - مما فهمت - على حلّها .

Chapter Six

On the interrelations of distances, motions, and times with respect to this topic, and an explanation that no first part belongs to them

(1) We say: When the distance can be potentially divided infinitely,[1] the motion along it (in the sense of *to traverse*) likewise can be divided infinitely. If the motion were indivisible, then the distance it [traverses] would either be indivisible as well—which is absurd—or divisible. If [the distance] were divisible [while the motion is not], then the motion from the beginning of [the distance] to the place where there is the division would be less than from the beginning to the end; but nothing is less than what is indivisible. Additionally, that motion [to where there is the division in the distance] would be a *part* of [that purportedly indivisible] motion that completely crosses the distance. Also, when the motion is divisible, the time corresponding with it is divisible—or, rather, the motion is divisible precisely because the time and distance are divisible. It also becomes clear that each one of these [namely, distance, motion, and time] is divisible from the fact that there exists fast and slow motion. [That] is because the distance that the slow motion traverses in a given time must be less than that which the fast motion traverses, and so the distance is divided. Also, the fast motion traverses that lesser [distance] in less time, and so the time is divisible, and (as you have already learned)[2] motion follows distance and time with respect to divisibility.

---

1. I have moved the phrase *allatī bi-maᶜná al-qaṭᶜ* (in the sense of *to traverse*), which appears here in **Y**'s edition, so that that it follows the phrase "the motion along it," which is how it appears in **Z**, **T**, and the Latin.

2. See 2.12.7 and 2.13.3–4.

<الفصل السادس>

في مناسبات المسافات والحركات
والأزمنة في هذا الشأن
وتبيين أنّه ليس لشيءٍ منها أول جزء

(١) فنقول: إذا كانت المسافة تنقسم إلى غير النهاية بالقوة فكذلك يجب أنْ تنقسم الحركة التي بمعنى القطع، معها إلى غير النهاية بالقوة. ولو كانت حركة لا تتجزأ؛ لكانت مسافتها إمّا غير متجزئة وهذا محال، أو متجزئة. ولو كانت متجزئة لكانت الحركة من مبدئها إلى موضع القسمة أقل من الحركة من مبدئها إلى منتهاها، ولا أقل من غير المتجزئ، ومع ذلك لكانت تلك الحركة جزءاً من الحركة التي بها استوفت المسافة. وإذا انقسمت الحركة انقسم بإزائها الزمان، بل إنّما تنقسم الحركة بسبب انقسام المسافة أو الزمان. ومن الموجود حركة سريعة وبطيئة، ومنها يتبين أنَّ كل واحدٍ من هذه ينقسم؛ فإنَّ المسافة التي تقطعها السريعة في زمانٍ يلزم أنْ تكون البطيئة تقطع أقلّ منها فتنقسم المسافة، والسريعة تقطع ذلك الأقل في زمانٍ أقل فينقسم الزمان، والحركة تتبع المسافة والزمان في الانقسام كما علمت.

(2) Motion, however, happens to have a certain kind of divisibility that does not correspond with anything in time—namely, that it is divided by dividing the mobile, where this seems best suited to motions other than local motion. [That] is because the parts of what undergoes local motion are either actual or potential. If the parts are, in fact, actual, then they are joined together either by way of continuity or contiguity; but how could there be [motion on the part of what is moved locally], for not one of them leaves its place? [That is] because, if [the parts] are continuous, they do not actually have a place, whereas if they are contiguous, they have a place; but they leave their place as a surface that is a part of the place of the whole, while not leaving the place[3] that surrounds them, and so they would not depart from their place and so would not have undergone motion. If they are potential parts, the improbability of motion resulting from them is all that more obvious, for how would the actual parts of a given motion be related to them? As for the rest of the motions, if they have actual parts, then one correctly says that part of the change changes the part. If they have potential parts, the motion likewise will have potential parts for, which if they were differentiated, there is, corresponding with each part of what changes, a change that is proper to it that is part of the whole's change. [That] is because, from *this* change that is in *this* part and *that* change that is in *that* part, there comes to be the sum total of the change of the whole, since that summed total is a total change, and the total change is a change. Now, every change is *of* something; but there is nothing that underlies these changes except the whole and the parts, and it is not *of* a part of a part. So [the change] is *of* the whole.

---

3. Reading *makān* with **Z** and **T** for **Y**'s *kamāl* (perfection) and the Latin *complete* (completely). Neither **Y** nor **Z** mentions the other's variant. If *perfection* is retained, the sense of the argument might be thus: since *motion*, according to Avicenna, is the actuality and *perfection* of the potential as such (see 2.1), when the perfection of what surrounds—that is, the place—is not left behind (and, consequently, a perfection remains, and there is no actualization of a potential), there cannot be a motion with respect to place.

(٢) لكن الحركة يعرض لها ضربٌ من الانقسام لا يطابقها فيه الزمان، وذلك هو انقسامها بانقسام المتحرك، ويشبه أنْ يكون هذا بغير الحركة المكانية أولى. فإنَّ أجزاء المتحرّك الحركة المكانية لا يخلو إمّا أنْ تكون أجزاء حاصلة بالفعل أو أجزاء بالقوة. فإنْ كانت أجزاء حاصلة بالفعل؛ فلا يخلو إمّا أن يكون اجتماعها على سبيل تماسّ أو اتصال، وكيف كانت، فإنَّ كلّ واحدٍ منها لا يفارق مكانه. لأنّها إنْ كانت متصلة فلا مكان لها بالفعل، وإنْ كانت متماسّة فلها مكان؛ لكنها تفارق من مكانها سطحاً هو جزء مكان الكل، ولا تفارق مكان المحيط بها، فلا تفارق مكانها، فلا تتحرك. وإنْ كانت الأجزاء بالقوة فبُعْد الحركة عنها أظهر، فكيف ينسب إليها أجزاء حركة بالفعل؟ وأمّا في سائر الحركات، فإنْ كان لها أجزاء بالفعل، صحّ أنْ يقال إنَّ جزء التغيّر يغيّر الجزء. وإنْ كان لها أجزاء بالقوة فللحركة أيضاً أجزاء بالقوة، لو فصلت لكان بإزاء كل جزء من المتغيّر تغيّر يخصه، هو جزء تغيّر الكل. فإنَّ من هذا التغيّر الذي في هذا الجزء، وذلك التغيّر الذي في ذلك الجزء، يحصل مجموع تغيّر الكل، إذ تلك الجملة المجتمعة جملة تغيّر، وجملة التغيّر تغيّر، وكل تغيّر فهو لشيء، ولا شيء يحمل هذه التغيّرات إلاّ الكل والأجزاء، وليس لجزء جزء؛ فهو للكل.

(3) Also, since every motion and change is in a time that is infinitely divisible, it would be absurd that the motion should have something that is the first that the mobile moves. That is because, if a certain motion is first, it inevitably is along a certain distance, where that distance is potentially divisible. Now, when [the distance along which there is the purported first motion] is divided, one of its two parts will be prior and the other posterior, and so the motion along that first part is first, but we stipulated that this [whole] motion was the first. This is a contradiction.

(4) The fact is that *first* with respect to motion and change, is understood in only one of three ways, one of which is [(1)] *first* in the sense of the limit—that is, that which corresponds with the first of the distance and its limit as well as the first of the time that is congruent with that motion and the limit of [that time]. This one, then, is first.[4] *First* has another sense [(2)]—namely, when the motion happens to bring about a division, whether actually or by supposition, then the prior part is the first of the parts of motion that is actual. It might also be thought that motion has a *first* in some other way [(3)]—that is, some have said that even if these bodies can be potentially divided *ad infinitum* it is not the case that, when they are divided, they preserve their forms and dispositions other than that of quantity. So [for example] the body reaches a certain limiting point that, should it be divided further, it would not, in fact, be water, air, or fire. They said that whether it is what undergoes motion or some spatial magnitude, when the spatial magnitude as such has some limiting point (which, in their opinion, is unsurpassable in smallness), the motion will also have some limiting point, which is the smallest existing motion. In that case, no separate motion exists that is smaller than it, even though, through an act of the estimative faculty,

4. Avicenna is probably punning here and so means something like, "This sense of *first* is the primary (first) sense of the term *first*."

(٣) ولمّا كان كل حركةٍ وكل تغيّرٍ فهو في زمانٍ ينقسم إلى غير النهاية، فمحالٌ أنْ يكون للحركة شيءٌ هو أول ما يحرّكه المتحرّك. وذلك لأنّه إنْ كان حركة هي أوّل حركة، فإنّها لا محالة في مسافة، وتلك المسافة منقسمة بالقوة، وإذا قُسّمت كان أحد جزئيها متقدماً والآخر متأخراً، فكانت الحركة في الجزء الأول هي أوّل حركة، وقد جُعلت هذه أوّل حركة، هذا خُلْف.

(٤) بل الأول في الحركة وفي التغيّر إنّما يفهم على أحد وجوهٍ ثلاثة: أحدها؛ الأول بمعنى الطرف، وهو الذي يوافق أول المسافة وطرفها، وأول الزمان المطابق لتلك الحركة وطرفه، فهذا أوّل. وأوّل بمعنى آخر؛ وهو أنّه إذا عرض للحركة تقسيم بالفعل أو بالفرض، كان الجزء المتقدّم أوّل أجزاء الحركة التي بالفعل. وقد يظنّ أنَّ للحركة أولاً على وجهٍ آخر، وهو أنّه قد قال بعضهم إنَّ هذه الأجسام، وإنْ كانت تنقسم إلى ما لا نهاية في القوة، فليست تنقسم حافظة لصورها وهيئاتها غير هيئة الكم، فإنَّ الجسم يبلغ حَدّاً لا يصحّ لو انقسم بعد أنْ يكون ماء أو هواء أو ناراً. قالوا؛ أو كان متحركاً أو مسافة؛ فإذا كان للمسافة - من حيث هي مسافة - حدٌّ عندهم لا يتعداه في الصغر، كان للحركة حدٌّ هو في الوجود أصغر الحركات، فلا توجد حركة مفردة أصغر منه. وإنْ كان قد يجوز أنْ

it might be possible for us to imagine something smaller than that—namely, half or some part of it, since that in itself is potentially divisible. It is just that that division will never pass into actuality in the sense of being separate and differentiated. (We'll discuss this later.)[5] In that case, the mobile would have a first motion during its motion, and that would be in[6] potency—namely, as that which is equivalent to the smallest motion.

(5) The *first* of the motions, in the sense of *limit,* is not a motion, and so nothing can be the first motion in that sense of *first.* What moves can be first in the second sense; however, its being first is hypothetical and accidental, not real. As for the third way, even if it is true that motion has something that is the smallest motion that can exist, it is so only in the sense that there is a certain motion that, in itself, [has] an actually separate beginning and ending. It is not [first] in the sense that it is some first of a total motion, of which that first is part and after which the total would have continued on. This division into parts, which is the focus of discussion, is by supposition, whereas that indivisible unity belonging to motion does not depend upon supposition, but existence [itself]. At best, one might say that the amount of that[7] motion, out of the total of all motion, rightly deserves to be posited as first, since no motion in existence is smaller than it, save by supposition. So the discussion comes to a standstill until we clarify this view.

(6) As for the *first* that would result in the motion by our dividing [the motion] corresponding with division of the distance (the division of which does not stop at some limiting point), there [simply] is no magnitude possessing a beginning and end such that it is indivisible at that

5. See 3.12.

6. Reading *fī al-qūwah* with **Z** and the Latin (*in*), which is also the same in meaning as **T**'s text *bi-l-qūwah;* Y's text simply reads *qūwah.*

7. **Y** has conflated the text *qadr tilka*, as it appears in **Z**, **T**, and the Latin (*quantitatem istius*), to *qudratak* (your power).

نتوهم ما هو أصغر من ذلك، وهو نصفه أو جزء منه، إذ كان ذلك يتجزأ في نفسه بالقوة، لكن ذلك التجزؤ لا يخرج إلى الفعل بتّةً خروجاً على معنى الأفراد والفصل، وسنتكلم في هذا بعد. فإنْ كان كذلك، فالمتحرّك يكون له في حركته أوّل حركةٍ وذلك في القوة، وهو ما يساوي الحركة التي هي أصغر الحركات.

(٥) فأوّل الحركات - بمعنى الطرف - ليس بحركة فلا يكون للشيء بمعنى ذلك الأول، أول ما يحرّك، وأمّا بالوجه الثاني فيكون له أول ما يحرّك، لكن أوّليته وضعية عرضية لا حقيقية. وأمّا الوجه الثالث؛ فهو وإنْ صحّ أنَّ للحركة شيئاً هو أصغر حركة يمكن أنْ توجد، فإنّما يصح على أنّها حركة بنفسها مفردة ابتداء بالفعل وانتهاء بالفعل، لا أنْ تكون هي أوْل جملة حركةٍ، ذلك الأول بعضها، وقد استمرت الجملة بعده. فإنَّ هذا التبعيض - الذي كلامنا فيه - هو بالفرض، وتلك الوحدة غير المنقسمة للحركة ليست بحسب الفرض، بل بحسب الوجود. اللّهم إلاّ أنْ يقول قائل إنَّ قدرتلك الحركة مستحقٌ في جملة كل حركةٍ أنْ يُفرض أولاً، إذ كان لاحركة أصغر منها في الوجود إلا بالفرض، فيقف الكلام إلى أن نوضح عن أمر هذا المذهب.

(٦) وأما الأول في الحركة الذي يكون بتقسيمنا إياها موازيا لقسمة المسافة التي لاتقف عند حدّ في القسمة فإنه لايكون مقدار ذو ابتداء وانتهاء غير منقسم الى ما يصحّ

which can be supposed to be *first*.[8] The same holds for whatever parallels the magnitude in that [respect], for it likewise will not stop at some limiting point that has some beginning and ending and [yet] cannot be so divided further. Consequently, no smallest motion could exist in a continuous motion so as to exist as a part of what is continuous. That is because the part in what is continuous is assumed to be actual only by assigning limiting points in one of the ways mentioned earlier.[9] Now, the assignment of limiting points cannot simply stop being possible. Only on account of an actual separation and discontinuity—in which case there would no longer be something continuous—could it come to stop, where this separation and discontinuity seem to terminate at limiting points that cannot be separated and discontinuous. If, however, it is possible to posit some division at them by assigning limiting points, then the division of the continuous, which does not involve separation in some way, is infinite, where these sorts of divisions of [the continuous] will be equivalent, none of them having some special status that the others do not. In that case, the smallest of motions is still divisible in this way, while perhaps being indivisible in another way. In other words, there is no motion, which actually passes from some starting point to some end point at which it is complete, that is actually the smallest. When this is the form, motion has no first part in this sense except as the limit, unless certain motions follow upon one another successively, not continuously, where the prior of which would have this description. As for what is continuous, no first part exists having this description, because no motion is found along [a continuum] that is in itself independent [and] discontinuous; rather, the parts of that motion are continuous

8. **Y** seems to have inadvertently omitted a number of lines through homeoteleuton, which appear in **Z**, **T**, and the Latin. They are *idh kāna lā ḥarakah aṣgar minhā fī al-wujūd illā bi-l-farḍ fa-yaqifu al-kalām ilá an nūḍiḥu ʿan amr hādhā al-madhhab. wa-ammā al-awwal fī al-ḥarkah alladhī yakūnu bi-taqsīminā iyyahā muwāziyan li-qismat al-masāfah allatī ʿinda ḥadd fī al-qismah fa-innahu lā yakūnu miqdār dhū ibtidāʾ wa-intihāʿ ghayr munqasim ilá mā yaṣiḥḥu an yafriḍa awwalan;* corresponding with "since no motion in existence is smaller than it, save by supposition. So the discussion comes to a standstill until we clarify this view. As for the *first* that would result in the motion by our dividing [the motion] corresponding with division of the distance (the division of which does not stop at some limiting point), there [simply] is no magnitude possessing a beginning and end such that it is indivisible at that which can be supposed to be *first*."

9. See 3.2.8.

أن يفـرض أوّلا وكذلك ما يحاذي المقـدار في ذلك؛ فإنّه أيضاً لا يقف عند حدّ يكون له ابتداء وانتهاء، ولا ينقسم هذا النحو من الانقسام. فإذا كان كذلك، كانت الحركة المتصلة لا يجـوز أنْ يوجد فيها ما هـو أصغر حركة، على النحو الذي توجد جزءاً في المتصل. وذلـك أنَّ الجزء في المتصل إنّما يفرض بالفعل بتعيـين الحدود على أحد الوجوه المذكورة. وليـس لتعيـين الحدود وقوف البتة فـي الاحتمال، إنّما الوقوف عسـى أنْ يكون للتفريق والتقطيـع بالفعل، وحينئذ لا يكون متصلاً البتة. ويشـبه أنْ يكون هذا التفريق والتقطيع يتناهى إلى حدودٍ لا يمكن تفريقها وتقطيعها. وإنْ أمكن فرض قسمة فيها بتعيين الحدود، فتجزئـة المتصل الذي يقـع لا على وجه التفريق، غير متناهٍ، وأصنـاف هذه التجزئة فيه متسـاوية، ليس بعضها أولى من بعض. فأصغر الحركات لا يعدم هذا النحو من التجزئة، عسـى أنّه يعدم التجزئة بنحو آخر؛ أي لا تكون حركة خارجة إلى الفعل عن مبدأ وإلى منتهى، يتمّ عنده بالفعل أصغر منها. وإذا كانت الصورة هذه، فلا يكون للحركة أوّل جزء بهـذا المعنى إلاّ الطرف، إلاّ أنْ تكـون حركات متتالية غير متصلة ومُتقدمها بهذه الصفة. وأمّـا في المتصل فلا يوجد جزء أوّل بهذه الصفة، لأنّه لا توجد فيه حركة منفردة منقطعة بنفسـها بل تكون أجزاء تلك الحركة متصلة بعضها ببعض. فلو كان في جملة تلك الحركة

with one another. So, if, in the totality of that motion, some motion were the first that something produces (in the sense that [the motion] is a part of what is continuous, no part in the continuous thing being smaller than it), then that part of the motion would undergo the division that removes continuity[10] (which is our focus, since we are supposing that the division of the entire motion to this first is such that it does not remove the continuity). Now, if this part of the motion were not susceptible to division in this way, the first motion would be wholly unextended, and so no distance would be covered at all. In that case, there would be no motion. When motion is infinitely divisible in such a way as to preserve the continuity, then whatever part you stipulate to be *first* (in the sense of *part*, not *limit*) will potentially have some other first [part]. The same holds for rest and the phenomenon called *coming to a rest,*[11] which either increases the motion's speed, if it is natural [motion], or decreases it, if it is not natural but forced, in both cases leading to rest. The same holds for the accidental things that accompany motion, such as separating, joining, drawing near, and shattering (which is a certain scattering by motion).[12] There is no period of time for arriving and becoming contiguous and what is similar to that, [so] being first is denied of them by way of absolute negation. (Later we'll explain that more clearly.)[13]

(7) To the question of whether that which has no parts (if it exists) can undergo motion, one finds in the works of the Peripatetics that that is absurd and that what is indivisible cannot undergo motion.[14] One thing that they rely on to illustrate that is that whatever moves, first moves some [distance] equivalent to itself, and then again something equivalent to itself, and so on until it exhausts the distance.[15] So, if what

---

10. Literally, "then that part of the motion would not undergo the division that does not remove the continuity."

11. See the Arabic, *Physics* 6.8, where *tawaqquf* translates *histamenon.*

12. Reading *bi-ḥarakah* with **Z**, **T** (*bi-l-ḥarakah*), and the Latin (*cum motu*) for **Y**'s *yuḥarrikuhu* (that moves it).

13. The reference may be to 3.7.2, where Avicenna discusses what is meant by *absolute negation.* Alternatively, the reference may be to 3.12, where he argues that there is a smallest motion that can exist in actuality and reality, even if it is not the smallest motion that can exist potentially.

14. See Aristotle, *Physics* 6.10.

15. See specifically Aristotle, *Physics* 6.10.241a6–14.

حركة هي أوّل ما يحركها الشـــيء ، وكانت بمعنى أنّها جزء من المتصل لا جزء في المتصل أصغــر منه ، لم يكن يعرض لذلك الجزء من الحركة الانقســام الذي لا يبطل الاتصال الذي كلامنا فيه ، إذ وضعنا أنَّ انقسام الحركة كلّها إلى هذا الأول انقسامٌ لم يبطل الاتصال . ولو كان هذا الجزء من الحركة لا يقبل هذا النوع من الانقسام ، لكان أوّل الحركة ليس فيه امتدادٌ بتّةً - فلم يكن على مســافة البتة - فلم تكن حركة . وإذا كانت الحركة تنقسم الانقسام الحافــظ للاتصال إلى غير النهاية ، فكل ما جعلته أولاً - بمعنى الجزء لا بمعنى الطرف - فله أوّل آخر بالقوة . وكذلك الســكون ، وكذلك الشـــيء الذي يســمى توقّفاً ؛ وهو تزيّد الحركة في السرعة إنْ كانت طبيعية ، أو في البُطْء إنْ كانت غير طبيعية بل قسرية ، متجهاً بالوجهين إلى الســكون . وكذلك الأمور العارضة مع الحركة ؛ كالمفارقة والمقارنة والمجاورة والانكســار الذي هو افتراق ما بحركة . وأمّا الموافاة والمماسّــة وما أشبه ذلك ، فلا زمان لها ، ونفي الأولية عنها هو على السلب المطلق ، وسنوضّح القول في ذلك بعد .

(٧) وأمّا هل يجوز أن يكون ما لا جزء له يتحرك - إنْ كان له وجود - فالموجود فــي كتب المشــائين أنَّ ذلك محال ، وأنّ ما لا يتجزأ لا يصــحّ أنْ يتحرك . والمعوّل لهم في إيضاح ذلك ؛ هو أنَّ كل متحركٍ فإنه يتحرك أوّلاً مثل نفســه ، وبعد ذلك أيضاً مثل نفسه ،

is indivisible undergoes motion, then the distance would be composed of indivisible parts, and the point would be a certain distance because it is what is first left behind. I don't find this argument satisfying at all. That is because this judgment does not extend to what moves essentially to the exclusion of what moves accidentally; rather, it is general to whatever has some position vis-à-vis something, whatever the position might be, and then steadily moves some distance away from it. If this [judgment], then, is not accidental to what stands in for the state of coinciding [with the distance equal to itself], it will not be accidental to what stands in for the place; and, if it is accidental to what stands in for the place, it will be accidental for that which stands in for the state of coinciding. So, if the point actually exists at the limit of some moving body, then, as a result of the [point's] accidental motion, it will describe a line over which it continuously had moved and with which it coincided. That line, however, would not be composed of points. Nor would it be said that, to the extent that that point coincides, it first coincides with something equivalent to itself, and, to the extent that it leaves, it first leaves something equivalent to itself, and, through coinciding again with something equivalent to itself, it makes [a line] follow in succession,[16] and so on until the line is finished. So, likewise, it would not be said of [that point] (were it independent) that it undergoes motion essentially and that, for instance, it has a place essentially and that it must, in fact, describe something like itself, one thing successively following after another. The fact is that this is not necessary. The motion has no first motion such that that would inevitably traverse something indivisible that is equivalent to itself; rather, its states of coinciding, at any instant

16. Reading *wa atlathu* with **T** and two of the MSS consulted by **Y** and **Z** for **Y**'s and **Z**'s *wa latlathu* (as voweled in **Y**), which I must confess to not understanding. The Latin *sequitur* (to follow) suggests the Arabic *talat* or perhaps *atlat,* as I have suggested.

وكذلك هَلُمَّ جرا ، حتى تفنى المسافة. فلو كان ما لا يتجزأ يتحرك؛ لكان تركيب المسافة مـن أجزاء ما لا يتجزأ ، ولكانت النقطة مسـافة لأنّها أوّل مـا تفارق ، وهذا الكلام ليس يقنعني بوجه. وذلك أنَّ هذا الحكم ليس يتناول المتحرّك بالذات دون المتحرك بالعرض ، بل هو عام لكل ما يكون موضوعاً أي وضع كان عند شـيء ، ثم يفارقه مستمراً على نسبةٍ مسـافية. فإنْ كان المستبدل للملاقاة لا يعرض له هذا ، فلا يعرض للمستبدل للمكان ، وإنْ عرض للمسـتبدل للمكان عَرَضَ للمستبدل للملاقاة. فإنْ كانت النقطة الموجودة بالفعل في طرف جسمٍ من الأجسام المتحرّكة ترسم بحركتها التي بالعرض خطاً ، تكون قد استمرت عليـه ملاقيةً له ، ولا يكـون ذلك الخطّ مؤلفاً من نقطة ، ولا يقـال إنَّ تلك النقطة أوْل ما لاقتْ لاقتْ مثل ذاتها ، وأوْل ما فارقتْ فارقتْ مثل ذاتها ، وَأَتلَته بملاقاة أخرى مثل ذلك ، وكذلـك حتى انتهى الخط. فكذلك لا يقال لها - لو أنّها كانت منفردة - تتحرك بذاتها ، ولها مثلاً مكان بذاتها ، ‹و› أنّها يجب أن تكون ترسـم بالفعل مثل ذاتها شيئاً بعد شيء علـى التتالي ، بل ليس هذا بواجب. ولا للحركة ، أوّل حركةٍ ، حتى يكون ذلك لا محالة

one cares to take, are with something equivalent to itself, where instants are not immediately between one another, but between them there is always a period of time, as we explained in responding [to the objection involving] the sphere's motion over the surface.[17] So, whenever you posit something coinciding with what is equivalent to itself, it has already traversed what does not correspond with itself—namely, the line. This proof is simply not necessary so as to be convincing.

(8) The proof that appears convincing to us is that whatever is moved essentially and essentially undergoes *corporeal* changes—not just because it changes—essentially has some position proper to it [call that thing *x*]. In that case, insofar as [*x*] separates the extremities of what surrounds it, should an indivisible point, similar to *x*, encounter *x*, either [that point] does not fully encounter *x* and, instead, a portion falls short of it, or that is not the case. If it has the former description [namely, a portion falls short], then, obviously, *x* is itself divisible. If it does not have this description, then, were a point to encounter *x*, it would correspond with *x* completely. Now, *x* has a distinct position, and whatever corresponds with what has a distinct position comes to have a distinct position [itself]; and so the point would have a distinct position separate from the position of the line. In that case, the line would terminate at a point before that one, and the same argument will arise about [that point]. In short, each point would come to have a distinct position and could be separate from the line, while the line terminates at some other point before them, which is absurd. So, from this, it is clearly obvious that the position of whatever is indivisible is not independently separate, but whatever is not such does not undergo the motions that in themselves are in place.

---

17. See 3.5.10.

قطعاً ممّا لا يتجزأ مثل ذاته، بل تكون ملاقاتها في كل آنٍ يفرض شيئاً مثل ذاته، ولآنات لا تتشافع، وبينها زمانٌ دائماً على ما أوضحناه في جواب حركة الكرة على السطح. فكلما فرضت ملاقية مثل ذاتها، تكون قد قطعت ما لا يطابق ذاتها وهو الخط، وهذه الحجّة ليست واجبة تقنع.

(٨) فيشبه أنْ تكون الحجة التي تقنعنا هي أنَّ كل متحرك بذاته، وكل متغيّر التغيّرات الجسمانية بذاته، لا لأجل أنّه يتغيّر، فله وضْعٌ بذاته يخصّه. فحينئذ لا يخلو إمّا أنْ يكون بحيث يفصل بين نهايات ما يحيط به، ويكون لو لقيته نقطة غير متجزئة مثله لم تستغرق ذاته لقاء، بل أصابت منه جانباً، أو لا يكون كذلك. فإنْ كان على هذه الصفة؛ فظاهرٌ أنَّ ذاته منقسمة، وإنْ لم يكن على هذه الصفة، كان بحيث لو لاقته نقطة طابقت ذاته بأسرها، وذاته لها وضْعٌ متميّز. وما طابق ذا وضع متميّز صار له وضْعٌ متميّز، فيكون للنقطة وضْعٌ متميّزٌ منفصلٌ عن وضْع الخط، فيكون الخط منتهياً دون تلك النقطة بنقطة، الكلام فيها هذا الكلام. وبالجملة تصير كل نقطة ذات وضْع متميّز، ولكل نقطةٍ انفصالٌ عن الخطّ، والخطّ ينتهي دونها بنقطةٍ أخرى، وهذا محال. فواضحٌ بيّنٌ من هذا أنَّ ما لا يتجزأ لا ينفصل وضعه منفرداً، وكل ما لم يكن كذلك، لم يتحرك الحركات التي بذاتها في المكان.

(9) The same holds for the state of other corporeal motions, and it follows that whatever changes so as to undergo corporeal[18] alteration and augmentation is also divisible. That is obvious in the case of augmentation, because it is an increase of some pre-existing thing. As for alteration, because the cause of alteration affects the encountered side of what is being altered before affecting the unencountered side, if [the cause] were to surround completely [what is being altered], it will affect what is near its exterior before affecting what is near its interior, since every case of change is divisible (with generation and corruption alone being indivisible). Some erroneously believe that there are cases of alteration that [occur] all at once, but that is because the event escapes perception on account of the brevity of time. The instantaneous occurrence of illumination all at once is, in fact, not an alteration primarily in bodies but, rather, is something that is a concomitant of surfaces in that they are visible, whereas it will become clear, concerning transparency in the air, that nothing at all happens to the air during the transparency; rather, what happens is only in the object of sight.[19] So, when the object of sight becomes visible as a result of light's shining upon it, then the air can convey [the light] to the body and so is called *transparent*. Because of this, when someone is deep in the shadows of a cave and the air between him and the object of sight is thick with shadows, but the object of sight is luminous, like a light is shining on it, the gloom of the air will not prevent one from perceiving it.

---

18. Adding *jismānīyah* with **Z**, **T**, and the Latin (*corporalibus*), which **Y** (inadvertently) omits.

19. See *Kitāb al-nafs* 3.1.

(٩) وكذلك حال الحركات الجسمانية الأخرى. ويلزم أنْ يكون كل متغيّر تغيّرات الاستحالة الجسمانية والنمو منقسماً. أمّا النمو فذلك ظاهر فيه؛ لأنّه ازدياد على أصل موجود. وأمّا الاستحالة الجسمانية فلأن تأثير المحيل في الجهة التي يلقاه المستحيل، أقدم من تأثيره في الجهة التي لا تلقاه. فإنْ كان مشتملاً عليه فتأثيره فيما يلي ظاهره أقدم من تأثيره فيما يلي غوره، إذ كان كل متغيّر منقسماً - وإنّما الكون والفساد هو الذي يكون غير منقسم. وأمّا الذي يظنّ في بعض الاستحالات أنّها تكون دفعة، فذلك لفوات الأمر الحسي لقصر زمانه. وأمّا الإضاءة دفعة فليس ذلك استحالة أولية في الأجسام، بل أمراً يلحق السطوح بأنْ تظهر. وأمّا الاشفاف في الهواء فسيتبين أن الهواء ليس يعرض له في الاشفاف شيء البتة، بل العارض إنّما هو في المرئي، فإذا صار المرئي بحيث تجوز رؤيته بإشراق الضوء عليه، أمكن الهواء إداءه إلى الجسم فسميَ مُشفّاً؛ ولهذا إذا ما كان الإنسان في كهفٍ بعيدٍ مظلم، وكان بينه وبين المرئي هواء مظلم جداً، وكان المرئي نيّراً أشرق عليه الضوء، لم تمنع ظلمة الهواء إدراكه.

Chapter Seven

The beginning of the discussion about the finitude and infinitude of bodies and people's opinions concerning that

(1) Let us now investigate the meaning of *infinite* and how it exists in natural bodies and their states. (For now, this is not the place to investigate things outside of natural philosophy—that is, to discuss whether there is an infinite with respect to number, power, or the like,[1] nor does anything in the following demonstrations touch those [topics].) [Instead,] our discussion should be about quantities possessing position and numbers that possess some ordered position, either in nature or in position, where our investigation into them will involve whether there is something infinite in them or whether this is absurd. Now, the first thing we should inquire into is what is to be understood by *infinite.* Afterwards, we must indicate the reasons prompting one to affirm an infinite in some way or other. We'll also mention the differences among the Ancients concerning it, followed by the truth that we ought to believe about it, and then we will refute the doubts surrounding it.

1. The proper place for such a discussion would seem to be the science of metaphysics, and while Avicenna has no appreciable discussion of the infinite in number in book 3 of his *Ilāhīyāt,* which is his most extended account of the philosophy of mathematics, he does have scattered, extended discussions of the infinite in book 6 (particularly chapters 2 & 4) where he discusses causes.

<الفصل السابع>

في ابتداء الكلام في تناهي الأجسام
ولا تناهيها وذكر ظنون الناس في ذلك

(١) فلننظر الآن معنى غير المتناهي كيف وجوده في الأجسـام الطبيعية وأحوالها؛ فإنَّ النظر في الأمور غير الطبيعية، وأنّها هل تكون غير متناهية في العدد أو في القوة، أو غير ذلك، فليس الكلام فيها لائقاً بهذا الموضع، ولا شـيء من هذه البراهين يتناول تلك. ويجب أن يكون كلامنا في الكميّات ذوات الوضع، وفي الأعداد التي هي ذوات الترتيب فـي الطبع أو في الوضع؛ وننظر من أمرها أنها هل يكون فيها ما لا نهاية له أو هذا محال. فـأول ما يجب أنْ نبحث عنه هو المفهوم مـن قولنا لا نهاية له. وبعد ذلك فيجب أنْ ندلّ على الأسباب الداعية إلى إثبات ما لا نهاية له على وجهٍ ما، ونذكر اختلاف القدماء في أمره، ثم نذكر الحق فيما يجب أنْ نعتقد فيه، ثم نبطل الشكوك في أمره.

(2) So we say that *infinite* may be spoken of literally and metaphorically. That which is spoken of literally may be spoken of either by way of absolute negation or not by way of absolute negation. That which is said [to be infinite] by way of absolute negation involves something of which what is concomitant with being finite is negated owing to [that thing's] having no quantity—as, for example, the way *point* is said to be infinite [that is, having no limit]. This is like when we say that sound is not seen because that which is concomitant with being seen—namely, color—is negated of it, since sound neither is a color nor has a color. [The literal infinite] that is not spoken of by way of absolute negation frequently is said of what, in fact, corresponds with the finite—that is, something whose nature or essence is of such a character that it can be finite, but then is not. This is said in two ways, one of which is that the species or nature of [the thing] is of such a character in itself that it can be finite, but the particular instance is not of such a character that it be [finite]. An example would be the infinite line (should there be one), for numerically one and the same line cannot be a subject of the finite and the infinite. Yet, in the opinion of those who posit an infinite line, the nature of line is susceptible to being finite. The doubt concerns only the infinite [line], in which case, if this line is infinite, it cannot be that the very same [line] at some other time is finite. This sense of *infinite* is what we want to inquire into—namely, that which whatever you take from it—and any of the things equal to that thing you took from it—you [always] find something outside of it.[2] The second [sense in which the literal infinite is not spoken of by way of absolute negation] is that it is of such a character that it accidentally has some endpoint,[3] but [that endpoint] does not exist in actuality. An example

---

2. Cf. Aristotle, *Physics* 3.6.206b33–207a2.

3. The terms *endpoint* here and *no endpoint* in the next sentence are translations of *nihāyah* and *lā nihāyah,* which are the same terms that were translated above as *finite* and *infinite.*

(٢) فنقــول : إنَّ مــا لا نهاية له يقال على الحقيقة وقد يقال على المجاز . فالذي يقال على الحقيقة فقد يقال على جهة الســلب المطلق ، وقد يقال لا على جهة السلب المطلق . والذي يقال على جهة السلب المطلق فهو أنْ يكون الشيء مسلوباً عنه المعنى الذي تلحقه النهاية ؛ بأنْ يكون لا كمّ له ؛ مثل ما يقال إنَّ النقطة لا نهاية لها ، وهذا كما نقول إنَّ الصوت لا يُرى ؛ لأنّه مســلوبٌ عنه المعنى الذي يلحقه أن يُــرى وهو اللون ، إذ ليس الصوت بلونٍ ولا ذا لون . وأمّا الذي يقال لا على جهة السلب ؛ فقد يقال لمقابلة التناهي بالحقيقة ، وهو أنْ يكون الشــيء من شــأن طبيعته وماهيته أنْ تكون له نهاية ثم ليست . وهذا يقال على وجهين : أحدهما على أنّه من شــأن نوعه وطبيعته أنْ تكون له نهاية ، لكنه ليس من شأنه بعينــه أنْ يكون له ذلك ، مثل الخط غيــر المتناهي لو كان ؛ فإنّه ليس يجوز أنْ يكون خطّ واحــد بالعدد موضوعاً للتناهي ولغير التناهي ، لكن طبيعة الخط قابلة لأن تكون متناهية عند مَنْ يضع خطّاً غير متناهٍ . إنّما الشــك في غير التناهي ، فإنْ هذا الخط غير المتناهي ليس من شأنه أنْ يكون هو بعينه وقتاً آخر متناهياً . وهذا المعنى ، من معنى غيرِ المتناهي هو الذي نريد أنْ نبحث عنه ، وهو الذي أي شيء أُخذت منه ، وأي أمثال أُخذت لذلك الشــيء منه ، وجدت شــيئاً خارجاً عنه . والثاني أنْ يكون من شأنه أنْ تعرض له نهاية ،

would be the circle, since it has no endpoint. I do not mean that the circumference does not limit the surface of the circle; rather, I mean only that the circumference [is without an actual endpoint], since there is no actual point in it at which the line terminates. Instead, it is continuous, with no division in it. Still, [the circumference] is of such a character that some point along it could be posited as that point that is [the circle's] limiting point. [That] is because within the circle there are as many potential points as you like having this description, [and they can] emerge into actuality either by severing [the continuity of the line] or by [one's merely] positing them, since there is no point that does not have this description (I mean as a limit of a line; and, again, there is no actual line here save the circumference). So these are the ways in which an infinite is spoken of literally. As for [the infinite] that is spoken of metaphorically, it is said of that whose endpoint cannot be reached by motion, nor is it delimited by motion. An example is the distance between Heaven and Earth, [which is said] to be infinite, even though it is finite. Also, [*infinite*] is said [metaphorically] in those cases where doing that is extremely difficult, even though it is possible, where being extremely difficult is likened to being impossible.[4] So these are the ways that *infinite* is understood.

(3) Now, our purpose in inquiring into the infinite concerns whether among bodies (whether through their magnitude or number) some are such that, anything you take from them, you always find something outside of them; for some people made the existence of that necessary.[5] There are a number of reasons for [believing] that. One of them is the truth of the claim that numbers can be added to or multiplied infinitely,

4. Literally, "the absence [of being possible]."

5. For the various views and arguments presented in this paragraph, see Aristotle, *Physics* 3.4.

لكنها غير موجودة بالفعل، مثل الدائرة فإنّها لا نهاية لها . لستُ أعني أنَّ سطح الدائرة غير محـــدود بحدّ هو المحيط، بل إنّما أعني المحيط، فإنّـــه ليس فيه نقطة بالفعل ينتهي عندها الخــط، بل هو متصل لا فصل فيه، لكنه من شـــأنه أنْ تُفرض فيه نقطة تكون تلك النقطة حدّاً لها . فإنَّ في الدائرة نقطاً بالقوة على هذه الصفة - كم شـــئت - تخرج بالفعل بقطع أو فرضٍ، إذ لا نقطة إلاّ وهي بهذه الصفة، أعني طرف خطٍ، ثم لا خطّ هناك بالفعل إلاّ المحيط. فهـــذه هي الوجوه التي يقال عليها لا نهاية بالحقيقة. وأمّا الذي يقال بالمجاز؛ فإنّه يقال لما لا يقدر على أنْ ينتهي ويحدّ بالحركة، كالطريق بين الأرض والسماء، إنّه لا نهاية له، وإن كان له نهاية. وقد يقال أيضاً لما يعْســر ذلك فيه - وإنْ كان ممكناً - تشـــبيهاً للعسير بالمعدوم، فهذه وجوه مفهوم لا نهاية.

(٣) وغرضنا أن نبحث عمّا لا نهاية له، من جهة أنّه هل يكون من الأجسام أجسامٌ هـــي بمقدارها أو بعددها؛ بحيث أي شـــيء أخذت منها دائماً وجدت شـــيئاً خارجاً عنه؛ فإنّه قد أوجب قومٌ وجود ذلك. والسبب في ذلك أمور منها صدق قول القائل إنَّ الأعـــداد تذهب في الازدياد والتضعيف إلى ما لا نهاية لـــه، أو أنّها لا تتناهى في ذلك.

or that there is no end to that. Consequently, something infinite belongs to them. The same holds for magnitudes with respect to divisibility. Another [reason] is what is supposed concerning time—namely, that it must extend infinitely into the past and future, [and] is not merely adding to some starting point in the finite [past] or some [finite] section. They say that, whenever time terminates at some first past [moment] or some last future [moment], its past must have some [moment] before and its future some [moment] after, saying that all of that is time, as we noted earlier.[6] Generation and corruption, which are supposedly uninterrupted, provide a further [reason], [for] there are those who suppose that [these] require infinite matter. So some of them made [the matter] one of the simple bodies, whether fire, air, or water. Others made it a body intermediate between two of [the simple] bodies, like those who made it the vapor intermediate between water and air. In short, they made it the body from which everything is believed to be generated. Others made [the matter] many infinite bodies, from which is combined a single body called a *mixture.* Still others made bodies infinitely many in number that are not joined together, but are separate and spread throughout an infinite void; [and] some of [these theorists] make [these bodies'] forms (which, in their view, are their shapes) infinite in species, while others stipulate that their species' forms are finite in number. They were driven to their belief that this is inevitable only because infinite generation requires that there be enough matter that it not run out. Some of them make the infinite a principle because there is the nature of the infinite, not because anything happens to be infinite.

---

6. See 2.11.4–5, where Avicenna discusses time's essential relation to *before* and *after;* and 2.12.1, where he argues that there cannot be an ultimate first or last moment of time, since there would either be a *before* it or an *after* it, respectively, and so a time.

فإذا كان كذلك، فقد وجد لها معنى أنّها لا تتناهى، وكذلك للمقادير في الانقسام. ومن ذلك ما يظنّ من أمر الزمان أنّه يلزم أنْ لا يتناهى فيما مضى ولا فيما يستقبل، امتداداً لا تضعيفاً فقط، مبتدأً من متناه، ولا قسمة فقط. قالوا: لأنّه كلّما انتهى الزمان إلى أوّل ماض أو آخر مستقبلٍ وجب أنْ يكون لماضيه قبْل ولمستقبله بعْد - أو على ما أشرنا إليه قبلُ - قالوا وذلك كله زمان. ومن ذلك أمر الكون والفساد، الذي يظنّ به أنّه أمرٌ غير منقطع، وهناك مَنْ يظنّ أنّه يجب أنْ تكون له مادة غير متناهية. فبعضٌ يجعلها جسماً من الأجسام البسيطة؛ ناراً أو هواء أو ماء، وبعضٌ يجعلها جسماً متوسطاً بين جسمين منها، كمَنْ يجعلها البخار المتوسط بين الماء والهواء، وبالجملة يجعلها الجسم الذي يعتقد أنّه يتكون منه كل شيء. ومنهم مَنْ يجعلها أجساماً كثيرة بلا نهاية يجتمع منها جسمٌ واحدٌ يسميه خليطاً. ومنهم مَنْ يجعلها أجساماً كثيرة بلا نهاية في العدد، لكنها ليست متلاقية بل منفصلة، مبثوثة في خلاء غير متناه. فمنهم مَنْ يجعل صورها التي عندهم أشكالها بلا نهاية في النوع، ومنهم مَنْ يجعل لأنواع صورها عدداً متناهياً. وإنّما ألجأهم إلى هذا ظنّهم أنّه لا بُدّ من ذلك، فإنّه يجب أنْ يكون للكون غير المتناهي مادة وافرة لا ينقطع إمدادها. ومن هؤلاء مَنْ يجعل غير المتناهي مبدأ؛ لأنّه طبيعة غير المتناهي، لا لأنّه شيء عرض له أنْ لا يتناهى.

(4) One of the ways in which the estimative faculty has led people to affirm that something is infinite is the imagination that every finite thing necessarily terminates at something similar to what is immediately observed. From that, it necessarily follows that every body terminates at a body and that the amassing and piling up[7] of bodies goes on infinitely. An imposition and judgment of the act of the estimative faculty plays a part in these ways [of proving the existence of the infinite], for the act of the estimative faculty does not impose upon a given thing some determinate limit; rather, the estimative faculty can always imagine something greater than it. So these are the ways prompting one to affirm that there is something infinite.

---

7. Reading *intiḍād* with **Z**, **T**, and the Latin (*ordinationem*) for **Y**'s *intiẓār* (expectation).

(٤) ومن الوجوه التي تدعو قوماً إلى توهم إثبات ما لا يتناهى، ما يتخيل من أنَّ كل متناهٍ فيلحقه أنْ يكون تناهيه إلى شيء على نحو المشاهدات، فيلحق من ذلك أنْ يكون كل جسم يتناهى إلى جسم، وأنْ يذهب ارتكام الأجسام وإنتضادها إلى غير النهاية. ومن هذه الوجوه مقتضى التوهم وحكمه؛ فإنَّ التوهم لا يضع لشيءٍ من الأشياء حدّاً يتعين عليه، بل دائماً للوهم أنْ يتوهم أزيد منه. فهذه هي الوجوه الداعية إلى إثبات ما لا يتناهى.

Chapter Eight

On the impossibility that either a body or magnitude or number in an ordered series is infinite, and that it is impossible that there be some infinite body that is moved either in its entirety or partially

(1) The first thing we say is that it is impossible that there exist as wholly actualized some unlimited magnitude, number, or [set of] numbered things having an ordered position either in nature or in position. That is because any infinite magnitude, as well as any of the infinity of numbered things possessing an order, would proceed toward an actual infinite either in [all of] their directions or in a single direction. If they are [infinite] in all of their directions, then let us posit a certain limiting point among them—such as a point in a line, or a line in a surface, or a surface in a body, or some unit in a numeric total and make it a limiting point—and our discussion will focus on it[1] inasmuch as we mark it off as a limiting point. Now, [starting] from it, we take some determinate part—for instance, AC, from AB, where [AB] is infinite in the direction of B. So, if some amount equal to CB were superimposed or laid next to AB (or you were to consider some other analogous relation between them), then either [CB] will proceed infinitely in the way AB does, or it will fall short of AB by an amount equal to AC. If, on the one hand, AB exactly corresponds with CB [in proceeding] infinitely, and CB is a part or portion of AB, then the part and the whole exactly correspond [with one another], which is a contradiction. If, on the other hand, CB falls short of AB in the direction of B and is less than it, then

1. Reading *natakallamu ʿalayhi* with **Z**, **T**, and the Latin (*loquamur*) for **Y**'s *nuʿallimu ʿalayhi* (we designate it).

<الفصل الثامن>

في أنه لا يمكن أن يكون جسمٌ أو مقدار
أو عددٌ ذو ترتيب غير متناهٍ.
وأنه لا يمكن أنْ يكون جسمٌ
يتحرك بكليته أو جزئيته غير متناه

(١) فنقول أوّلاً إنّه من المستحيل أنْ يكون مقدارٌ أو عددٌ أو معدوداتٌ لها ترتيب في الطبع أو في الوضْع، حاصلاً موجوداً بالفعل، غير ذي نهاية: وذلك لأنّ كل مقدار غير متناهٍ، وكل معدودات ذوات ترتيب لا نهاية لها، إمّا أنْ يكون ذهابها إلى ما لا نهاية له بالفعل في جهاتها، أو في جهةٍ واحدة. فإنْ كانت في جهاتها كلّها، فلنا أنْ نفرض حدّاً فيها كنقطةٍ في خط، أو خطٍ في سطح، أو سطح في جسم، أو واحد في جملة عددٍ ونجعله حدّاً، ونتكلّم عليه من حيث نحدّه حدّاً، ونأخذ منه جزءاً محدوداً مثلاً (آ ج) من (آ ب) غير المتناهي منه من جهة (ب) - فلا يخلو إمّا أنْ يكون (آ ب) لو أُطبق عليه مساوٍ لـ(ج ب) أو حوذي به، أو اعتبرت مناسبة بينهما؛ أن يكون ذاهباً فيما لا نهاية مذهب (آ ب) أو يقصر عن (آ ب) بمساوٍ لـ(آ ج). فإنْ كان (آ ب) مطابقاً لـ(ج ب) إلى غير النهاية و(ج ب) جزء أو بعض من (آ ب)، فالكل والبعض متطابقان، هذا خُلْف. وإنْ كان يقصر (ج ب) من جهة (آ ب) في جهة (ب) وينقص عنه فـ(ج ب)

CB is finite and AB exceeds it by the finite [amount] AC, in which case AB is finite; but it was infinite. So it becomes evidently clear from this that the existence of an actual infinite with respect to magnitudes and ordered numbers is impossible.

(2) Let us take up the other topic and say that no infinite body is subject to motion. That is because motion is intellectually understood only in one of two ways: either involving an exchange of place, or not. Now, it would be impossible for the infinite body to move so as to exchange place. On the one hand, if [the body] is infinite in every direction, there is no vacant place such that there [could] be an exchange. On the other hand, if it is infinite in one direction to the exclusion of another, then perhaps we can conceptualize some empty space; but, when [the purportedly infinite body] moves toward [that empty space], it must either vacate the opposite side or not. Now, if it does not vacate [the opposite side], there is no motion but, rather, increase and augmentation. If it does move and vacates [the opposite side], then the infinite side was finite. Furthermore, this motion cannot be natural or forced. It is not natural because [what moves] naturally is that which seeks its natural place,[2] and, as we have already concluded earlier,[3] every natural place is a limiting point. Now, every limiting point is something delimited, but what has no limit neither moves toward nor joins with what is delimited. As for forced [motion], we shall soon show that what is infinite is not subject to force.[4] Also, forced [motion] is away from the natural place, and so, when there is no natural [motion], neither is there forced [motion].

---

2. Literally, its "where."
3. See 2.8.9–10.
4. See 3.10.

متنــاهٍ ، و(آ ب) يفضل عليه بــ(آ ج) المتناهي . فـ(آ ب) متناهٍ ، وقد كان غير متناهٍ . فبيّن من هذا بياناً واضحاً أنَّ وجود ما لا يتناهى بالفعل في المقادير والأعداد المرتبة مستحيل .

(٢) ولنبتــدىء في نمطٍ آخر ونقول : إنَّه لا يكون جرْم لا نهاية له متحركاً ؛ وذلك أنَّ الحركة لا تعقل إلاّ على أحد وجهين : حركة يكون فيها اســتبدال مكان ، وحركة لا يكون فيها اســتبدال مكان . فأمّا الحركة التي يكون فيها استبدال مكانٍ فذلك ممّا يستحيل على الجرْم غير المتناهي ، أمّا إنْ كان غير متناهٍ من جميع الجهات ، فلأنّه لا يخلو عنه مكان حتى يســتبدله ، وأمّا إنْ كان غير متناهٍ من جهةٍ دون جهــة ، فربّما أمكن أنْ نتصور عنه فراغ . لكنــه إذا انتقل إليه لم يخلُ إمّا أنْ يخلى عن الجهة المقابلة لها أو لا يخلى ، فإنْ لم يخلُ فما انتقــل ، لكنه ربا ونما ؛ وإنّ انتقل وأُخلي فالجهة غير المتناهية متناهية . وأيضاً هذه الحركة لا يجــوز أنْ تكون طبيعية ولا قســرية . أمّا أنها لا تكــون طبيعية فلأنَّ الطبيعي هو الذي يطلــب أيْناً طبيعياً ، وكل أيْن - كما قد فرغنا منــه قبل - حدٌّ ، وكل حدّ فهو محدود والمحدود لا ينتقل إليه ما لا حدّ له ، ولا ينحاز إليه . وأمّا القســري ، فإنّا سنبيّن عن قريبٍ أنَّ ما لا يتناهى لا ينقســر . وأيضاً فإنَّ القســري يكون إلى خلاف الأين الطبيعي ، فإذا لم يكنِ طبيعي لم يكن قسري .

(3) Moreover, how could the simple body or its like be finite in one direction but infinite in another when its nature is identical throughout? In this case, either the boundary that splits it must be something that its [own] nature imposes on it, or it is some accidental force (that is, something outside of nature) that has taken hold of it. If it is something that its nature imposes, and [if] its nature is simple and identical throughout, then the effect it produces as a result of its nature must not vary so as to delimit one side while not delimiting another. If it is by force [that the simple body is finite in one direction but not another], then the nature of this body requires that it be infinite. In that case, it might have been that some agent just so happened to mark off [the simple body's] boundary and split it so as to make it finite, and so the infinite part of it would exist but would be delimited and split off from it. In that case, it would not terminate at some empty space or void, but [it] would terminate at what was split off, which is of its [same] genus and nature, and so, again, it would have no place into which it could undergo this manner of motion. Alternatively, [the agent] might have delimited it without separating certain things from it but, rather, with respect to making its quantity one that has a boundary in one direction but not another, such as when the finite body is accidentally made smaller when there is condensing and larger when there is rarefaction. In this case, however, this body would be of such a character that it could be [both] finite and infinite, where that would be through the effect of some active influence. Later, when we show that the body is not acted upon in this way, whether by a finite or an infinite active influence, we'll expose what is false about that.[5] As for the composite [body], it would be impossible that one area be infinite and another finite. [That] is because, were our

5. See 3.10.

(٣) وأيضاً فإنّه كيف يكون الجسم البسيط، وما يجري مجراه، متناهياً من جهة، وغير متناهٍ من جهة، وطبيعته متشابهة؟ فلا يخلو إمّا أنْ يكون الحدّ القاطع له أمراً تقتضيه طبيعته، أو يكون إنّما عرض له قسرٌ وأمرٌ، خارج عن الطبع، قد أدركه. فإنْ كان مقتضى طبيعته – وطبيعته متشابهة بسيطة – فمن الواجب أنْ لا يختلف تأثيره عن طبيعته حتى يتحدّد منه جانب ولا يتحدّد منه جانب. وإنْ كان بالقسر فتكون طبيعة هذا الجسم توجب أنْ يكون غير متناه. فإمّا أنْ يكون قد عَرَضَ أنَّ حادّاً حدّه وقاطعاً قطعه فجعله متناهياً؛ فيكون غير المتناهي منه موجوداً لكنه حُدّ دونه وقُطع عنه، فلا يكون تناهيه إلى فضاء أو خلاء. ولكن تناهيه إلى مقطوع من جنسه وطبيعته، فلا يكون له أيضاً مكانٌ يتحرك إليه هذا النوع من الحركة. وإمّا أنْ يكون حدّده من غير أنْ أَبان منه أشياء، بل من جهة أنّه جعل كمّه كمّاً ذا حدّ في جهةٍ دون جهة، كما لعارضٍ أنْ يجعل كمّ الجسم المتناهي أقلّ عند التكاثف وأكثر عند التخلخل. فيكون حينئذ من شأن هذا الجسم أن يقبل تناهياً وغير تناهٍ، وذلك بتأثير مؤثّرٍ، وذلك ممّا سنوضّح بطلانه بعد؛ حيث نبيّن أنَّ الجسم لا ينفعل هذا النحو عن مؤثّرٍ متناهٍ أو غير متناه. وأمّا المركب فلا يجوز أنْ يكون غير متناهٍ من جهة؛ ومتناهياً من جهةٍ، فإنّا لو توهمنا كل واحد من أجزائه قد تحرّك

estimative faculty to imagine that each one of its parts has moved to the finite area, then either the whole has ended up moving out of the infinite side (but that is absurd) or [else] [the whole] has not moved from there, but some parts have moved but not others (but this contradicts what was posited). So this is [what results] when you believe that the motion [of an infinite body can occur] through exchange of place.

(4) The other motion, which is not through exchange of place, is circular motion. Now, it must be such that either the rotation is completed or it is not altogether completed. In the case where the rotation is completed, there arises the impossibility of circular motion in something infinite that we mentioned in the chapter on the void.[6] If the rotation is not completed, then its completion is either impossible or not. If it is not [impossible], there is no absurdity in positing it and no absurdity necessarily follows from [positing] it; but, as we noted,[7] an absurdity does necessarily follow from it. If [the completion of the rotation] is impossible, then some posited part of it can move through a certain arc but cannot move through another arc, where what is moved, the distance it covers (if it does), the arc, and all the states are exactly identical. It is impossible, however, that this should be the case, for it is impossible that there should be two things that are formally identical [and yet] one is possible and the other is impossible. From this it becomes clear that an infinite body simply cannot undergo circular motion; and, equally, a finite body within an infinite body cannot undergo it, just as we made clear in the chapter on the void.[8]

---

6. See 2.8.8.
7. Ibid.
8. Ibid.

إلـى جهة التناهي ؛ لم يخلُ إمّـا أنْ يحصل للكل انتقال من الجانب غير المتناهي - وذلك محـال ، وإمّا أنْ لا يكون له انتقال مـن هناك ، فتكون بعض الأجزاء تحركت دون بعض ، وهذا خلاف ما فُرض . فهذا إذا جعلت الحركة باستبدال المكان .

(٤) وأمّا الحركة الأخرى التي لا يستبدل بها المكان فهي المستديرة ، فلا يخلو إمّا أن تتمّ الدورة وإمّا أنْ لا تتمّ البتة . فإنَّ تتمّ الدورة عرض ما قلناه في باب الخلاء من استحالة الاستدارة فـي أمرٍ غير متنـاه . وإنْ لم يتممّ الدورة فلا يخلو إمّـا أنْ يكون تتميم الدورة مسـتحيلاً أو لا يكون . فإنْ لم يكن ، كان فرضه غير محالٍ ولا يلزم منه محال . لكنه يلزم منه كما قلنا محال . وإنْ كان تتميم الدورة مستحيلاً ؛ فيكون - لجزءٍ منه - مفروض أنْ يتحرك قوساً ، ولا يكون له أنْ يتحرك قوساً آخر والمتحرك والمسافة فيه ، إنْ كان ، والقوس والأحوال كلها متشـابهة . وهذا مستحيل أنْ يكون ؛ فمن المستحيل أنْ يكون أمران متفقا الصورة لأمرٍ واحد ؛ أحدهما جائزاً والآخر مسـتحيلاً . فبيّنٌ من هذا أنَّ الحركة المستديرة ممّا لا تعرض البتة للجسم غير المتناهي ، وأيضاً لا تعرض لجسمٍ متناهٍ في جسمٍ غير متناهٍ ، على نحو ما أوضحنا في باب الخلاء .

(5) Now, it is said[9] that, if [an infinite body] were to move circularly, it would have a circular shape; also, its two radii would both be infinite, in which case what is infinite would be doubled; also, then, the interval between the line extending from the center that is posited as moving and the stationary line toward which or away from which the [first line] moves would become infinite, and, moreover, [that interval] would be traversed in a finite period of time [all of] which would be absurd.

(6) All of this, however, is something that I don't truly understand to the point of feeling confident in its soundness. That is because it has not been demonstrated to me during the course of their argument that whatever is moved circularly must have a circular shape or that what is infinite in one direction cannot be doubled. If they prove this by showing that what is infinite is not susceptible to increase, as well as why it is not so susceptible, and [if] thereafter [they] busy themselves with talk about the circle, then they have unduly burdened themselves with something they do not need. [That] is because showing that that is not susceptible to increase is sufficient and does not require that they use as a middle term the *half* and its double with regard to halving the diameter. It also might just be that the half is nothing but what is delimited, and the same holds for the double.[10] As for the report about the interval, it does not seem necessary to me that that interval between the two lines should become

9. Cf. Aristotle, *De caelo* 1.5.

10. Here, Avicenna is relying on one of the points he made in par. 1 above: that, if there were a magnitude infinite in all directions, we could still posit some limiting point within it. For example, imagine a line AB, which extends infinitely toward both A and B. Now, posit some point C within AB. Both CA and CB would be infinite, even though either is only "half" of the whole AB. Similarly, AB would be "double" either CA or CB. Avicenna's point, then, is that there is no demonstrative absurdity in assuming infinities of different sizes.

(٥) وأمّا الذي يقال إنّه لو كان يتحرك على الاستدارة لكان له شكلٌ مستدير، أو كان نصف قطريه كلاهما لا نهاية له، فيتضاعف ما لا نهاية له، ولكان البُعْد بين الخط المتحرك المفروض خارجاً عن المركز، والخط الساكن المنتقل إليه أو عنه يصير غير متناهٍ، ثم يلزم أنْ يقطع في زمانٍ متناهٍ، وذلك محال.

(٦) فجميع ذلك ممّاً لم أفهمه حق الفهم حتى أومن بصحته. وذلك أنّه لم يتبرهن لي، في تعليمهم، أنَّ كل متحركٍ على الاستدارة يجب أنْ يكون له شكلٌ مستديرٌ، ولم يتبرهن لي من تعليمهم أنَّ ما لا نهاية له في جهةٍ لا ضعف له. فإنْ بيّنوا هذا بإبانةٍ أنَّ ما لا يتناهى لا يقبل الزيادة، وبينوا أيضاً أنّه لِمَ لا يقبل الزيادة، ثم اشتغلوا بحديث الدائرة، فقد تكلفوا شططاً لا يلزمهم تكلّفه. فإنَّ إبانتهم أنَّ ذلك لا يقبل الزيادة يكفيهم، وغير محوج إياهم إلى أنْ يوسّطوا أمر النصف والضعف فيه من جهة تنصيف القطر، وعسى أيضاً أنْ لا يكون النصف إلاّ المحدود وكذلك الضعف. وأمّا حديث البُعْد، فإنّه ليس يجب عندي أنَّ ذلك

infinite at all. How could it be, when two lines enclose it?[11] Even if that were true [namely, that the interval is infinite], I would do away with mentioning that it would be traversed in a finite period of time and, rather, raise a quick contraction—namely, [that the interval] is infinite, but [that] two lines limit it, which is a contradiction. The reason it is not necessary is because when the interval is always increased, it is not the case that an infinite interval must occur there; rather, the increase will proceed infinitely, where every increase will involve one finite [amount] being added to another, in which case the interval will be finite. This is just like what you learned concerning number—namely, that [number] is susceptible to infinite addition, and yet, any number that occurs is finite without some number actually being infinite, since any given number in an infinite sequence exceeds some earlier number [in that sequence] only by some finite [number].

(7) So this is what I think; but perhaps someone else has some well-established way to prove that. At any rate, should someone want to prove that an infinite interval occurs, then the proof that they mention does not get it right as it should, and I wouldn't bet that anyone else will get it right. The fact is that they ought to say this: Let us posit a certain interval between two opposite points on two lines extending infinitely. Now, let us connect the [points] by a line that is a chord of the intersecting angle. So, because the extension of the two lines, which is infinite, is proportional to the increase of the interval [that is, the length of the chord], the increases to that interval are infinite. [Those increases] can also exist together equally, because the increases that are below will actually be

11. In Aristotle's original objection, the claim was that the interval—for example, the arc between the lines AB and AC in the diagram below—would become infinitely large if the two lines extended infinitely. Avicenna's point is that, no matter how far one extends the lines—even infinitely, the interval represented by the arc BC is always going to be limited by the lines AB and AC and so, strictly speaking, will be finite.

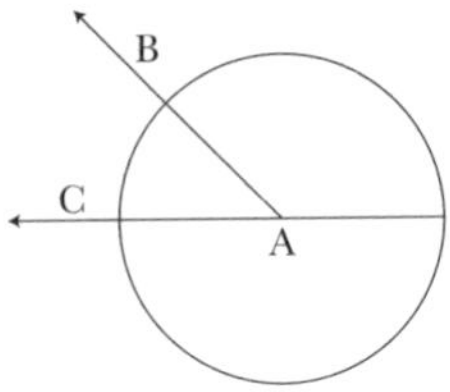

البُعْد بين الخطّين يصير البتة بلا نهاية؛ وكيف ويحيط به الخطّان الخارجان؟ ولو صحّ ذلك لا ستغنيتُ عن ذكر قطْع في زمانٍ متناه، بل كنتُ أُقيم خُلْفاً عن قريب، وهو أنّه غير متناهٍ ويحدّه خطّان، وهذا خُلْفٌ. وأمّا أنه لِمَ ليس يجب ذلك؛ فلأنَّه ليس إذا كان البُعْد دائماً يزيد يجب أنْ يحصل هناك بُعْدٌ غير متناهٍ، بل يكون التزيُّد ذاهباً إلى غير النهاية، وكل زيادة فهي بمتناهٍ على متناهٍ، فكل بُعْدٍ يكون متناهياً. وهذا كما تعرفه في أمر العدد، أنّه يقبل الزيادة إلى غير النهاية، ويكون كل عددٍ يحصل متناهياً، ولا يتحصل عددٌ لا نهاية له؛ لأنّه لا يزيد عددٌ في النظام غير المتناهي على عددٍ قبله إلاّ بمتناهٍ.

(٧) فهذا ما عندي، وعسى أنْ يكون عند غيري وجهٌ محقّقٌ لبيان ذلك. فإنه اشتهى أحدٌ أنْ يبيّن أنّه لا بُدّ من بُعْدٍ غير متناهٍ يقع، فليس طريق البيان ما يقولون ممّا لَمْ يحصّل فيه على وجهه، ولا نقدّر أنَّ غيرنا يحصله، بل يجب أنْ يقولوا هكذا؛ لنفرض بُعْداً بين نقطتين من الخطّين الذاهبين إلى غير النهاية متقابلتين، ولنصل بينهما بخطٍ يكون وتراً لزاوية التقاطع. فلأنَّ ذهاب الخطّين في زيادة البُعْد هو إلى غير النهاية، فإنَّ الزيادات على ذلك البُعْد موجودة بغير نهاية. ويمكن أنْ توجد متساوية؛ لأنَّ الزيادات التي توجد

joined to those that are above. For instance, the [amount that] the second increases the first will belong to the third, together with any other increase. So the infinite increases must actually exist in one of the intervals, and that is because the increases actually exist, and every actual increase will exist and so will belong to a certain one [of the intervals]. In that case, it necessarily follows that some interval will exist in which there is an actual infinity of equal increases. So that interval would increase the first finite [interval] by an infinite [amount], in which case there would be an infinite interval. Still, when this way is preferred, the contradiction obviously does not require [the introduction] of motion, because this infinite can exist only between two lines, in which case it is finite and infinite, which is absurd.

(8) We further note what is said about the parts of the infinite having to be at rest at every location while they are moved toward every location because every location belongs to it naturally.[12] This, however, is also something that I could not independently verify nor understand, for it is not necessary that when a single thing has [several] locations, each one of which belongs to it by nature, it must rest in and move away from each one of them. Indeed, instances of these locations, in any one of which the body might by chance occur, are between [the limits of] the whole of its universal location, where it naturally comes to rest and does not flee. A case in point is one of the parts of air in the whole of the region of air, and one of the parts of earth in the whole of the region of earth. Were this not the case, there would be no natural rest and motion, for the region [as a whole] always exceeds that which the parts [taken individually] occupy. Perhaps, however, there is some explanation of this way [of arguing against an infinite body] that I do not understand.

---

12. The argument seems to refer to that of Aristotle's *Physics* 3.5.205a12ff.

مـــا تحت تجتمع بالفعل فيما هو فوق، مثـــلًا أنَّ زيادة الثاني على الأول موجودة للثالث مع زيـــادة أخرى، فيجب أنْ تكون الزيادات غير متناهية موجودة بالفعل في بُعْدٍ من الأبعاد، وذلك لأنَّ الزيادات بالفعل موجودة وكل زيادةٍ بالفعل موجودة فهي توجد لواحد. فيلزم أنْ يكون بُعْدٌ موجودٌ فيه زيادات غير متناهية بالفعل متساوية، فيكون ذلك البُعْد زائداً على المتناهي الأول بما لا نهاية له، فيكون بُعْداً غير متناه. لكنه إذا فَضُل على هذا الوجه كان الخُلْف ظاهراً لا يحتاج فيه إلى الحركة، وذلك لأنَّ هذا غير المتناهي لا يمكن أنْ يوجد إلاّ بين الخطّين؛ فيكون متناهياً وغير متناهٍ، هذا محال.

(٨) ونقول أيضاً إنَّ ما يقال من أجزاء غير المتناهي يجب أنْ يســـكن في كل موضع وَيتحرّك إلى كل موضع، لأنَّ كل موضع له طبيعي، فهذا أيضاً ما لَمْ أتحقّقه ولَمْ أفهمه. فإنّه ليس يجب إذا كان لشيءٍ واحدٍ مواضع كل واحدٍ منها له بالطبع أنْ يلزمه، أنْ يسكن في كل واحد منها وأنْ يتحرك عن كل واحدٍ منها. فإنَّ أمثال هذه المواضع - أيّها اتفق للجسم الحصـــول فيه - من بين جملة الموضع الكلّي له، وقف بطبعه فيه ولم يهرب، كحال جزءٍ مـــن أجزاء الهواء في جملة حيّز الهواء، وجزء من أجزاء الأرض في جملة حيّز الأرض. ولولا هذا لما كان ســـكونٌ ولا حركةٌ بالطبع؛ فإنَّ الحيّز دائماً يفضل على مُشْغل الأجزاء، فعسى أنْ يكون لهذا وجه بيانٍ لم أفهمه.

(9) It is true, however, that that body's parts would not have a natural motion. That is because, on the one hand, should the body be infinite in every direction, there would be no location for its parts to seek by moving that is different [from the one] at the time the motion began. On the other hand, if it is [infinite] in one direction to the exclusion of another, such that the part is moved when it is outside the limiting point that is in the delimited direction, then the part inevitably will be moved to some place that is sought by nature. Still, that which the part seeks must be the very [place] that the whole seeks, whereas the whole will not naturally seek a place, since there is no place for it either generically like itself or unlike itself. By *generically like itself* I mean that there would be some surface similar to its surface, while by *generically unlike itself* [I mean] that there would be some surface dissimilar in its nature from its surface, just as, for us, [the surface] belonging to air [differs] from the surface of fire. So, when the nature of the whole does not seek some place that is proper to it and specified,[13] then neither does the part seek some place, because the region of the whole is homogeneous throughout [and, instead, the part] rests at any location where it, by chance, happens to be, and there is no region outside the region of the whole. That is [so] unless we stipulate that the whole is finite in one direction, in which case the region of the whole would necessarily be that which the part seeks—namely, that [region] where the whole is at rest—and so this region would either be some interval or something that is surrounded; but the doctrine of the interval is vacuous,[14] and what is infinite cannot be surrounded. Perhaps the part would then seek the whole by means of its natural motion so as

13. Reading *yataᶜayyanu* with two MSS consulted by **Y**, **Z**, **T**, and the Latin (*specificat*) for **Y**'s preferred *yataᶜallaqu* (to be associated), which, if retained, should include the preposition *bi* as part of its syntactical regimen, but which is, in fact, absent.

14. See 2.7.

(٩) وأمّا أنّه لا تكون لأجزاء ذلك الجسم حركة طبيعية فذلك صحيحٌ، لأنّه لا يخلو إمّا أنْ يكون الجسم غير متناهٍ في جميع الجهات، فلا يكون موضع مطلوباً لأجزائه بالحركة مخالفاً لمبدأ الحركة؛ وإنْ كان في جهة دون جهة حتى يكون الجزء يتحرك إذا كان خارجاً عن الحدّ الذي في الجهة المحدودة، فلا محالة أنَّ الجزء يتحرك إلى مكانٍ يطلبه بالطبع. ولكن الذي يطلبه الجزء يجب أنْ يكون هو بعينه الذي يطلبه الكل؛ والكل لا يطلب مكاناً بالطبع، إذ لا مكان له مجانسٌ ولا غير مجانسٍ؛ أعني بالمجانس أنْ يكون سطح شبيه بسطحه، وغير المجانس أنْ يكون سطح غير شبيه بسطحه في طبيعته؛ كما للهواء عندنا من سطح النار. فإذا كانت طبيعة الكل لا تطلب مكاناً ولا يختص لها ولا يتعّين، فطبيعة الجزء أيضاً لا تطلب مكاناً، لأنَّ حيّز الكل الذي هو متشابه يسكن في أي موضع اتفق، ولا حيّز خارجاً عن حيّز الكل، اللّهم إلاّ أنْ نجعل الكل متناهياً في جهة. فيجب حينئذ أنْ يكون حيّز الكل هو الذي يطلبه الجزء وهو الذي يسكن فيه الكل، فيُرى أنْ هذا الحيّز بُعْدٌ أو محيطٌ، والقول بالبُعْد باطل، ولا محيط لغير المتناهي. فعسى أنْ يكون

to come into contact with it, the first [place] of which would be along the nearest of the Heavens. This, however, is not the situation with natural bodies, which will, at some point, become clear to you from what we will teach you.[15] So, then, the part will not seek some place by nature; and whatever does not seek some place by nature does not move by nature, for the erroneous view that natural motion is not toward the natural place but toward the universal [place] or the like is something whose falseness has been explained to you.[16] So, from this, you know that bodies whose parts have a natural motion in delimited directions, the number of which is denotable, are all finite, and it is all that more obvious in the case of the body to which that [namely, having a natural motion in delimited directions] belongs entirely.

(10) We additionally say that bodies cannot be delimited in magnitudes while unlimited in number, for they must either be contiguous with one another or distinct and scattered with respect to place. On the one hand, if they are distinct from one another, then, were the estimative faculty to imagine them [all] contiguous with one another, the volume [occupied] by all of them in every direction would become smaller and closer to the center than the volume of what surrounds. In that case, the [occupied] volume would be finite and fall short of the initial volume by the amount that [the bodies] cover, up to the point where they [all become] contiguous with one another. So the initial volume would have been finite. Thus, the number of [bodies] in the finite volume containing them is finite, because the actual parts exiting in what is delimited is numerically delimited. From this you also know that it is impossible for a rectilinear motion to proceed infinitely, since you have learned that intervals are finite[17] and the preceding [showed] you that directions are finite. Also, [you know] that it would be impossible for some motion to

15. The reference would appear to be to 4.10.2–4.
16. The reference appears to be to 2.9.10.
17. See pars. 6–7.

الجزء يطلب الكل بحركته الطبيعية حتى يتصل به، وأوّلاه على أقرب السماوات، وليس الحال في الأجسام الطبيعية. هذا قد يتضح لك مما نعلّمه إياك. فإذن الجزء لا يطلب مكاناً بالطبع، وما لا يطلب مكاناً بالطبع فهو لا يتحرك بالطبع: فإنَّ الذي يُظنّ أنَّ أمر الحركة بالطبع هو إلى غير المكان الطبيعي، بل إلى الكلّية أو غير ذلك، أمرٌ تبين لك بطلانه. فتعلم من هذا أنَّ الأجسام التي لأجزائها حركة طبيعية إلى الجهات المحدودة العدد المشار إليها كلّها متناهية، فالجسم الذي ذلك لكليته أظهر.

(١٠) ونقول أيضاً إنّه لا يجوز أنْ تكون الأجسام محدودة المقادير غير محدودة العدد، فإنّها لا تخلو إمّا أنْ تكون متماسّة أو تكون متباينة مبثوثة في المكان. فإنْ كانت متباينة، فلو توهمناها متماسّة متلاقية صار حجم جملتها - من جميع الجهات - أصغر وأقرب إلى الوسط من حجم ما يحويها، فتكون متناهية الحجم وقاصرة عن الحجم الأول بمقدار ما قطعت من مقامها إلى التماس، فيكون الحجم الأول أيضاً متناهياً، فيكون عدد الموجود منها في حجم متناهٍ منها متناهياً؛ لأنَّ الأجزاء الموجودة بالفعل في كل محدودٍ محدودة بالعدد. ومن هذا تعلم أنّه لا يجوز أنْ تكون حركة ذاهبة إلى غير نهايةٍ في الاستقامة، إذ قد علمتَ تناهي الأبعاد، وسلف لك تناهي الجهات، وأنه يستحيل

move downward, for instance, when *down* is not delimited. The same would hold in the case of *upward.* For when *downward* can be delimited,[18] then its opposite must be able to be delimited; and, likewise, if *upward* can be delimited, then its opposite must be able to be delimited. Otherwise, neither it nor its opposite would exist. In that case, downward would have no opposite, and so downward would not be downward, since downward is such relative to upward.

(11) Also impossible is the claim of those who make the infinite *qua* infinite an element and principle, but not that *qua* some other nature, like [that] of water or of air, where that nature is accidentally infinite.[19] The proof that this claim is impossible is that that which is infinite is either divisible or not. If, on the one hand, it is indivisible, then it would not, in fact, be infinite[20] in proceeding in some direction, but [would be infinite] only by way of negation, just in the way that the point is said to be *infinite*—but this is not what they were getting at. Instead, they intended *infinite* in the sense that we can take whatever we want. On the other hand, if it is divisible, while not being divisible into some other nature (since there would be no nature other than the nature of the infinite *qua* infinite), then every part should have the nature of the whole, and the part that is surrounded and delimited by the division should also be infinite, which is absurd. So, from what we said, it has become clear that no body exists as infinite, nor is there any infinite body that is moved naturally, nor is there any elemental body that has some influence whose effect is infinite. Likewise, there are no numbers having a natural ordered position that are actually infinite.

---

18. **Y** has omitted by homeoteleuton the lines *ka-dhālika ḥāl al-ʿalūw. Fa-idha kāna al-sufl mutaḥaddid,* which occur in **Z**, **T**, and the Latin, corresponding with "The same would hold in the case of upward. For when downward can be delimited. . . ."

19. Aristotle attributes this position to the Pythagoreans; see Aristotle, *Physics* 3.4.203a4–8.

20. Reading *ghayr mutanāhin* with both **Z**, **T**, and the Latin (*infinitum*) for **Y** *mutanāhin* (finite).

أنْ تكون حركة إلى السُفْل مثلاً، والسُفْل غير متحدد، وكذلك حال العلو. فإذا كان السفل متحددا فمقابله لا محالة متحدّد. وكذلك إنْ كان العلو متحدّداً؛ فمقابله لا محالة متحدّد، وإلاّ لم يكن موجوداً؛ ولم يكن مقابلاً، فلم يكن للسُفْل مقابل، فلم يكن السُفْل سُفْلاً، لأنَّ السُفْل سُفْلٌ بالقياس إلى العلو.

(١١) ومن الكلام المستحيل قول مَنْ جعل غيرِ المتناهي، من حيث هو غير متناهٍ، أُسْطُقُسّاً ومبدأ. وليس ذلك من حيث هو طبيعة أُخرى، كمائيةٍ أو هوائيةٍ تلك الطبيعة يعرض لها أنْ لا تتناهى. والدليل على استحالة هذا القول أنَّ هذا الذي هو غير متناهٍ، إمّا أنْ يكون منقسماً أو يكون غير منقسم. فإنْ كان غير منقسم فليس هو غير متناهٍ من الجهة التي يذهب إليها، بل على سبيل السلب؛ كما يقال للنقطة إنّها غير متناهية، وليس إلى هذا يذهبون، بل يريدونه غير متناه، ليكون لنا أنْ نأخذ منه ما شئنا. وإنْ كان منقسماً، وليس ينقسم إلى طبيعةٍ أُخرى، إذ ليست هناك طبيعة غير طبيعة ما لا نهاية، من حيث هو لا نهاية، فيجب أنْ يكون كل جزءٍ في طبع الكل، وأنْ يكون الجزء المحاط المحدود بالقسمة منه أيضاً غير متناهٍ، وهذا محال. فقد وضح ممّا قلناه إنّه لا وجود لجسم غير متناهٍ، ولجسم متحركٍ بالطبع غير متناهٍ، ولجسم أُسْطُقسيّ مؤثّر متأثر غير متناهٍ، وكذلك لا أعداد لها ترتيبٌ بالطبع غير متناهية بالفعل.

(12) It remains, then, that we consider some other way that the infinite might exist in bodies and whether, in fact, it does or does not—namely, in the case of their augmentation. So we say that some of the Ancients believed that, just as one affirms of the body that it can be progressively divided without there having to be some minimal limiting point of which nothing is smaller, so, likewise, that holds on the part of largeness. [That] is because, just as those divisions do not occur together actually, but only as one after another such that they do not terminate at some limiting point of which none is smaller, so, likewise, the situation is the same concerning largeness. They said: Indeed, while it is impossible for an infinitely large body to exist actually, it is not impossible for it to proceed toward it, just like in the case of adding numbers.

(13) Since we maintain that in one sense this school of thought is right and in another it is not, let us then inquire into how it is and how it is not right. The sense in which this school of thought is right is because, in the estimative faculty, you can[21] imagine that a finite body is divided without stopping, and you can imagine yourself continually taking a part from what is divided and adding it to some other part or body so that [the other part or body] becomes bigger than it was. Moreover, you can take a part from what remains that is smaller than the remainder and add it to the first increase, and then continue doing that, so that each subsequent addition is smaller than the first. Now, the body increased by those additions will not reach the point at which the sum of the additions that make up [the body] equal all of the divided body. This manner of addition does not make the body reach every [degree of] largeness that

21. Reading *li-ka* with **Z**, **T**, and the Latin (*tu*) for **Y**'s (inadvertent) *dhālika* (the demonstrative "that").

(١٢) فبقي أنْ نتأمل نحواً آخر من وجود ما لا يناهى في الأجسام؛ أنّه هل هو ممّا يصحّ أو لا، وذلك حال نموّها، فنقول: قد ظنَّ، بعض المتقدمين أنَّه كما أنَّ للجسم أنْ يمعن ذاهباً في الانقسام من غير أنْ يقتضي حدّاً في الصغر لا أصغر منه؛ كذلك له ذلك في جانب العِظم. فإنّه كما أنَّ هذا الانقسام ليس يحصل بالفعل معاً، ولكن يحصل شيئاً بعد شيء، فلا ينتهي إلى حدّ لا أصغر منه، كذلك الحال في العِظم. قال: فإنّه، وإنْ استحال وجود عظمٍ للجسم غير متناهٍ بالفعل، فليس يستحيل السلوك إليه، كما الحال في تزايد الأعداد.

(١٣) فلننظر في هذا المذهب؛ ولنتأمل كيف يصحّ وكيف لا يصحّ. فنقول: إنّه يصحّ من وجهٍ ولا يصحّ من وجه. أمّا الوجه الذي به يصحّ هذا المذهب، فذلك لأنَّ لك – في التوهم – أنْ تقسم جسماً متناهياً قسمة لا يقف، ولك في التوهم أنْ لا تزال تأخذ جزءاً من المقسوم وتضيفه إلى جزء آخر أو جسم آخر، فيصير أكبر ممّا كان. ثم تأخذ جزءاً آخر من الباقي، أصغر من الباقي، وتضيفه إلى زيادةٍ أُولى، فلا يزال يزداد ذلك زيادة كل تالٍ منها يكون أصغر من الأول، ولا يبلغ الجسم المزيد عليه تلك الزيادات أنْ يساوي جملة

can by chance be; rather, [the body] will have a delimiting point that it will never reach, let alone exceed.[22] The manner of addition, however, by which the body could be augmented until every limiting point in largeness arises or is exceeded is impossible and is not analogous to smallness, for division does not need something outside of the body, whereas [this is not so for] augmentation and increase. [That is because] augmentation and increase, on the one hand, might occur through joining some matter to the original stock, where this would require that the materials of the bodies be infinite. On the other hand, [augmentation and increase] might occur though rarefaction and spreading out without stopping. This is impossible, because everything that undergoes rarefaction does so in some void or plenum, all of which is finite, as you have already learned.[23] Also, specifically, the void has no existence. Again, [it is also impossible] because there cannot be some motion, which necessarily entails some direction, unless it has some limiting point.

---

22. An example of Avicenna's point is given in Aristotle's *Physics* 3.6.206b4ff, where, for example, one imagines some magnitude ABC that is divided in half. Then BC, for instance, is divided again in half, and, half of BC is added to AB. One then continues in this fashion, such that progressively smaller divisions of BC are made and those divisions are, in turn, added to AB. In this manner, AB can be infinitely augmented, but it never actually becomes as large as the initial magnitude ABC, let alone larger.

23. See par. 3.

الزيادات التي يحصل منه، جميع الجســم المقسوم. وهذا الضرب من الزيادة لا يبلغ بالجسم كل عظــم اتفق، بل له حــدّ لا ينتهي إليه البتة، فضلاً عن أنْ يزيد عليه. وأمّا الضرب من الزيادة، التي من شأنها أنْ تنمّي الجسم حتى توافي كل حدّ في العِظم أو يزيد عليه؛ فذلك متعذرٌ وليس على قياس الصغر، فإنَّ القسمة لا تحتاج إلى شيءٍ خارج عن الجسم. والنمو والتزيّد يكون إمّا بمادةٍ تنضمّ إلى الأصل، وهذا يوجب أنْ تكون موادّ الأجسام بلا نهاية، وإمّا بتخلخلٍ وانبســاطٍ لا يقف، وهذا يســتحيل لأنّه يحتاج كل متخلخلٍ أنْ يتخلخل في حيّــز خلاءً أو ملاء؛ وكل ذلك متناهٍ كما قد عُلم، والخلاء خاصة لا وجود له، ولأنّه لا يجوز أنْ تكون حركة تقتضي جهة إلاّ ولها حدّ.

Chapter Nine

An explanation of the way that the infinite does and does not enter into existence, and a refutation of the arguments of those who defend the existence of an actual infinite

(1) Since all of this has become clear, we are ready to learn how the infinite can exist with respect to divisibility and numerical increase. So we say that *infinite* sometimes includes within its meaning [(1)] the things that that [term] describes, while at other times we mean by it [(2)] the very reality of what is infinite. It is just like when we speak of twenty cubits[1] and sometimes mean by it the timber that is twenty cubits [in length] and at other times the nature of this quantity. We likewise speak of this very nature that it is infinite, by which we mean that it is such that anything you take from it, you will find, without repetition, some part of it existing outside. When we say that, we mean that [the infinite] never reaches some limiting point at which it stops, so as to terminate at it. So, therefore, it is not later limited—that is, it does not reach some limit that cuts it short.

(2) As for those things belonging to the natures we mentioned that are said to be infinite, it is correct for us to say that they exist potentially, although not altogether, but each one [individually]. So each one of the things whose number is infinite exists potentially one after another; but the whole [infinite] *qua* a whole does not exist, either potentially or actually, except accidentally owing to its parts (if something like that can

1. Approximately 30 U.S. feet, or 9 meters.

<الفصل التاسع>

في تبيين كيفية دخول ما لا يتناهى في الوجود وغير دخوله فيه، وفي نقض حجج مَنْ قال بوجود ما لا يتناهى بالفعل

(١) وإذ قد تبين هذا كله، فبالحري أنْ نعلم أنّه كيف يمكن أنْ يكون لما لا يتناهى في انقسام الجزء وفي تزيّد العدد، وفيما يجري مجرى ذلك وجود؛ فنقول إنَّ قولنا ما لا نهاية له، تارة يتناول الأمور التي توصف بذلك، وتارة نعني بها نفس حقيقة غير المتناهي، كما إذا قلنا هو عشرون ذراعاً؛ فتارة نعني به الخشبة التي هي عشرون ذراعاً، وتارة نعني به طبيعة هذه الكميّة. وأيضاً نقول لنفس هذه الطبيعة أنّها لا تتناهى، ونعني بذلك أنّها بحيث أيّ شيءٍ منها أخذتَ وجدتَ منه موجوداً من خارج من غير تكرير. ونقول ذلك ونعني به أنّها لم تصل عند حدّ تقف عليه فتتناهى عنده؛ فإذن هي غير متناهية بعد، أي غير واصلة إلى نهاية الموقف.

(٢) فأمّا الأمور التي يقال لها أنّها غير متناهية من الطبائع التي ذكرناها، فصحيح أنْ نقول إنّها موجودةٌ في القوة لا في الجملة، بل كل واحد. فتكون الأمور التي لا نهاية لعددها كل واحدٍ واحدٍ منها موجوداً بالقوّة؛ والكل بما هو كل غير موجود، لا بالقوة

be said). Concerning the nature of the infinite itself, the first sense is not something that belongs to these things, either potentially or actually. That is because, if it exists, either it is a certain accident belonging to something else (but we have explained that there cannot be something that is accidentally infinite),[2] or, insofar as it is infinite, it is in itself a subsisting nature or even principle that actually exists (a view held by a certain group, that we have already refuted).[3] The second sense, however, is something that actually always exists, for we always find divisibility to be actually such that it does not terminate at some limiting point beyond which there is no [other] limiting point that [can] come to exist. So you have now learned how something is potentially infinite and how it is actually infinite, and how it is neither in potency nor in actuality. Now, what is actual about it is not free of a certain potential nature. [That] is because the sense of that [namely, what is actual about the infinite] is that it does not lead to the disappearance of the potential's nature, but, instead, the nature of the potential is always preserved in it. So the permanence and reality of what is infinite is associated with a potential existence, and so it is associated with the nature of matter rather than the nature of form, which is in actuality.[4] Now, the whole either is a form or possesses a form, and so what is infinite is not a whole. From the things that we have explained, you know that what is infinite has a privative nature and does not encompass everything, as some have supposed,[5] but, instead, is [itself] encompassed by the form because it is the power of the material.

---

2. See 3.8.11.

3. Ibid.

4. Reading *bi-fiʿl* with **Y** and the Latin (*in actu*). In **Z** and **T**, the preposition *bi* is omitted, which gives the sense that the form simply is the actuality.

5. Cf. Aristotle, *Physics* 3.4.203b10–13, where he identifies this position with that of the majority of earlier natural philosophers.

ولا بالفعل إلاّ بالعرض من جهة أجزائه، إنْ كان قد يقال مثل ذلك. وأمّا طبيعة ما لا نهاية له نفسها؛ فالمعنى الأول منه غير موجود لهذه الأشياء لا بالقوة ولا بالفعل، وذلك لأنّه إنْ كان موجوداً فإمّا أنْ يكون عارضاً لشيءٍ آخر - وقد بيّنا أنّه لا يجوز أنْ يكون شيء عرض له أنْ يكون بلا نهاية - وإمّا أنْ يكون بنفسه طبيعة قائمة من حيث هو لا نهاية، هو الموجود بالفعل أو المبدأ أيضاً، على ما يراه قومٌ، وقد أبطلناه. والمعنى الثاني موجودٌ بالفعل دائماً، فإنَّ الانقسام دائماً نجده بالفعل لم يتناه إلى حدّ لا حدّ بعده في حدوث الوجود. فقد علمتَ أنَّ ما لا نهاية له كيف هو بالقوة وكيف هو بالفعل، وكيف هو لا بالقوة ولا بالفعل، والذي منه بالفعل فغير خالٍ عن طبيعة ما بالقوة، فإنَّ معنى ذلك أنَّه لم يتناه إلى زوال طبيعة القوة، بل طبيعة القوة محفوظة فيه دائماً. فيكون ما لا نهاية له، ثباته وحقيقته، متعلقة بوجودٍ ما بالقوة، فهو متعلّق بطبيعة المادة دون طبيعة الصورة التي هي بالفعل - والكل صورة أو ذو صورة - فما لا نهاية له ليس بكل. وتعلم من الأشياء التي بيّناها أنَّ ما لا نهاية له طبيعة عدمية، وليس هو محيطاً بكل شيء؛ كما ظنَّ بعضهم، بل هو محاطٌ بالصورة لأنَّه قوة الهيولى.

(3) Should someone say that being infinitely divisible is a property that follows upon quantity (which is a form), the response is that *divisibility* is said in two ways. One of them is [in the sense of literally] *becoming separated* and *discontinuity,* which follows upon quantity, owing to the preparedness of the matter. The other sense of *divisibility* is that it is in the nature of the thing that one thing in it is *posited* as different from some other thing and so on without coming to an end, which essentially follows upon magnitude. Now, the first [sense] inevitably involves some motion, whereas the second does not require motion. The first is true divisibility—namely, what changes the state of the thing—whereas the second is only a function of the estimative faculty. Now, the magnitude does not essentially receive the first [that is, true divisibility] at all, because that which receives something must remain together with that which is received. When that [true divisibility] happens, the initial magnitude ceases to exist, for the initial magnitude is nothing but that determinate continuity, not something that that determinate continuity is in. In fact, the magnitude, as you have learned repeatedly,[6] is the continuity itself, not some continuous thing resulting from a continuity in it. So, when the continuity is fragmented [into two parts], the initial magnitude is done away with, and two different magnitudes come to be; and two other delimited continuous things come to be actual after having been potential. [Otherwise], if the two were actual, then, in a single continuous thing, there would be two actual continuous things *ad infinitum.*

---

6. See, for example, 3.2, pars. 8-10.

(٣) فإنْ قال قائل: إنَّ الانقسام غير المتناهي خاصة يلحق الكميّة وهي صورة؛ فالجواب أنَّ الإنقسام يقال على وجهين: أحدها الافتراق والانقطاع، وهذا يلحق الكمّ، لأجل استعداد المادة، والآخر الانقسام بمعنى أنَّ في طبيعة الشيء أنْ يُفرض فيه شيءٌ غير الآخر، ولا يزال كذلك، وهذا يلحق المقدار لذاته. والأول لا بُدّ فيه من حركة، والثاني لا يحتاج إلى الحركة، والأول هو الإنقسام الحقيقي؛ وهو الذي يغيّر من حال الشيء، وأمّا هذا الثاني فهو أمرٌ موهوم. والأول لا يقبله المقدار لذاته البتة، لأنَّ القابل يجب أنْ يبقى مع المقبول: وذلك إذا عرض أبطل وجود المقدار الأول؛ فإنَّ المقدار الأول لم يكن إلاّ ذلك الاتصال المعيّن، ليس شيئاً فيه ذلك الاتصال المعيّن. فإنَّ المقدار - كما علمته مراراً - هو نفس الاتصال، ليس الشيء المتصل باتصالٍ فيه. فإنّه إذا عرض الإنفصال المفكّك أبطلَ المقدار الأول وأحدث مقدارين آخرين، وحدث متصلان محدودان آخران بالفعل بعد أنْ كانا بالقوة. ولو كانا بالفعل لكان في متصلٍ واحدٍ متصلان بالفعل بلا نهاية.

(4) Now, no one denies that the matter receives the divisibility that it does precisely because of the quantity belonging to it. [Still], there are people who apparently believe that the material has some form (namely, [the form of] corporeality) that disposes [the material] to always being divisible in the sense of becoming separated, while some other form prevents that or, at least, does not make it incumbent when it occurs. For example, they say that when the body is perpetually divided, no flesh will remain and, in fact, the [form] of flesh will cease, while [the form of] corporeality will remain. We'll need to consider this. Now, again, when we say that the quantitative form disposes the matter to the divisibility that is proper to it, that preparedness need not belong to a form, for it is not the case that what acts must do so in itself. Likewise, that form does not[7] have to remain together with the emergence of that which disposes it to act, for motion is that which brings the body to its natural rest and disposes it to it, but it does not remain together with that [rest]. [That] is because that activity of [the motion] is to produce a disposition, and so it must exist together with the production of the disposition. The same holds for the activity of the quantitative [form] that produces a disposition, whereas the division results from something else.

(5) The way that the infinite exists has now been made known. It occurs in numbers in being multiplied, whereas they are finite on the part of the unity. In magnitude, it occurs in being halved and decreased, whereas it is finite on the part of being multiplied. [That follows], since its being halved *qua* magnitude is a halving of it insofar as there is a number whose first part is the unit, whereas the unit is the principle [that is, the starting point] of numbers, since [numbers] begin from a unit and become two. Infinite divisibility belongs to motion because of the magnitude over which it [crosses].

7. Correcting **Y**'s *L-Dh-Ā* (?) to *lā,* following **Z**, **T**, and the Latin (*nec*).

(٤) ولا ينكر أنْ يكون الانقسام الذي تقبله المادة؛ إنّما تقبله بسبب وجود الكمّ لها، ويشبه أنْ يكون الناس يرون أنَّ للهيولى صورة تهيّئها للانقسام الدائم المفرّق وهو الجسمية، وصورة أخرى تمنع ذلك، أو لا تثبت عليه إذا وقع. كما يقولون إنَّ الجسم إذا قسّم دائماً فإنّه لا يبقى لحماً، بل تبطل اللحمية وتبقى الجسمية، وهذا يجب أنْ ننظر فيه. ثم ليس إذا قلنا إنَّ الصورة الكميّة تُهيىء المادة للانقسام الذي يخصّ المادة؛ وجبَ أن يكون ذلك الاستعداد للصورة، فليس ما يفعل فعلاً يجب أنْ يكون في نفسه يفعل، ولا أيضاً يجب أنْ تكون تلك الصورة باقية مع خروج ما تُهيّئه له إلى الفعل. فإنَّ الحركة هي التي تقرّب الجسم من السكون الطبيعي وتُهيّئه له، ولا تبقى مع ذلك؛ لأنَّ فعلها هو التهيئة، فيجب أنْ توجد مع التهيئة، وكذلك فعل الكميّة المُهيّئة. وأمّا القسمة فهي عن شيءٍ آخر.

(٥) فقد علم نحو وجود ما لا يتناهى؛ فالعدد يعرض له ذلك في التضعيف، ويتناهى من تلقاء الوحدة. والمقدار يعرض له ذلك في التنصيف والنقصان، ويتناهى من قبل التضعيف؛ إذ كان تنصيفه - من حيث هو مقدار - تضعيفاً له من حيث هو عددٌ أوّله واحد؛ والواحد مبدأ عددٍ، فإنّه يبتدىء من واحد ويصير اثنين. والحركة يعرض لها الإنقسام غير المتناهي بسبب المقدار الذي هي عليه.

(6) As for time, there is the preparedness of the division that the estimative faculty imagines in it, and so [the division] belongs to it precisely, inasmuch as it is a magnitude, and essentially, whereas what is actually determinate belongs to it on account of motion. Now, there is a difference between, on the one hand, what occurs actually and, on the other hand, the product of the estimative faculty and the preparedness. [That] is because magnitudes are essentially subjects, owing to the fact that the infinite divisibility produced by the estimative faculty and their having a preparedness is accidental to them, while the emergence of that into act is on account of something else. Wherever it is said that that [namely, infinite divisibility and being so prepared] accidentally belongs to time because of its motion, we mean the accidental thing that actually occurs as one thing after another infinitely. The nature of the preparedness belongs to time inasmuch as [time] is a magnitude; the motion does not provide it with that, but, rather, it makes the time to exist in such a way that it entails that preparedness. It is like when one counts, for example, [and] by counting (or in some other way) makes ten exist. In that case, he does not make [ten] so that it is even, but, rather, he makes it to exist, and its existence entails that it is even. As for motion *qua* traversal, just as being infinitely divisible is accidental to it, so, likewise, being multiplied and increasing are accidental to it. Since the property of finitude, as well as the absence thereof, do not follow upon the motion merely because it is essentially a quantity, they follow upon it because of some other quantity. Now, it is not the case that they follow on it because of the quantity of distance, since the distance is finite. So, therefore, they follow upon it because of some yet other quantity, which is time. So motion is the cause of time's existence, while time is the cause of motion's being either a finite or infinite magnitude. The mover, however, is the

(٦) وأمّا الزمان فإنَّ استعداد الموهوم من القسمة فيه، فإنّما يعرض له من حيث هو مقدار ولذاته. وأمّا المعيّن بالفعل فيعرض له بسبب الحركة، وفرقٌ بين الواقع بالفعل، وبين الموهوم والاستعداد، فإنَّ المقادير موضوعة بذاتها لأنْ تعرض لها القسمة الوهمية إلى غير نهاية ومستعدة لها. وأمّا خروج ذلك إلى الفعل فيكون بسبب شيء آخر؛ وحيث يقال إنَّ الزمان يعرض له ذلك بسبب الحركة، فنعني العارض الذي يوقع بالفعل شيئاً بعد شيءٍ بلا نهاية. وأمّا طبيعة الاستعداد فهي للزمان من حيث هو مقدار، والحركة لا تفيده ذلك، بل توجد الزمان وهو على نحوٍ من الوجود يلزمه ذلك الاستعداد. كما أنَّ العادّ مثلاً إذا أوجد بالتعديد، أو بعمل آخر، عشرة فليس هو الذي يجعله زوجاً، بل يوجده ويلزم وجوده أنْ يكون هو زوجاً. أمّا الحركة، من حيث هي قَطْعٌ، فإنّها كما يعرض لها أنْ لا تتناهى في القسمة، كذلك يعرض لها أنْ لا تتناهى في التضعيف والزيادة. وإذ خاصّية التناهي وعدم التناهي، ليس إنّما تلحق الحركة بسبب كميّةٍ لذاتها، فتلحقها بسبب كميّةٍ أخرى. وليس تلحقها بسبب كميّة المسافة، إذ المسافة متناهية؛ فتلحقها إذن بسبب الكميّة الأخرى التي هي الزمان. فالحركة علّة لوجود الزمان، والزمان علّة لكون الحركة

cause of the motion's existence and so is the primary cause of time's existence as well as the cause of the motion's persistence, which is a first perfection. So, given that [the mover] persists, there will be an increase in the extension of its quantity, which is time. [The mover] is in no way the cause of time's *being prepared* to extend infinitely; but it is a cause of time's *extending infinitely* so that the motion will go on infinitely, for that [preparedness] belongs to time essentially, just as it did in the case of divisibility. Still, the actual existence of this formal aspect of the time is on account of the mover by means of the motion, just as its actually being divided is on account of some external agent that divides [it].

(7) So motion is the cause of this accident's belonging to time, while time is the cause of this accident's belonging to motion, except one is in one way and the other is in another. As for motion, after the motive cause it is a cause of this accident's really belonging to time, since the mover does not render motion discontinuous but is continuous with it, whereas time is a cause of motion's possessing an infinite magnitude, for time is a cause of motion's being measured. So, when it accidentally happens to be infinite primarily through that requirement of the motion, and the time makes it to exist in that way, then it is accidentally said of the motion by means of [the time]—not primarily, but, rather, owing to the fact that something accidental to it, which is time, is like that. So the motion accidentally makes itself to be like that. In other words, it makes something accidental to it to be like that; and, on account of that accidental thing, that [namely, being infinite] is said of [the motion]. That is something that occurs frequently, for many things [for instance, *y*] exists as *x*, where *x* has some primary description, and, due to that, [*y*] has that description as a secondary description and by secondary intention, where [the description] is not primary. So this is what we have to say concerning the manner in which the infinite exists.

متناهية المقدار أو غير متناهية، والمحرّك علّة لوجود الحركة، فهو علّة أُولى لوجود الزمان، وعلّـة لثبات الحركة التي هي كمالٌ أوّل، فيتبع ثباته إزدياد امتداد كميّته التي هي الزمان، وليس بعلّةٍ بوجهٍ لكون الزمان مسـتعداً لأنْ يمتد إلى لا نهاية. وعلّة كون الزمان ممتداً بلا نهاية حتى تصير الحركة بلا نهاية، فإنّ ذلك للزمان لذاته؛ كما كان في الانقسام أيضاً. لكن وجود هذا المعنى بالفعل للزمان؛ فهو بسـبب المحرّك وبوسـاطة الحركة، كما كان وجود الإنقسام له بالفعل بسبب شيء من خارجٍ قاسم.

(٧) فالحركة سببٌ لوجود هذا العارض للزمان، والزمان سببٌ لوجود هذا العارض للحركـة، لكن هذا بوجهٍ وذلك بوجه. أمّا الحركة فهي علّة بعد العلّة المحركة لوجود هذا العـارض للزمان بالحقيقة؛ إذا كان المحرّك لا يقطع الحركـة بل يصلها. وأمّا الزمان فهو علّة لكون الحركة ذات مقدارٍ غير متناه، فالزمان علّة لتقدّر الحركة؛ فإذا عرض له أنْ لا يتناهى عروضـاً أولياً بإيجاب الحركة ذلك، وإيجاده الزمان على ذلك، عرض بوسـاطته انْ قيل علـى الحركة ليس عروضاً أولياً، بل لأجل أنّ عارضـه الذي هو الزمان كذلك. فالحركة جعلت نفسها بالعرض كذلك؛ أي جعلت عارضها كذلك، ولأجل العارض يقال لها ذلك. وذلك ممّا يكون كثيراً؛ فإنّ كثيراً من الأشياء يوجد أمراً، لذلك الأمر صفة أولية، ويكون له من جهة ذلك تلك الصفة صفة ثانية وبالقصد الثاني، وليست أوّلية. فهذا ما نقوله في تحقيق كيفية وجود غير المتناهي.

(8) Concerning the proofs mentioned to establish [the existence of an infinite] that involved multiplying, division, generation and corruption, time, and the like,[8] know that we do not require that the infinite exist except[9] in the way we noted. What they say about every finite thing's terminating only in something else[10] is not sound. [That] is because, when it is further agreed that a single thing was finite and its endpoint is at something else, then it is finite *and* something that encounters another. Now, insofar as it is finite, it merely has an endpoint, where that is the sense of *being finite,* whereas, insofar as it is something encountering another, it terminates at something else; but its terminating at another is something that the encounter requires, not something that its being finite requires. [That] is because its being finite requires only that it possess an endpoint, while its terminating at something else means something additional to being finite. So, were it necessary that every finite thing encounter something else, either generically like itself or not, then perhaps their account would be valid, and every body would terminate at a body. As it stands, however, it is not necessary that every finite thing must encounter something generically like itself such that the body inevitably encounters a body; for you know that motion terminates at rest, which is either simply a privation or a contrary. As for the account of the operation of the estimative faculty,[11] that is something granted; but from that it does not necessarily follow that existing things are infinite in reality, but only in the act of the estimative faculty.

8. See 3.7.3.

9. Reading *ᶜalá ghayr* with **Z**, **T** and the Latin (*nisi*) for **Y**'s simple *ᶜalá* (according to).

10. See 3.7.4.

11. See 3.7.4.

(٨) فأمّا الحججُ المقولة في إثباته؛ فما قيل فيها من أمر التضعيف وأمر القسمة وأمر الكون والفساد والزمان وغير ذلك، فمعلومٌ أنّه لا نوجب لغير المتناهي وجوداً على غير النحو الذي نقوله. وأمّا ما قالوه من أمر أنَّ كل متناهٍ فإنّما يتناهى إلى شيءٍ آخر؛ فإنّه ليس بمسلّم، لأنَّه إذا اتفق أيضاً أنْ كان شيءٌ واحدٌ متناهياً ونهايته عند شيء آخر فهو متناهٍ وملاقٍ، ومن حيث هو متناهٍ فله نهاية فقط، ومعنى أنّه متناهٍ هو ذلك. وأمّا من حيث هو ملاقٍ فنهايته عند شيء آخر، فتكون نهايته عند شيء آخر أمراً تقتضيه الملاقاة وليس هو مقتضى تناهيه، فإنَّ مقتضى تناهيه هو أنَّه ذو نهاية فقط، وأمّا أنَّ نهايته عند شيء آخر فهو معني أزيد من معناه. فلو كان كل متناهٍ يلزمه أنْ يكون ملاقياً لشيء من جنسه أو غير جنسه، كان ربّما يصحّ قولهم، وكان كل جسم يتناهى إلى جسم، ولكن ليس يجب أنْ يكون كل متناهٍ ملاقياً لجنسه حتى يلاقي الجسم - لا محالة - جسماً؛ فأنت تعلم أنَّ الحركة تتناهى إلى السكون، وهو عدم فقط أو ضد. وأمّا حديث التوهم فليكن ذلك مسلّماً، لكن لا يلزم من ذلك أنَّ الموجودات لا تتناهى في الوجود، بل أنَّ الموجودات لا تتناهى في التوهم.

Chapter Ten

That bodies are finite with respect to influencing and being influenced

(1) We say that there cannot be an infinite body that either temporally acts on or temporally is acted upon by another body. It is impossible that one body act upon another in that way, because the body that is acted upon must be either finite or infinite. On the one hand, if it is finite, it is undoubtedly the nature that brings about the acting and being acted on between each one of them, not because it is finite or infinite. Now, if it is on account of the natures of the two that the patient is acted upon by the agent, then a part of one of them—namely, the patient—will tend to be acted upon by a part of the other. So, when, in a given period of time, some part of the infinite acts upon the finite (or some part thereof), then the ratio of that time to the time that it takes for the [whole, rather than a part of] the infinite to do that is proportional to the ratio of the power of what is finite [namely, a part of the infinite] to the power of what is infinite [namely, the whole of the infinite]. So [for example] the more massive the bodies are, the greater their power becomes, and the more effectively they act, and the time it takes for them [to act] decreases. It follows from that, however, that the action of something infinite should take no time at all; but it was assumed that it did take some time. If, on

<الفصل العاشر>

في أنَّ الأجسام متناهية من حيث التأثير والتأثّر

(١) ونقول، إنَّه لا يجوز أنْ يكون جسمٌ فاعلٌ في جسمٍ، أو منفعلٌ عن جسمٍ فعلاً وانفعالاً زمانياً؛ وهو غير متناه. أمّا أنّه لا يجوز أنْ يكون جسمٌ فاعل في جسم كذلك. فلأنَّ ذلك الجسم المنفعل لا يخلو إمّا أنْ يكون متناهياً أو يكون غير متناه. فإنْ كان متناهياً، فلا شك أنْ الفعل والانفعال يجري بينهما للطبيعة. كل واحدٍ منهما، لا لأنَّه متناهٍ أو غير متناه. فإنْ كان انفعال المنفعل عن الفاعل لطبيعتهما؛ فمن شأن جزءٍ من أحدهما - الذي هو المنفعل - أنْ ينفعل عن جزءٍ من الآخر، فإذا فعل جزء من غير المتناهي في المتناهي، أو في جزءٍ منه، في زمان، فتكون نسبة ذلك الزمان إلى زمان الذي يفعل غير المتناهي في ذلك بعينه، كنسبة قوة غير المتناهي إلى قوة المتناهي. فإنَّ الأجسام كلما كانت أعظم صارت قوتها أشدّ، وكانت أفعل، وزمانها أقصر. فيجب من ذلك أنْ يكون فعل غير المتناهي لا في زمان، وقد فُرض في زمان. وإنْ كان ذلك المنفعل

the other hand, the patient is infinite, then the ratio of some part of its being acted upon to the whole of its being acted upon is proportional to the ratio of two periods of time, in which case every part of it should be acted upon in no time. Now, the smaller part of it will be acted upon more quickly than a larger part is acted upon, since smallness of size is one of the factors that determines speed—and so something would occur more quickly than what comes to be in no time. Equally, when we assume that some part of the patient [call it *x*] is acted upon in no time, either what is adjacent to *x* [call it *y*] is acted upon together with *x*'s being acted upon (in which case all would be acted upon in no time) or *y* is acted upon afterward. In that case, let us assume another part [*z*] that is after *y*—but, then, that part [*z*] is either acted upon together with *y* (in which case what we already mentioned happens) or *z* is acted upon after *y* in no time (in which case instants would immediately follow upon one another, and the truth opposes this).[1]

(2) Since you have discovered this about acting, you are in a position to discover the counterpart of that about being acted upon. So, from this it is known that the elements—some of which temporally act upon others and in which, whenever they become more massive, there is an increase in power—are all finite. Now, one cannot object that the power of bodies is their forms and [that] forms do not become stronger and weaker. That is because, even though they do not become stronger in their substance, their influence does become increasingly stronger. I mean [that], although the form that is in *this* fire is neither stronger nor weaker in *this* fire and one like it, it is more powerful in twice the amount of fire; and also, a clod of earth that is twice as large [as another] is

1. See 2.11.6 and 3.4.3 for Avicenna's argument as to why instants—and, more generally, any indivisibles—cannot be adjacent to or immediately follow one another.

غيـر متناهٍ، فإنَّ نسـبة إنفعال جزءٍ منه إلى إنفعال الكلّ كنسـبة الزمانين، فيجب أنْ يقع إنفعـال كل جزء منه لا في زمان، ويكون إنفعال الجزء الأصغر من ذلك أسـرع من انفعال الجزء الأكبر، إذا كان الصغر مقتضياً للسرعة، فيكون شيء أسرع من الكائن لا في زمان. وأيضاً إذا فرضنا للمنفعل جزءاً فانفعل لا في زمان؛ فلا يخلو إمّا أنْ يقع انفعالٌ ما يليه مع إنفعالـه، فيكون إنفعال الجميع واقعاً لا في زمـان، وإمّا أنْ يقع بعده. فلنفرض جزءاً آخر بعده، فلا يخلو إمّا أنْ يكون ذلك الجزء انفعل معه فيعرض ما قلناه، أو انفعل بعده أيضاً لا في زمان، فتكون الآنات تتتالى، والحقّ يمنع هذا.

(٢) وإذ قـد عرفت هـذا من جهة الفعل، فلـك أنْ تعرف مقابـل ذلك من جهة الإنفعال؛ فمعلومٌ من هذا أنَّ الأُسْطُقسّات التي يفعل بعضها في بعض فعلاً زمانياً، وتكون كلّما عظمتْ إزدادت قوة، كلّها متناهية. وليس لقائلٍ أنْ يقول إنَّ قوة الأجسـام صورها؛ والصور لا تشـتد ولا تضعف؛ وذلك لأنّها - وإنْ كانت لا تشـتد في جوهرها - فإنّها تشـتد في تأثيرها في الزيادة؛ أعني أنَّه وإنْ كان لا يجوز أنْ تكون الصورة التي في هذه النار تشـتد وتضعف لا في هذه النار ولا في مثلها، فإنّها في ضعف النار تكون أقوى،

heavier. This does not mean that there is a greater increase *qua* substance. Instead, there is only an increase of the influence inasmuch as the form acts through certain accidents that do become stronger and weaker together with an increase in the number of forms, as well as [the accident's] becoming weaker as a result of the [increased] magnitude. (This kind of increase in the form is different from increasing the generable thing through[2] intensification, which you will learn about later.)[3] From this you also know that there is no infinitely strong power in any of the bodies to produce either forced or natural motion—as, for instance, the inclination [linked with] heaviness or lightness—for that would require that it act in no time. It is impossible, however, that there be some motion that takes no time. It would have to occur in no time precisely because the stronger the power becomes, the shorter the duration; and so, when [the power] is infinitely strong, [the duration] would become infinitely small.

(3) At this point, we need to consider the finitude and infinitude of powers; but before that, we note that there are differences between one power and another. These include [(1)] the variations in speeds, [(2)] the varying lengths of continual duration, and [(3)] the number of times [the power can] perform the action. An example of (1) is that the more powerful of two archers is the one who shoots the arrow so that it more quickly covers some designated distance. An example of (2) is that the more powerful archer is the one who, all things being equal, shoots the arrow so that it hangs in the air the longer period of time. Finally, an example of (3) is that the more powerful archer is the one who is capable of shooting one arrow after another the greater number of times. Since

2. Reading *bi* with **Z**, **T**, and the Latin (*per*) for **Y**'s *fī* (in, or with respect to).

3. The reference would seem to be to *Fī al-kawn wa-l-fasād* 6, where Avicenna discusses the difference between generation and alteration and uses the language of *ishtidād* (intensification) and *ḍuʿf* (weakening), or, in Latin, *intensio* and *remissio* respectively.

وفي ضعف المدرة تكون أثقل ، وليس هذا المعنى زيادة الشدّة في الجوهر ، بل في زيادة الأثر على أنَّ الصورة تفعل بأعراضٍ تشتد وتضعف مع تكثّر الصور وتضعُّفها تبعاً للمقدار ، وهذا نوعٌ من الزائد في الصور غير الزائد الكائن بالاشتداد ، وأنت تعلم هذا بعد . ومن هذه الأشياء تعلم أنّه لا يكون في جسم من الأجسام قوة على التحريك القسري أو الطبيعي غير متناهية الشدّة، كالميل الثقيل أو الخفيف؛ فإنَّ ذلك يوجب وقوع فعله لا في زمان، ويستحيل أنْ تكون حركة لا في زمان، وإنّما يجب أنْ يقع لا في زمان، لأنَّه كلّما اشتدت القوة قصرت المدّة، فإذا لم تتناه في الاشتداد بلغت من الصغر ما لا نهاية له.

(٣) فيجب أيضاً أن ننظر في حالِ القوى وتناهيها ولا تناهيها ، وقبل ذلك نقول إنَّ القوة يقع بينها وبين قوة أخرى تفاوتٌ في أُمور منها : سرعة ما تفعله وبطؤه، ومنها طول مدة استبقاء ما تفعله وقصرها ، ومنها كثرة عدّة ما تفعله وقلّتها . مثال الأول: إنَّ أشدّ الراميين قوة هو أسرعهما في الرمي لمسافةٍ معينة قطعاً ، ومثال الثاني أنَّ أشدّ الراميين قوة هو أطولهما زمان نفوذ الرمية في الجو مع تساوي المعاني الأُخر ، ومثال الثالث إنَّ أشدّ الراميين قوة هو أكثرهما قدرة على رمي بعد رمي . وإذا كان التفاوت يقع من هذه الوجوه،

these are the ways that [powers] differ, increase will occur in these ways. Also, in these ways there will be a greatest increase. So what goes on infinitely will occur in these ways, since powers in themselves have no quantity. Its quantity is only accidental, either relative to the thing in which there is the power or the thing over which there is a power. Now, the thing in which there is the power is always finite, since bodies are finite, whereas, if they were infinite, the power due to them would be infinite. Thus, it remains that a power is finite or infinite only relative to the quantity of that over which there is the power. So, when that thing can be infinite in the way that something can be infinite, the power relative to it will be infinite. Assuming, then, that there were some body that has power in one of the three ways and is infinite, let us consider whether its power would also be infinite with respect to that one of the three [ways in which powers differ].

(4) We say that if the more massive the body is, the more abundant and the greater the power is with respect to whichever of the three it is related, then, when it is infinite, its power must be infinite. Now, you know that two movers or agents (whatever the action might be) collectively have more power than one of them. [That] is because, collectively, they inevitably have the power over what one alone has power, as well as something more beyond that, since they have a power beyond the power of the one alone. Thus, the power of the more massive [body] is larger and more intense. So, necessarily, whenever [a body] becomes more massive, the power becomes greater and increases more. Now, the power of that which is infinitely massive must likewise increase infinitely with respect to that thing to which the power is related. Now,

فالتزايد يقع على هذه الوجوه، والأزيد يقع على هذه الوجوه. فالذاهب في الزيادة إلى غير النهاية يقع على هذه الوجوه، لأنّ القوة في نفسها لا كميّة لها، وإنّما كميّتها بالعرض؛ إما بالقياس إلى الشيء الذي فيه القوة، وإمّا بالقياس إلى الشيء الذي عليه القوة. والشيء الذي فيه القوة يكون أبداً متناهياً؛ إذ الأجسام متناهية، ولو كانت غير متناهية لكانت القوة تكون بسببها غير متناهية. فبقي أنْ تكون القوة إنّما هي متناهية وغير متناهية بالقياس إلى كميّة ما عليه القوة. فإذا كان ذلك الشيء جائزاً فيه أنْ يكون غير متناهٍ على نحو الجواز الذي لغير المتناهي، كانت القوة بالقياس إليه غير متناهية. فلننظر أنّه هل يجب لو كان جسم يقوى على أمرٍ من الثلاثة - وكان غير متناهٍ - أنْ تكون قوته أيضاً غير متناهية بالقياس إلى ذلك الأمر من الأمور الثلاثة.

(٤) فنقول: إنّه إنْ كان يجب أنْ يكون الجسم الأعظم أوفر قوة وأكبر في الأمر المقيس إليه من الأمور الثلاثة، فيجب إذا كان غير متناهٍ أن تكون قوته غير متناهية. وأنت تعلم أنَّ جملة محركين وفاعلين اثنين - أي فعلٍ كان - أكثر من قوة أحدهما، فإنَّ الجملة تقوى على ما يقوى عليه الواحد وعلى أمرٍ خارج عن ذلك لا محالة؛ إذ لها قوة خارجة عن قوة الواحد، فلذلك قوة الأعظم أكثر وأشدّ، فيجب أنْ يكون كلّما صار أعظم صارت القوة أكثر وأزيد. والذي يذهب إلى غير نهايةٍ في العِظم، فكذلك قوته تزداد إلى غير نهاية في الأمر المقيس إليه القوة، ولو كان المقيس إليه القوة متناهياً لكان لقوة

should that thing to which the power is related be finite, the power of some part of the body would stand in some ratio to a part of that over which there is the infinite power. So, when one part of the patient and one part of the agent are taken multiple times until [all the parts of] the finite patient are exhausted, where some finite collection of parts taken from the infinite body will be paired up with [the finite patient], then the ratio of one part of that which has the power to the powers of all those finite parts is proportional to the ratio of part of the patient to the whole of the patient. That, however, will be proportional to the ratio of the power of part of the supposedly infinite body to the power of the whole of the infinite body. In that case, a power of some finite part of the infinitely powerful body is equivalent to the whole of the body, which should exceed it by its power existing in the parts beyond that body. This is a contradiction, for [the whole] must be greater than [the part] in accordance with the ratio, but perhaps collectively it requires some greater power above that which the ratio necessitates. Clearly, then, if there were an infinitely massive body, the power would be infinite relative to what is subject to the power; but, since an infinite body is impossible, it is impossible that there be an infinite power of this sort.

(5) Let us now consider whether an infinite power can exist in a body that is not infinite, and whether there can exist an infinite[4] power relative to the speed of the action. We say that this does not exist. Otherwise, with respect to the speed, its action would take place in no time. Every speed, however, takes some time, because *speed* is a certain distance traveled (or the equivalent of some distance),[5] the whole of which is

4. Reading *ghayr mutanāhīyah* with **Z**, **T**, and the Latin (*infinita*) for **Y**'s *mutanāhīyah* (finite).

5. The "equivalent of distance" is rotations of a spinning object that does not leave its place, and so, strictly speaking, covers no distance. For Avicenna, traveling a certain distance would be motion with respect to the category of place, but, as he argued in 2.3.13–16, there is also motion with respect to the category of position, as is observed in the rotations of the celestial spheres.

جزءٍ ما من الجسم نسبة إلى جزءٍ ما من الذي عليه القوة غير متناهية. فإذا ضوعف من المنفعل جزءٌ ومن الفاعل جزءٌ إلى أنْ يفنى المنفعل المتناهي، وتحصل بإزائه من الجسم غير المتناهي جملة أجزاء متناهية، فكانت نسبة قوة الجزء الواحد من ذي القوة، إلى قوى جميع تلك الأجزاء المتناهية؛ كنسبة الجزء من المنفعل إلى جميع المنفعل، وذلك كنسبة قوة الجزء من الجرم المفروض غير متناهٍ إلى قوة جميع غير المتناهي، فتكون قوة جزء متناهٍ من هذا الجسم القوي غير المتناهي، مساوية لقوة الجسم كلّه الذي يفضل عليه بقوته الموجودة في الأجزاء غير المتناهية الخارجة عن ذلك الجسم، هذا خُلْف؛ فالواجب أنْ يكون أزيد منه بحسب النسبة، بل ربّما أوجب الاجتماع اشتداد قوةٍ فوق الذي توجبه النسبة. فبيّنٌ أنّه لو كان جسمٌ غير متناهي العِظم؛ لكان غير متناهي القوة بالقياس إلى المقوى عليه. ولمّا لم يجز أنْ يكون جسمٌ غير متناهٍ، لم يجز أنْ تكون قوة غير متناهية من هذا القبيل.

(٥) فلننظر هل يجوز أنْ توجد قوة غير متناهية لا في جسم غير متناه ولننظر هل يمكن وجود قوة غير متناهية بالقياس إلى سرعة الفعل؟ فنقول: إنَّ هذا لا يوجد؛ وإلاّ لكان فعلها في السرعة واقعاً لا في زمان، وكل سرعةٍ في زمان؛ لأنَّ كل سرعة هي قطْعٌ لمسافةٍ أو نظير مسافة، وكل ذلك في زمان. فلو كانت حركة لا نهاية لها في السرعة،

[divided] by time. So, were there an infinitely fast motion, there would be an infinitely small time; but this, as we know, is absurd. In short, speed applies only to things that exist in time. As for things that occur at an instant, [terms like] *fast* and *slow* are not said of them. Someone might say that infinite power acts at an instant, whereas the rest of the powers take time to act. So let us just posit that speed does not apply to the action of an infinite power. The response to that is that, in this chapter, we are concerned with instances[6] of local motions, which require that some distance be traversed and with respect to which the speeds vary. Now, [such a traversal] would be impossible unless it took some time, since it is impossible to traverse some distance in an instant unless the instant has been divided to correspond with the division of the distance. The same holds for whatever is analogous to local motions that vary in speed, because the occurrence of that necessarily requires time. So, [even] if something can take place at an instant and in time, our present discussion is not about it. Instead, [we are concerned] about things that do vary in speed and whose occurrence is never free of time and, indeed, as their power increases, their time decreases. So, if something pertaining to them occurs as a result of an infinite power, it will either take place at an instant—but that is absurd, since distance and its analogues are not traversed in an instant—or it takes some period of time. In that case, it would have a certain ratio to a given period of time that it takes for an action that results from a finite[7] power to occur. So it comes back to the fact that the ratio of the two times is proportional to the ratio of the two powers; and so the power that is infinite with respect to the proper object of its power will stand in a certain ratio to something finite that is finite[8] with respect to the proper object of its power.

---

6. Reading *amthāl* with two of the MSS consulted by **Y**, **Z**, and **T** for **Y**'s preferred *ithbāt* (establishing). There is nothing in the Latin that corresponds with either of these, although the sense of the Latin is simply "we are concerned with local motions."

7. Reading *mutanāhīyah* with **Z** and **T** for **Y**'s *ghayr mutanāhīyah* and the Latin's *infinita*.

8. Omitting **Y**'s *lā*, as in **Z**, **T**, and the Latin (*finitum*); if retained, the translation's "finite" should be "infinite."

لـكان زمانٌ لا نهاية له في القِصَر، وهذا محالٌ كما نعلم. وبالجملة إنّما تعتبر السـرعة في الأمور التي لها وجودٌ في زمان، وأمّا الأمور الواقعة في الآن فلا يقال فيها سرعة وبُطْء. فـإنْ قـال قائل إنَّ القوة غير المتناهية تفعل في آن، وسـائر القوى تفعل في زمان. فلنضع القوة غير المتناهية – على أنْ يكون فعلها لا سـرعة فيه. فالجواب عن ذلك إنّا نعتبر في هذا الباب إمثال الحركات المكانية التي توجب قطع مسافةٍ ما، وتختلف فيها في السرعة والبُطْء، ولا يمكن إلاّ في زمان، إذ لا يمكن قطع مسـافةٍ في آنٍ وإلاّ لانقسـم الآن بإزاء انقسام المسافة، وكذلك ما يجري مجرى الحركات المكانية ممّا يقع فيه سرعةٌ وبُطْء لضرورة حاجه وقوع ذلك إلى زمان. فإنْ كان شيءٌ يحتمل أنْ يقع في آنٍ وأنْ يقع في زمانٍ فليس كلامنا الآن فيه؛ بل كلامنا في الأمور التي تختلف بالسرعة والبُطء، ولا تخلو في وقوعها عن زمانٍ، فإنَّها كما تشـتدّ قوتها يقصر زمانها. فإنْ كان شـيء منها واقعاً عن قوةٍ غير متناهيـة كان إمّا في آن – وذلك محال – لأنَّ المسـافة وأمثالها لا تقطع في آن، أو في زمانٍ فتكون له نسـبة مّا إلى زمان فعلٍ واقعٍ عن قوة متناهية، فيعود إلى أنْ تصير نسـبة الزمان إلى الزمان كنسبة القوة إلى القوّة، فيصير للقوة التي لا تتناهى ما تقوى عليه، نسبة إلى المتناهية التي يتناهى ما تقوى عليه.

(6) Hence, if there is an infinite power, then that over which it has power will be in one of the two other ways [in which powers differ]—I mean duration and multiplicity [that is, the length and number of times the power can perform the actions]. Let us consider, then, whether this power that is infinite with respect to the proper object of its power—whether [with respect to] duration or multiplicity—can exist in some body such that, by dividing the body, [that power itself] would be accidentally divided. [First], however, *multiplicity* might [refer to] the multiplicity [of events] that follow one another successively from some determinate starting point according to some determinate order of position, which exactly mirrors duration; or it might [refer to] the multiplicity that is a mixture of different events in various orderings. For now, we need to set aside an inquiry into the infinite power that involves mixed multiplicity, since our discussion is not about it. Let us focus on power that involves a multiplicity that continuously follows a single order that parallels the duration, and then investigate whether there can be in bodies some power that involves multiplicity with this description and infinite duration.

(7) We say that that is impossible because this body can inevitably be divided into parts, and, with it, the power is also divided. Now, a part of this power must, at some given instant, either have or not have the [same] power to do what the whole has the power to do with respect to multiplicity and duration. [If the part has the same power as the whole], the power in both of them to do that thing would be [exactly] the same, and so the whole would not exceed the part in what it has the power to do; but this absurd. [If the part does not have the same power as the whole], either [the part] has a power over something generically like [what the whole has power over], or it simply does not. Now, it would be absurd that it not have power over something generically similar, for

(٦) فـإذن إنْ كانت قوة غير متناهية، فيكون ما تقوى عليه أحد الأمرين الآخرين؛ أعني المدّة والكثرة. فلننظر؛ هل يمكن أنْ يكون لهذه القوّة التي لا تتناهى ما تقوى عليه؛ كثرة أو مدة وجودٍ في جسـم حتى يعرض لها انقسـامٌ بانقسام الجسم لكن الكثرة إمّا كثرة متوالية من مبدأ محدود على ترتيب محدود يحاذي المدّة، وإمّا كثرة مختلطة من أشـياء مختلفـة، وفي تراتيب مختلفة. فيجب أنْ نترك الآن النظـر في القوة على كثرةٍ مختلطةٍ غير متناهية - فلا كلام لنا فيها - ولنبحث عن قوةٍ على كثرةٍ متصلةٍ من ترتيبٍ واحد، محاذية للمدة. فلننظر؛ هل يجوز أنْ يكون في الأجسام قوة على كثرةٍ بهذه الصفة، وعلى مدةٍ غير متناهية؟

(٧) فنقـول إنَّ ذلك لا يمكن، لأنَّ هذا الجسـم لا محالـة يتجزأ وتتجزأ معه القوة، وجـزء هـذه القوة لا يخلو إمّا أنْ يقوى على ما يقوى عليـه الكل في الكثرة والمدّة في آنٍ معين، فيكون المقوى عليه فيهما جميعاً في القوة شيئاً واحداً، فيكون لا فضل للكل على الجـزء في المقوى عليه، وهذا محال، وإمّا أنْ يكون لا يقوى عليه، فحينئذ إمّا أنْ يقوى على شـيءٍ من جنسـه أو لا يقوى على شيءٍ من جنسه البتة. ومحالٌ أنْ لا يقوى على

the power pervades the body that has it. So the part will have a power generically like that of the whole, and the object of the [part's] power will be of that genus that belongs to the whole. So, either both of them have the power, for example, to move one and the same thing, or the part has a power over something smaller than that thing. If, on the one hand, [that over which they have power] is one and the same, and the sum total of the power (the multiplicity and duration of which is infinite at a given instant) is over that which each one of them [is over], then they are equal with respect to that which they have the power to do; but this is absurd. On the other hand, if the part has the power to move something smaller, and the whole likewise has the power over that smaller thing, then either they are equal at a given instant in that over which there is power with respect to multiplicity and duration—which is absurd—or [the power of the part] is less and falls short [of the whole]. Now, when that over which the part has power falls short, its falling short with respect to its continuation will not be at the instant at which we assumed it began, but at some other limit. So, when, with respect to its being infinite, it falls short of the infinite, then, in that respect [namely, being infinite], the infinite exceeds it; but, to the extent that something exceeds it in a given respect, it is finite in that respect. Hence, the posited part is of a finite power relative to the duration of the action. The whole of the finite body, however, is related to the posited part by some determinate ratio, and so the power that is in the whole is related to it by some determinate ratio, where this ratio is relative to what the power does. So that over which the whole has power is related by some determinate ratio to that over which the part has power; but, then, the period of time that the whole [takes to perform the action related to its power] would be determinate [that is, limited], and the same would hold for the number of times [it performs that action].

شـــيءٍ من جنســـه. فإنَّ القوة تكون سارية في الجســم ذي القوة، فيكون للجزء قوة من جنس قوة الكل، ومقوى عليه من ذلك الجنس الذي للكل. فلا يخلو إمّا أنْ يكون، مثلاً، المقوى عليه الذي يُحرّكانه شـيئاً واحداً، أو يكون ما يقوى عليه الجزء أصغر من ذلك. فـإنْ كان شـيئاً واحداً، وكان جميع ما في القوة، ممّـا لا نهاية له، كثرة ومدّة من آنٍ معيّنٍ يقوى عليه كل واحد منهما، فهما سواء في المقوى عليه، وهذا محال. وإنْ كان ما يقوى الجـزء على تحريكه أصغر، والكل أيضاً يقوى علـى ذلك الأصغر، فإمَّا أنْ يكون المقوى عليه في الكثرة والمدّة من آنٍ معيّنٍ فيهما سواء - وذلك محال، أو يكون أقلَّ وأنقص، وإذا كان مـا يقوى عليه الجزء أنقص، لم يكن نقصانه في اتصاله من الآن الذي فرضنا الاعتبار منـه، بل من الطرف الآخر. فإذا نقص عـن غير المتناهي - في جهة كونه غير متناه - زاد غيـر المتناهـي عليه في تلك الجهة، وما زاد عليه شـيء في جهةٍ فهو متناهٍ في تلك الجهة. فيكون إذن الجزء المفروض متناهي القوة بالقياس إلى مدة الفعل، لكن جملة الجسم المتناهي تناسـب الجزء المفروض مناسبة محدودة، فالقوة التي في الجملة تناسبها مناسبة محدودة، . وهذه المناسـبة بالقياس إلى المقوى عليه، فالمقوى عليه الذي للجملة يناسب المقوى عليه الذي للجزء مناسـبة محدودة، فزمـان الجملة أيضاً محدود وكذلك عدده.

(8) Now, the discussion of these suppositions is just like the one involving the suppositions posited about the makeup of the plenum and the void.[9] That is because we do not need to consider whether these ratios actually exist. Instead, we say, in the way that geometers use suppositions, that, whenever the supposition that something stand in some ratio necessarily entails this status, it is finite. In short, it is not the nature of the power that prevents that, but the nature of things that don't exist. So we say that, if things were to exist in some given way, to that extent their nature would require such and such, whereas, if this power were an infinite power in a finite body, it would not be such that, were the things to exist as such, their nature would require such and such; and [yet] that is necessary for it to be.

(9) From this it is clear that, within a finite body, there cannot be an infinite power relative to the duration and the number of like-ordered events previously mentioned, whereas, relative to the number of mixed events, the case is perhaps not as straightforward, and so the same proof cannot be used for it. That is because the number of nonexistent future events is not necessarily finite, when [those events] fall short of some other number. It is possible, then, that future[10] events are infinite, and yet some fall short of others. An example is infinite motions that are faster and infinite motions that are slower, for the faster rotations will inevitably be greater than the slower ones. Likewise, the infinitely many multiples of ten are less than the infinitely many multiples of one, while [the multiples of ten] are greater than the infinitely many multiples of one hundred and one thousand.

9. See 2.8.12.

10. As a result of dittography **Y** has repeated the phrase *fa-yajūzu an takuna fī al-mustaqbal idhā kānat anqaṣ min ʿiddah ukhrá an takuna mutanāhīyah* (when future events are less than some other number that it is finite). The phrase is not repeated in **Z**, **T**, or the Latin.

(٨) والكلام في هذه التقديرات كالكلام في التقديرات التي فرضناها في قوام الملاء والخلاء، وذلك لأنّا لسنا نحتاج إلى اعتبار وجود هذه المناسبات بالفعل، بل نقول إنَّ ما تقدير مناسبته يوجب هذا الحكم فهو متناه، على التقديرات التي يفعلها المهندسون، وبالجملة ليس العائق في ذلك من طبيعة القوة، ولكن من طبيعة الأُمور التي ليست توجد. فنحن نقول إنَّ هذه القوة - بحيث لو كانت الأمور توجد على نحوٍما - لكان‹ت› طباعها توجب كذا وكذا، ولو كانت قوة غير متناهيةٍ في جسم متناهٍ لما كانت تكون؛ بحيث لو كانت الأمور توجد كذا ‹لكانت› طباعها توجب كذا وكذا، وذلك أوجب لها أن تكون.

(٩) بيّنٌ من هذا أنّه لا يجوز أنْ تكون في جسم متناهٍ قوة غير متناهية بالقياس إلى المدّة والعدّة المنتظمة المذكورة. وأمّا بالقياس إلى العدّة المختلطة فعسى الأمر أنْ يشكل فيه، فلا يمكن استعمال هذا البيان بعينه فيها؛ وذلك لأنّه لا يلزم أنْ تكون العدّة المعدومة التي في المستقبل إذا كانت أنقص من عدّةٍ أُخرى أن تكون متناهية. فيجوز أن تكون في المستقبل أُمور بلا نهاية، ولكن بعضها أنقص من بعض، كحركاتٍ بلا نهايةٍ هي أسرع، وحركات بلا نهاية هي أبطأ، فإنَّ دورات الأسرع لا محالة أكثر من دورات الأبطأ، وكذلك العشرات غير المتناهية أقل من الوحدات غير المتناهية وأكثر من المئين والألوف غير المتناهية.

(10) With respect to the time reaching up to now, a given period of time beginning now that is less than the infinite [time] beginning now can only be finite. Still, when there is that which has power over infinitely many mixed events for all of their ordered series, then it may have power over one of their ordered series, whether beginning at some designated unit or instant. When, however, the body does not have power over even a single infinite ordered series, it likewise does not have power over some mix of varying ordered series, and it is clear from what we [already] said that it does not have power over an infinite ordered series. (In this science, however, we do not discover why [that] is impossible in the case where the whole multiplicity of events completely lacks any order or that multiplicity is a certain genus in which there is no order.)[11] So it has become clear that a body cannot have an infinite power with respect to strength, duration, or number of times [the action is performed]. Should someone say that, in the celestial sphere closest to us, there is the power to rotate fire forcibly without discontinuity and that it is corporeal, we say, initially, that (as you will learn in its proper place)[12] that motion is an accidental motion because it is being moved by what is [itself] being moved; but, nevertheless, [the rotation of fire] perpetually results[13] from the motive cause belonging to the celestial sphere by means of the celestial sphere's motion.

(11) Now, we ourselves are not opposed to some infinite power moving a body and, by means of [that body], moving another one infinitely many times, when the infinite power is not something residing in either of the two bodies. We are opposed only to an infinite power's being in a body that moves that body or some other one. As for when [the power]

11. See, for instance, *Ilāhīyāt* 8.1, where Avicenna argues that the order of existing things must terminate at a First Principle of all existence.

12. See, for instance, *Ilāhīyāt* 9.5.

13. Reading *ʿan* with three of the older MSS consulted by **Y**, **Z**, **T**, and the Latin (*per*) for **Y**'s preferred *ghayr* (different from).

(١٠) وأمّا في الزمان المتصل من الآن، فلا يجوز أن يكون زمانٌ يُعتبر من آنٍ أقل من غير المتناهي المبتدىء من الآن، إلاّ متناهياً ولكنه إذا كان ما يقوى على كثرات مختلطة غير متناهيةٍ، ‹ل› كلِ ترتيبٍ منها، فقد يقوى على ترتيبٍ واحدٍ منها، مبتدئاً من وحدةٍ معيّنةٍ أو آنٍ معيّن. فإذا كان الجسم لا يقوى على ترتيبٍ واحدٍ غير متناهٍ؛ فكذلك لا يقوى على خلْطٍ من ترتيباتٍ مختلفة – وأمّا أنّها لا تقوى على ترتيبٍ غير متناه فذلك بيّنٌ بما قلناه، وأمّا إذا كان كل كثرةٍ منها غير منتظمة في ترتيب، أو تكون الكثرة جنساً واحداً لا ترتيب فيه، فلا يتبيّن لنا من هذا العلم امتناعه. فقد بان أنّه يستحيل أنْ يكون لجسم قوة بلا نهاية في الشدّة وفي المدّة وفي العدّة. فإنْ قال قائل إنَّ القوة التي في الفلك الأقربٍ إلينا تقوى على تحريك النار على الدور قسراً من غير انقطاعٍ وهي جسمانية، فنقول: أولاً، إن تلك الحركة – كما ستعلمه في موضعه – حركة بالعرض لتحرّك ما المتحرّك به فيه، ومع ذلك فهو عن السبب المحرّك للفلك دائماً بتوسط حركة الفلك.

(١١) ونحن لا نمنع أنْ تكون قوة غير متناهية تحرّك جسماً، وتحرّك بتوسطه شيئاً آخر حركاتٍ غير متناهية، ولا تكون القوة غير المتناهية مستقرة في أحد الجسمين. إنّما نمنع أنْ تكون قوة غير متناهية هي في جسمٍ تُحرّك ذلك الجسم أو جسماً آخر؛ فأمّا إنْ

is not in a body but it moves some body, and that body, on account of its being moved by [that power], moves another body infinitely, that is indisputably found to be the case. [That] is because nothing prevents there being an infinite power, wholly unmixed with bodies, that moves some body; then, owing to [that moving body], a number of large interconnected bodies are moved; and, from them, there is produced a system involving numbers of continuously generated things. What we are talking about precisely is the infinite power that is the foundation and principle of the system of an infinitely ordered series, whether [in] duration or number with respect to generation, or continuous motion, whether mediately or immediately; for we are quite adamant that that principle is never in a body.

(12) Someone might say that it is not impossible for the body to have some power over that which entails the existence of that [first] body, and [that] thereafter that [first] body is of a character that it perpetually continues on, and [that] then, from it, there perpetually results that production of motion or that number. The answer to this is that it is impossible on account of what we have [already] explained. The fact is that it necessarily follows from what we explained that none of the bodies has a power by which it perpetually acts on what is contiguous with it; rather, the power of every body is one by which it acts upon what is contiguous with it so as to produce discontinuous motions by bringing [the body] closer or farther away. In no body can there be some power—whose action is one, continuous, and homogeneous—that perpetually continues together with the continuation of the body. Instead, the power of

كانت لا في جسم، وتُحرّك جسماً ، ويحرّك ذلك الجسم - بسبب تحرّكه عنها - جسماً آخر حركة غير متناهية؛ فذلك ممّا هو موجود وليس عليه كلام. فإنّه لا مانع أنْ تكون قوة غير متناهية، بريئة عن مخالطة الأجسام، تحرّك جسماً فتتحرّك له أجسام كبيرة ملتحمة بــه، ويتولّــد عنها نظام في أعداد متكونة لا تنقطع - إنّمـا كلامنا في القوة غير المتناهية التــي هي أصل ومبدأ لنظام الترتيب غير المتناهي، مدّة كان أو عدّة في التكوّن أو حركة متصلة، وكان بواسطة أو بغير واسطة. فإنّا نجزم أنَّ ذلك المبدأ لا يكون في جسم دائماً .

(١٢) فإنْ قال قائل : إنّه ليس من المستحيل أنْ يكون للجسم قوة على ما يلزم وجود ذلك الجسم، ثم يكون ذلك الجسم ممّا من شأنه أنْ يبقى دائماً ، فيصدر عنه ذلك التحريك أو ذلك العدد دائماً - فالجواب عن هذا أنَّ ذلك من المستحيل لما بيّناه، بل يلزم لما بيّناه أنْ لا يكون لجســم من الأجسام قوة يفعل بها فيما يماسّه دائماً ؛ بل قوة كل جسم قوة يفعل بها فيما يماســه تحريكاً منقطعاً من تقريب وتبعيد . ولا جسم من الأجسام يمكن أنْ تكون فيه قوة تبقى دائماً مع بقاء الجســم؛ يكون فعلها واحداً مستمراً متشابهاً ، بل يجب أنْ تكون

the body must be one from which there results only some action that must itself be finite. Even if the body does perpetually continue, it will be something that, for instance, pushes or pulls or transmits, or something analogous.

(13) Should someone say that it is a matter of observation that if earth were to continue [to exist] perpetually and nothing accidental were to happen to it, then, as a result of its nature, it would be found continually resting in its natural place, we would say [that] rest is a privation of some action, not an action. Additionally, the perpetual continuation of earth and bodies subject to generation and corruption (as well as the continuation of their powers) is something whose alteration we will explain later.[14]

(14) One might also object that this infinite power can belong only to the whole body, so [that] when the body is divided, [the power] passes away. In that case, none of that infinite power will belong to the part. So the part will not have power over some part of that over which the whole has power, because the substrate of this power is the whole. It is just like the powers that exist in composite bodies after mixture, where [those powers] are not found in any of the underlying elements that were mixed together in it, [or] similar to rowers of a ship, for one of them [alone] would not move it at all. We say that the situation is not as you have supposed. [That] is because, even if the power belongs to the body in virtue of the combining of the parts and its mixture, it nonetheless permeates the whole of [the body], unless a given power

---

14. See *Fī al-kawn wa-l-fasād* 6. Alternatively, the text's *istiḥālah* might more naturally be read as "impossibility." In that case, the sense of the text would be, "Still, we will explain later that it would be impossible for earth and bodies subject to generation and corruption to continue forever, and the same will hold for their powers." Since, in the next chapter, Avicenna will argue for the eternity of the world, this reading seems unlikely; however, he may merely mean here that the individual instances of generable and corruptible things must ultimately pass away, even though the species does not (see *Ilāhīyāt* 6.5).

قوة الجسم قوة إنّما يصدر عنها فعلٌ يقتضي نفسه التناهي، وإنْ بقي الجسم دائماً فيكون مثلاً دافعاً أو جاذباً أو محيلاً أو شيئاً مما يجري هذا المجرى.

(١٣) فإنْ قال قائل إنّا نشاهد الأرض لو بقيت دائماً ولم يعرض لها عارضٌ لكان يوجد عن قوتها سكونٌ متصل في مكانها الطبيعي، فنقول: أمّا السكون فعدم فعلٍ لا فعل، ومع ذلك فبقاء الأرض والأجرام القابلة للكون والفساد دائماً، وبقاء قواها كذلك، ممّا سنبيّن استحالته.

(١٤) ثم لقائل أنْ يقول: إنّه يجوز أنْ تكون هذه القوة غير المتناهية إنّما توجد لجملة الجسم، فإذا قُسّم الجسم بطلت فلم يوجد من تلك القوة المتناهية شيء للجزء، فلم يقوَ الجزء على شيءٍ ممّا يقوى عليه الكل، لأنَّ محل هذه القوة الكل؛ كما يوجد من القوى في الأجسام المركبة بعد المزاج، ولا تكون موجودة لشيءٍ من الأركان التي امتزجت عنه، كما أنَّ المحركِين للسفينة فإنَّ الواحد منهم لا يحرّكها البتة. فنقول إنَّ الأمر ليس على ما قدّرت، فإنَّ القوة - وإنْ كانت للجسم بحال اجتماع أجزائه وبحال مزاجه - فإنّها مع ذلك تكون سارية في جملته، وإلاّ لكانت قوّة لبعض تلك الجملة دون الكل. فإذا كانت

belongs to some of the whole but not all. So, when it does permeate the whole of it, then some of the power belongs to part of [the whole]. So, once there is the mixture, the simple [body] does harbor the power permeating the whole that occurs after the mixture, whereas, when taken alone, it does not. Now, our supposing that the body has some part does not necessarily force us to take that part with the condition that it [actually] be cut off and separated [from the whole], such that one could then object that the separated part does not harbor any of the power. Instead, it is enough that we designate some part of it as it is so that we can recognize, in the way that was supposed, the state of that which results from that part and from the power that is in it. Also, even though one of the rowers of the ship cannot move the whole ship, he inevitably can move something smaller than it, which necessarily follows from what we have said.

(15) One might also object that the incorporeal mover of infinite power that moves some body must either provide a certain motion or provide some power by which [the body] is moved. On the one hand, if it provides some power, it would have provided an infinite power to the body. In that case, it would follow that [that power] is divisible, necessarily entailing what we have mentioned. On the other hand, if it provides only motion while not providing it with some inherent desire or inclination, [the body] would be forcibly moved; but you believe that forced motion is not perpetual. The answer is that if it provides a certain inclination, then, even though the inclination is the proximate

ســارية في جملته، كان لبعضها بعض القوة، فيكون البسيط إذن في حال المزاج؛ حاملاً للقوة الحاصلة بعد المزاج السّارية في الكل، وإنّما لا يحملها في حال الانفراد. وليس يجب أنْ يكون فرضنا للجسم بعضاً يلجئنا إلى أنْ نأخذ ذلك البعض بشرط قطْعه وإبانته، حتى يكون للقائل أنْ يقول إنَّ البعض المباين لا يحمل من القوة شيئاً، بل يكفينا أن نعيّن بعضاً منه وهو بحاله، فنتعرّف حال ما يصدر عن ذلك البعض وعن القوة التي فيه وحدها؛ التعرّف المفروغ منه على ســبيل التقدير. والمحرّكون للســفينة فإنَّ الواحد منهم - إنْ لم يمكنه أنْ يحرك كل السفينة - فيمكنه أنْ يحرك أصغر منها لا محالة، ويلزم ما قلناه.

(١٥) ولقائــل أنْ يقول فالمحــرّك ‹غير› المتناهي القوة غير الجســماني الذي يحرّك جسماً؛ لا يخلو إمّا أنْ يفيد حركة، وإمّا أنْ يفيد قوة بها يتحرك، فإنْ أفاد قوة؛ فقد أفاد قوة غير متناهية للجســم فيلزمها أنْ تنقسم ويوجب ما ذكرتم. وإنْ أفاد حركة فقط؛ ولم يفد شــوقاً غريزياً وميلاً لها فهو قَسْــرٌ، وعندكم أنَّ القسري لا يدوم، فالجواب أنّه إنْ أفاد ميلاً؛ فإنَّ الميل - وإنْ كان مبدأ قريباً للحركة - فليس مبدأ قريباً لها من حيث هي غير

principle of the motion, it is not its proximate principle insofar as it is infinite, but insofar as it is that motion. So inclination alone is not such that there results from it actions of an infinite power; rather, [they] are the results of the influence of that which makes [the inclination] to continue perpetually and by which it is perpetual, whereas, in itself, it is finite with respect to that over which it has power (if it can have power over it). Also, even if [the incorporeal mover of infinite power] does not provide an inclination, the motion won't be forced, as they reckoned, since forced motion is that which is contrary to the natural inclination in the thing, whatever it might be. So, when that which was provided with the motion has no inclination, but only motion, then [the motion] is not forced. It has thus become clear that it is simply impossible for there to be some power in a body that, of itself, necessitates infinitely many things.

(16) Again, one might say that your professed demonstration deals only with an infinite power that moves some foreign body that is external to it, whereas it does not deal with infinite powers internal to the bodies that they move, for you cannot say that the entire power moves something smaller, which we assume that part of the power does move. [That] is because part of the power moves that in which it is and the entire power moves that in which it is, but at any given moment that entire power won't be moving what the part moves, because it is not in it; but, in that case, the account is not driven into the contradiction. The response is to recall what we stipulated at the onset of the account[15]—[namely, that] this depends upon a hypothetical, conditional proposition based upon a supposition, not upon an existence claim.

---

15. See par. 8.

متناهيــة ، بل من حيث هي تلك الحركــة - فالميل وحده ليس بحيث تصدر عنه الأفعال غير المتناهية القوة ، بل عن تأثير من مســتبقيه على الدوم ويدوم به ، وهو في ذاته متناهي المقوي عليه إنْ كان له مقوي عليه - وإنْ لم يفد ميلاً فليســت الحركة بقســريةٍ أيضاً كما حسبوا ، إذ القسرية هي التي تخالف الميل الطبيعي في الشيء ما كان ، فإذا لم يكن ميلٌ لما أُفيد من الحركة ، بل الحركة لم تكن بالقســر ، فقد اتضح أنّه من المستحيل أنْ تكون قوة في الجسم هي التي تقتضي لذاتها أُموراً بلا نهاية .

(١٦) ولقائــل أنْ يقول : إنَّ البرهان الذي ادعيتم إنّما قام على قوةٍ غير متناهيةٍ تحرّك جســماً غريباً خارجاً عنها ، ولم يقم على قوةٍ غير متناهية تحرّك الجســم الذي هي فيه . فإنّه ليس لكم أنْ تقولوا إنَّ جميع القوة تحرّك الشــيء الأصغر الذي فرضنا أنَّ بعض القوة تحرّكه ، لأنَّ بعض القوّة تحرّك ما هو فيه ، وجميع القوة تحرّك ما هو فيه ، وليس جميع القوة محــرّكاً - في وقت مــن الأوقات - لما يحرّكه الجزء لأنّه ليس فيه . وإذا كان كذلك ، لم ينسق الكلام إلى الخُلْف ، فيكون الجواب عنه أنْ تتذكر ما اشترطناه من حديث في اعتبار هذا على حسب قضيةٍ شرطيةٍ متصلةٍ تقديرية ، لا بحسب الوجود .

(17) We have now examined this topic of investigation in the right way, explaining it in a way different from the simple-minded one that those who prattle on about the sciences use and who take infinite power as if it were something infinite in itself and derive some contradiction inasmuch as that [the power] will be necessarily multiplied or doubled or have some other ratio. They just don't understand that power in itself is neither finite nor infinite; rather, what *infinite powers* means is that that which is correlated with [that power], as that over which there is the power, is potentially, not actually, infinite, and that the potentially infinite might happen to become bigger and smaller. Indeed, there are many things, each one of which is in the class[16] of infinite, in whose case the infinite [can be multiplied by] two, three, four, or even more than that, where these might belong to a single genus as well as different genera. So it is not impossible to multiply the potentially infinite. So it is not impossible to multiply the power that is over that which is not impossible. The fact is that one ought to come around to what we have proven. Having explained that, let us consider whether there can be an infinite number of continuous motions and instances of generation, and, if they are infinite, whether they have a temporal beginning that is an [extreme] limit before which there is no before.

16. Reading *tabaqa* with four of the MSS consulted by **Y**, with **Z**, and with the Latin (*ordine*) for **Y**'s and **T**'s *tabīʿah* (nature).

(١٧) وإذ قد فتشنا عن هذا البحث حقّ التفتيش، وبيّناه على غير الوجه السخيف الذي يذكره مَنْ يخرّف في العلوم، ويأخذ القوة غير المتناهية كأنها في نفسها شيء غير متناهٍ ويُخرج خُلْفاً بأنّها يلزم أنْ تتضعّف أو تتنصّف، أو يكون لها نسبة أخرى؛ ولا يعلم أنَّ القوة في نفسها لا متناهية ولا غير متناهية، بل معنى قوة غير متناهية أنَّ مقابلها من المقوي عليه غير متناهٍ في القوة لا بالفعل، وأنَّ غير المتناهي في القوة قد يعرض له ما يصير أكثر وأقل. وأنْ تكون أشياء كثيرة، كل واحد منها في طبقة غير متناهية، فيكون غير المتناهي مرتين وثلاثة وأربعة وأكثر من ذلك، ويكون ذلك من جنسٍ واحدٍ ومن أجناسٍ مختلفة، فلا يستحيل تضعيف غير المتناهي في القوة، فلا يستحيل تضعيف القوة التي هي قوة على ما لا يستحيل، بل يجب أن يُحامَ حول ما بيّناه. وإذ بيّنا ذلك، فلننظر هل من الممكن أنْ تكون حركاتٌ وأكوانٌ متصلة بلا نهاية؟ وهل - إنْ كانت بلا نهاية - فلها بداية زمانية هي طرف لم يكن قبلها قبل؟

Chapter Eleven

That nothing precedes motion and time save the being of the Creator (may He be exalted) and that neither of the two has a first [moment] of its being

(1) Let us investigate whether motion can begin at some moment of time before which there was nothing, or whether motion is an atemporal creation, where every limit of time has some before and, in fact, [only] the being of the Creator is before anything. We say that, before the existence of any nonexisting thing, it was something whose existence was possible. So the possibility of its existence existed before its existence. [That] is because, if the possibility of existence were not to exist, it would be nonexistent and there would be no possibility of its existence, in which case its existence would be impossible. So the possibility of [a thing's] existence exists before [the thing] exists. Now, the possible existence of what exists must be some determinate thing, not merely nonexistence, for how many nonexisting things are different from [the thing's] possible existence?[1] So [possible existence] is either a substance subsisting in itself or some factor existing in something. If it were something subsisting in itself, which is neither in a substrate nor in a subject, then, as such, it would not be a correlative; however, as possible existence, it is correlative

1. Avicenna's point is that the *possibility to be x* cannot simply be identified with *not being x;* so, for example, while elephants, pinecones, stones, and galaxies are all *not* human, none, as such, is *possibly* a human.

<الفصل الحادي عشر>

في أنه ليس للحركة والزمان
شيء يتقدم عليهما إلاّ ذات الباريء تعالى
وأنهما لا أول لهما من ذاتهما

(١) فلننظر أنّه هل يمكن أنْ تبتديء الحركة من وقتٍ ما من الزمان لم يكن له قبل، أو الحركة إبداعية؟ وكل طرفٍ من الزمان فله قبل، وأنَّ ذات الباري ‹هي› قبل كل شـــيء. فنقول: إنَّ كل معدوم فإنّه قبل وجوده جائز الوجود، فجواز وجوده موجود قبل وجوده. فإنّــه لو لم يكن موجــوداً أنّه جائز الوجود، كان معدوماً أنّــه جائز الوجود، وكان ليس بجائز الوجود، فكان ممتنع الوجود. فجواز الوجود موجود قبل الوجود، وجواز الوجود للموجود أمرٌ محصل لا محالة، ليس هو نفس العدم، فكم من معدوم غير جائز الوجود، فهو إمّا جوهرٌ قائم بنفســه، وإمَّا أمرٌ هو موجود في شـــيء. ولو كان أمراً قائماً بنفســه لا فــي محلٍ ولا في موضوعٍ لكان - من حيـث هو كذلك - هو غير مضافٍ لكنه من حيث هو جواز وجودٍ هو مضاف إلى شـــيءٍ، ومعقول بالقياس، فليس هو جوهراً قائماً

to something and intelligible by the comparison. So it is not a substance subsisting essentially. Perhaps instead it is a certain relation or accident belonging to a substance. Now, possible existence cannot be a substance that has a relation, because that relation is associated with something assumed to be nonexistent. Also, it is impossible that[2] that relation be simply a certain association,[3] however it might by chance be; rather, it is a determinate association, where that association is not determinate save only in that it is possible. Therefore, possibility is itself the relation, not some substance upon which a relation other than possibility follows, where the two, taken together, would be possibility. Also, its existence in reality is not in that whose existence is possible while it is still nonexistent, for the existing description is not some accident of what does not exist. It is also not a description of the efficient principle so as to be a power, for the power or possibility to bring something into existence is not [the same as] possible existence. Thus, one rightly says that power over what is impossible and over what, in itself, does not possibly exist are absurd. That, however, is not [the same as] our saying that the power over that which cannot bring something into existence or the possibility to bring into existence that which cannot bring something into existence are absurd. The first of the two sayings leads to a concept different from the second. Again, if the first speaker conveys something that is not laughable, while the second speaker conveys something that is laughable (namely, when he says "that which cannot be brought into existence cannot be brought into existence"), then the laughable statement is not the same as that of the one who says "that whose existence is not possible in itself cannot be brought into existence by another." Indeed, this is a true and well-received statement used in making inferences.[4] It is also for

2. Reading *lā yumkinu an takūna* with **Z** and **T** for **Y**'s preferred *yumkinu an lā takūna* (it is possible that it is not).

3. The term "association" here is translating the Arabic *nisba,* which I have frequently rendered "relation." In the present context, however, I have reserved "relation" for the Arabic *iḍāfa,* which is the technical Arabic vocabulary for the Aristotelian category of relation (*pros ti*).

4. Alternatively, *qiyās* may be a reference to Avicenna's syllogistic; see, for instance, *Kitāb al-qiyās* 3.4, where he speaks about the definition of the possible (*mumkin*).

بذاته؛ بل عسى أنْ يكون إضافة ما وعرضاً ما لجوهر. ولا يجوز أنْ يكون جواز الوجود جوهراً له إضافة؛ لأنَّ تلك الإضافة تكون نسـبة إلى الشيء المفروض معدوماً، ولا يمكن أنْ تكون تلك الإضافة نسبة مطلقة كيف اتفقت، بل نسبة معيّنة؛ لا تتعين تلك النسبة إلاّ بأنها جواز فقط، فيكون إذن الجواز نفس الإضافة لا جوهراً تلزمه إضافةٌ هي غير الجواز؛ ومجموعهمـا هو الجواز. وليس وجوده بالحقيقة فيما يجوز وجوده وهو معدوم بعد؛ فإنَّ الصفـة الموجودة لا تعرض لمعـدوم، ولا هو صفة للمبدأ الفاعلي حتى تكون هي القدرة، فإنّ القدرة على الإيجاد أو جواز الإيجاد، ليس هو جواز الوجود. ولذلك يصحّ أنْ يقول القائل إنَّ القدرة على الممتنع محال، وعلى ما ليس في نفسـه جائز الوجود محال، وليس يكون ذلك هو قولنا إنَّ القدرة على ما ليس جائز الإيجاد محال، أو جواز إيجاد ما ليس بجائز الإيجاد محال. فإنَّ الأول من القولين يؤدي مفهوماً غير مفهوم القول الثاني: فإنْ قال قائلٌ القولَ الأول يفيد معنى غير هذرٍ، وقائل القول الثاني يفيد هذراً؛ أي إذا قال إنَّ ما لا يجوز إيجاده لا يجوز إيجاده فإنَّ قوله قولُ هذر لا كقول مَنْ يقول: إنَّ ما لا يجوز وجوده في نفسـه لا يجوز إيجاده عن غيره. فإنَّ هذا قول صحيح مسـتعمل في القياس مقبول.

that reason that theoreticians inquire into whether things are possible or not, so as to judge whether their being brought into existence is possible, whereas it is impossible for them to inquire into whether their being brought into existence is possible or not in order to infer from that and so learn whether their being brought into existence is possible or not.[5]

(2) It remains, then, that the possibility of existence—that is, the potential to exist—subsists in a substance other than the mover and its power. The substance in which there is the possibility of motion's existence is that which is of the character to be moved. From this, it is obvious that what has not been moved but is of the character to be moved precedes the beginning of its motion's existence. So, when that thing exists but is not moved, then there cannot [presently] exist the motive cause or states and conditions on account of which the mover moves the mobile, but thereafter they will exist. In that case, there will be a change of state before that motion, for the motion and whatever [previously] did not exist and then does [call it *x*] has some cause that necessitates its existence after its nonexistence. If there were no [such cause], [*x*'s] nonexistence would be no more fitting than its existence, nor would one of the two [states (namely, existence or nonexistence)] be essentially preferred over the other. So one [state] must be preferred because of some factor. Now, if that factor equally confers and does not confer a preference for that existence that results from it over the nonexistence, then the situation remains the same. The factor, instead, must be something with respect to which the preference for existence over nonexistence is selectively determined, where the selective determination is either such as to

5. While Avicenna's point here is not as clear as one might wish, he does make it more clearly at *Ilāhīyāt* 4.2, where he again argues that what is possible cannot simply be reduced to the power of the agent to do something. That is because, if these were equivalent terms, then one should be able to replace one term with another *salva veritate*—that is, in a way that is truth preserving. Avicenna begins by observing that no agent has the power over that which, in itself, is impossible. In that case, consider, for example, the meaningful statement "God's omnipotence means that God has the power to do everything possible," and replace "possible" with "the power of an agent to do something." The seemingly meaningful statement then becomes the vacuous tautology, "God's omnipotence means that God has the power to do everything that God has the power to do." Conversely, consider the empirically false claim, "I [(that is, some finite agent)] have the power to do everything possible." Under the present interpretation of *possible,* that false statement would turn out to be true, since I do have the power to do everything that I have the power to do.

ولذلــك فإنَّ الناظرين ينظــرون في الأُمور ؛ هل هي جائزة الوجود حتى يحكموا أنّها جائز إيجادها ، أو هل هي غير جائزة الوجود حتى يحكموا أنّها غير جائز إيجادها ، ويستحيل أنْ ينظروا أنّها هل هي جائز إيجادها أو غير جائز إيجادها ؛ ليتعرفا من ذلك على ســبيل الإنتاج أنّها جائز إيجادها أو غير جائز إيجادها .

(٢) فبقــي أنْ يكون جواز الوجــود ؛ وهو القوة على الوجــود ، قائماً في جوهر غير المحرّك وغير قدرته ، والجوهر الذي فيه جواز وجود الحركة ، هو الذي من شــأنه أنْ يتحرك . وظاهرٌ من هذا أنَّ الذي لم يتحرك – ومن شــأنه أنْ يتحرك يسبق ابتداء وجود حركتــه . فإذا كان ذلك الشــيء موجوداً ولا يتحرك ، وجــب أنْ لا تكون العِلّة المحرّكة ، أو الأحوال والشــرائط التي لأجلها يصدر التحريك مــن المحرّك في المتحرّك موجودات ثم وجدت ، فيكون قد تغيّر حالٌ قبل تلك الحركة . فإنَّ الحركة ، وكل ما لم يكن ثم كان ، فله علّــة توجب وجوده بعد عدمه ، ولولاها لم يكن عدمه ليس بأولى من وجوده ، ولا يتميّز لــه أحد الأمرين لذاته ، فيجب أنْ يتميّــز لأمرٍ ، وذلك الأمر إن كان يميّز ذلك الوجود عنه عن العدم ولا يميّزه ســواء – كان الأمر بحاله ؛ بــل يجب أنْ يكون الأمر يترجّح فيه تمييز الوجــود عن العدم . والترجُّح إمّا أنْ يكون ترجّحاً يوجب ، أو ترجّحاً لا يبلغ أنْ يوجب ،

necessitate [the effect] or such as not to reach the level of necessitating [the effect], in which case the same issue would remain. The fact is that it must inevitably necessitate [the effect]. Whatever the case, there must some selectively determining or necessitating cause that has come to be, and the discussion concerning its coming to be is the same as the former one. So its coming to be has an infinite number of essentially ordered natural causes either [all] existing together or existing successively. If they exist together, something impossible would have existed [namely, there would exist an actual infinity]. If they exist in succession, then either each one of them endures for a period of time, or instants follow successively upon one another. If, on the one hand, they endure for a period of time, then one motion would be after another, such that there will be no discontinuity in between, and before the first motion there will be a motion. The motion, however, would be eternal; but we stipulated that it had a starting point, which is a contradiction. If each one remains [only] an instant, they follow successively upon each other without any intervening period of time; and that also is absurd. Clearly, then, when something that [previously] did not exist comes to be in a body, the cause of that thing has then come to have a certain relation to that body that [previously] did not exist, where that relation is one of existing after not existing (whether belonging to an entity or some state). [That relation] will be either as some motion that necessitates some proximity, remoteness, juxtaposition, or some variation thereof, or as the coming to be of some motive power that [previously] did not exist,[6] or as some temporal act of volition. The coming to be of all of that will have some continuous cause, one thing following another. That is possible, however, only through some motion that temporally orders one

6. **Y** has (inadvertently) omitted the line *wa-imma ḥudūth qūwa muḥarrika lam takun* (or as the coming to be of some motive power that [previously] did not exist), which appears in **Z** and **T**.

فيكـــون الكلام بحاله، بل يجب - لا محالـــة - أنْ يوجب، وعلى كل حال فيجب أنْ يكون سببٌ مرجّح أو موجبٌ قد حدث، والكلام في حدوثه ذلك الكلام بعينه. فإمّا أنْ يكون لحدوثه أسبابٌ ذات ترتيب بالطبع لا نهاية لها موجودة معاً، أو موجودة على التتالي، فـــإنْ كانت موجودة معاً فقد وجد المحال، وإنْ كانت موجودة على التتالي، فإمّا أن يكون كل واحدٍ منهـــا يبقى زماناً أو تتالـــى الآنات، فإنْ بقيت زماناً كانت حركة بعد حركة على التشافع لا تنقطع، وكان قبل الحركة الأُولى حركة، وكانت الحركات قديمة؛ وقد جعلنا لها مبدأ، هذا خُلْف. وإنْ بقيت آناتٌ؛ فتتالت الآنات بلا توسط زمانٍ، وذلك أيضاً محال. فتبين أنّه إذا حدث في جســـم أمرٌ لم يكن فقد حصل لعِلّة ذلك الأمر إلى الجســـم نسبة لم تكن، وتلك النســـبة نسبة وجودٍ بعد عدم لذاتٍ أو لحالٍ؛ إمّا حركة توجب قرباً أو بُعْداً أو مـــوازاة، أو خلافها، وأما حدوث قـــوة محركة لم تكن، وإمّا إرادة حادثة. وكل ذلك فلحدوثه ســـببٌ على الاتصال، شيئاً بعد شـــيء. وذلك لا يمكن إلاّ بحركةٍ تنظّم الزمان

thing after another and that preserves the continuity, because it is impossible for instants to follow one another successively and because, if there is no motion undergoing transition from one thing to another, the causes and effects would necessarily occur simultaneously. Indeed, if the temporally originated cause (whether a necessitating or selectively determining one) has a fixed existence, then it necessitates and selectively determines [the effect] either through its nature or owing to some accidental factor belonging to it. If that is due to its nature—and, moreover, that which is its cause is distinct from it (even if it is something accidental)—it is not a cause owing to itself, but only together with that accident. So, necessarily, if [the cause] has a fixed existence, the effect must be simultaneous with it without delay, whereas, if it is something temporally originated that is renewed and not fixed, then the original issue arises once again. So, when the causes or conditions by which causes are causes have a fixed existence (whether temporally originated or not), then the existence is not brought to completion by something coming to be through them alone at some time. [That] is because, if what has a fixed [existence] always exists, then what it necessitates [as its effect] will not be delayed such as to be something coming to be at some time, whereas, if it comes to be [a cause] at some time, there is another cause of its being a temporally originated cause. Therefore, among the causes and conditions of causes, there must be some cause having an unfixed existence. Indeed, its existence involves exchange and transition from certain things to others, where this is nothing but motion or time. Now, time in itself does not act as a [cause], whereas motion produces proximity and remoteness. So [motion] is, in a certain way, a causal explanation, since it brings the cause into proximity.

شيئاً بعد شيء ، وتحفظ الاتصال - لامتناع تتالي الآنات ولأنه إنْ لم تكن حركة تنقل أمراً إلى أمر ، وجب أنْ تقع العلل والمعلولات معاً . فإنَّ السـبب الحادث ، الموجب أو المرجّح ، إنْ كان قـار الوجود فإنّه إمّـا أنْ يكون بطبيعته يوجب ويرجّح ، أو يكون لأمرٍ يعرض له . فـإن كان ذلك لطبيعته - ثم تميّز عنه ما هـو علّته - وإنْ كان لعارضٍ ، فليس هو لذاته علّـة بل مع ذلك العارض ، فيجب إنْ كانت قارة الوجود أنْ يجب معها المعلول بلا تأخر ، وإنْ كانـت حادثة متجدّدة غير قارّة لـزم بعينه الكلام الأول . فإذا كانت العِلل أو الأحوال التـي بها العلـل عِللاً قارّة الوجود - حادثة أو غير حادثة - لم يتم للحادث بها وحدها وجـود ، فإنَّ القارّ إنْ كان دائماً ، كان موجبـه لا يتأخر فيصير حادثاً ، وإنْ كان حادثاً ، كان لكونه علّة حادثة؛ عِلّة أُخرى . فيجب إذن أنْ يكون في العِلل وأحوال العِلل عِلّة غير قـارّة الوجود ، بل وجودها على التبدّل وعلـى النقل من أُمور إلى أُمور ، وليس هذا غير الحركة أو الزمان ، والزمان في نفسـه لا يفعل فعلها ، والحركة تقرّب وتُبعد ، فتكون سـبباً وعِلّة بوجهٍ ما ، إذْ تُقرّب العِلّة .

(3) It has become clear that if, whenever we assume that motion has some starting point with this description, some motion is before it, then motion, considered absolutely, does not have a starting point but involves an atemporal creation, where nothing is before it save the being of the Creator, preceding essentially [and] not temporally. How could [anything] be before it except the being of the Creator when we have shown that it is impossible that time in itself have some first instant or first thing that precedes it except the being of the Creator?[7] Thus, motion has no temporal beginning, but exists in the manner of an atemporal creation, where nothing precedes it save the being of the Creator.

(4) Someone cannot say: You have made motion something existing necessarily, whereas what exists necessarily does not need something that makes it exist. The answer is that there are two ways that something exists necessarily. One of them is to exist necessarily absolutely and through itself, while the other is to exist necessarily conditionally and through another. An example would be angles that equal up to two right angles. That is not necessary absolutely, but it is necessary when the figure is a triangle. Similarly, daytime is necessary with the rising of the Sun (for it is necessary through a cause), whereas neither daytime nor the rising of the Sun is necessary in itself. We ourselves required that the motion's existence be eternal if it is supposed that motion has a beginning in some way other than as an atemporal creation, and that [other way] is absurd. So this [necessity] is conditional. We did not require that [motion] exist necessarily of itself, nor is it the case that, when something is thought to have necessary existence as something transmitted to it and conditionally, it has been deemed to have that of itself. So our saying that motion is necessary does not prevent that

7. See 2.12.1.

(٣) فقد بانَ أنّه إنْ كان؛ كلما فرضنا للحركة مبدأ بهذه الصفة، كان قبلها حركة، فلا يكون للحركة المطلقة مبدأ إلاّ الإبداع، ولا قبلها شيءٌ إلاّ ذات المبدع؛ قبْلية بالذات لا بالزمان. وكيف يكون قبلها إلاّ ذات المبدع، وقد منعنا أنْ يكون للزمان في نفسه آنٌ أول يتقدم عليه، أو شيء أوّل، إلاّ ذات المبدع؟ فلذلك لا يكون للحركة ابتداء زماني إلاّ على جهة الإبداع، ولا شيء يتقدم عليها إلاّ ذات المبدع.

(٤) وليس لقائلٍ أنْ يقول إنكم قد جعلتم الحركة واجبة الوجود، وواجب الوجود لا يحتاج إلى موجد؛ فالجواب: إنَّ واجب الوجود على نحوين؛ أحدهما واجب الوجود مطلقاً ولذاته، والآخر واجب الوجود بشرطٍ وبغيره، مثل كون الزوايا مساوية لقائمتين، وذلك ليس واجباً مطلقاً، بل واجبٌ إذا كان الشكل مثلثاً. وكذلك وجوب النهار مع طلوع الشمس، فإنّه واجبٌ بعِلّة؛ وليس وجوب النهار ولا طلوع الشمس واجباً بذاته. ونحن أوجبنا قدم وجود الحركة - إنْ فُرض للحركة ابتداء - لا على نحو الإبداع، وذلك محال، فهذا بشرط، ولم نوجب لها وجوب الوجود لذاته، وليس إذا جُعل للشيء وجوب وجودٍ مرسلاً أو عند شرط، فقد جُعل له ذلك لذاته. فقولنا إنَّه يجب أنْ تكون

necessity from resulting from some beginning [principle], nor would our saying that motion necessarily emanates always from some mover (should we claim as much) require that the motion be thought to exist necessarily of itself. The fact is that, when we say that it is impossible that there be a motion, it is as if we said that it is impossible that there not be some mover who has caused the motion. [That] is because, when we say that it is impossible that there be some motion that comes to be in time unless there was some motion before that time, it is as if we said that it is impossible that there be[8] some mover causing motion in time unless some mover had caused motion before it itself or anything else.

(5) One might object, saying: The possibility you appropriate for God's power is like His creating a creation before every creation and a motion before every motion, like someone who wants to appropriate a certain possibility [for God], in that—if God does create a creation before creating—it would be possible for Him, in just the way that you made motion have no beginning. Now, this requires that you maintain that there have been an infinite number of motions in the past. In that case, the motions up to the Flood would be less, while those up to our time would be greater; but, undoubtedly, being less than infinite is to be finite, and so what is infinite would be finite. Again, the existence of the last motion would depend upon the existence of an infinite number of motions, whereas that whose existence depends upon an infinite does not exist. Moreover, you would have made an actually infinite number of motions exist, since each one of the motions would have inevitably existed actually. Finally, when every motion comes to be in time, the motions as a whole and their set come to be in time.

---

8. **Y** seems to have inadvertently omitted the lines here corresponding with the translation "there not be some mover who has caused the motion. [That] is because, when we say that it is impossible that there be some motion that comes to be in time unless there was some motion before that time, it is as if we are saying that it is impossible that there is. . . ." The Arabic appears in **Z** and **T** and reads: *muḥrrik ḥarraka fa-innā idhā qulnā lā yumkinu an takūna ḥaraka taḥduthu fī al-zamān illā wa-qad kāna.*

حركة، لا يمنع أنْ يكون ذلك الوجوب عن مبدأ، ولا قولنا إنّه يجب أنْ تكون الحركة دائمة الفيضـان عن محرّك - لو قلناه - توجب أنْ تجعل الحركة واجبة الوجود لذاتها. بل إذا قلنا لا يمكن أنْ لا تكون حركة، نكون كأنا نقول لا يمكن أنْ لا يكون محرّك حرّك. فإنا إذا قلنا: لا يمكن أن تكون حركة تحدث في الزمان إلا وقد كان في القبْل لذلك الزمان حركة، نكـون كأنا قلنـا لا يمكن أنْ يكون محرّك حرّك في الزمان إلاّ ويكون قد حرّك محرّك قبله هو أو غيره.

(٥) فإنْ قال قائلٌ تجويزكم في قدرة اللّه أنْ يكون كأن يخلق قبل كل خلْقٍ خلْقاً وقبل كل حركةٍ حركةَ، كمَنْ شـاء تجويز بأنْ يكون اللّه جائزاً عليه - إنْ كان يخلق خلقاً قبل خلق - على وجهٍ جعلتم الحركة لا بداية لها؛ وهذا يوجب أنْ تقولوا بوجود حركاتٍ بلا نهاية في الماضي، فتكون الحركات التي إلى الطوفان أقل، والتي إلى زماننا أكثر، ولا شك فـي كون الأقل ممّا لا نهاية لـه متناهياً؛ فيكون ما ليس له نهاية متناهياً. وأيضاً فإنَّ الحركة الأخيرة يكون وجودها موقوفاً على وجود حركات بلا نهاية، وما توقّف وجوده على ما لا يتناهى لا يوجد. وأيضاً فإنكم تكونون قد أوجدتم بالفعل ما لا نهاية له في الحركات، إذ كل حركـة منهـا فقد وجد بالفعل لا محالة، وأيضاً فإنّه إذا كانت كل حركة حادثة، فكل الحركات وجملتها حادث.

(6) The first skeptical doubt is [again] that, when we posit the [infinite number of] motions, God (powerful and glorious is He) would have already created [an infinite number of motions] now. The response to it is that, when [those motions] are considered now, they have absolutely no existence and, instead, are nonexistent. So, when they are said to be infinite, it is not as some actually existing infinite quantity, but, rather, as whatever number our estimative faculty imagines to belong to the motions, we find a number that was before it. Since [the motions] are nonexistent, then, necessarily, either it can be said of nonexistent things that they are *more* and *less* and *finite* and *infinite,* or it cannot. On the one hand, if it is not possible, then the objection disappears. If, on the other hand, it is possible, then, necessarily, there can be an infinite number of nonexisting things at the same time, with some of them being less than others—just like nonexistent future things, such as eclipses of the Moon, for they will be less than the periodic rotations of the Moon. Also, the orbital periods of a number of celestial spheres will be less than the orbital periods of one certain celestial sphere,[9] and those [motions counting forward] from the time of the Flood will be more than those [counting forward] from our time. Yet they will be infinite. (There is a group[10] that believes that nonexisting things are entities occurring distinct from one another and [that] one class of them, such as black and white, is numerically infinite.) Now, if its being such is not said about each one of these nonexisting things that are in the future (since they are nonexistent), then do not say it about each one of the nonexisting things that are in the past. If *each one* is predicated of what is future while not being required of the whole set, then, in the same way, let it be said about the past while not being required of some set.

---

9. For instance, Mercury, Venus and the Sun make their apparent rotations around the Earth in one year, Mars in two years, Jupiter in twelve, while it takes Saturn nearly twenty-nine and one-half years to complete an orbit.

10. Robert Wisnovksy suggests that such a view was prominent among the Muᶜtazilites; see *Avicenna's Metaphysics in Context* (Ithaca, NY: Cornell University Press, 2003), 148.

(٦) فالجواب عن التشـــكّك الأول أنَّ تلك الحركات إذا فرضناها قد خلقها اللّه عزَّ وجلَّ، فإنّها إذا اعتبرت من الآن كان لا وجود لها البتة، بل معدومة. فإذا قيل لها إنّها غير متناهية، فليس على أنَّ لها كمّاً حاصلاً غير متناهٍ، بل على أنّا أي عددٍ للحركات توهمناه وجدنـا قبله عدّة كانت. وإذ هي معدومة فـلا يخلو إمّا أنْ يجوز أنْ يقال في المعدومات إنّها أكثر وأقل، ومتناهية وغير متناهية أو لا يجوز. فإنْ لم يجز فقد زال الاعتراض، وإنْ جوّز فسيجوز ضرورة أنَّ المعدومات بلا نهاية معاً؛ وأنَّ بعضها أقل من بعض، كالمعدومات في المســتقبل التي هي كســوفات القمر، فإنّها أقل من دورات القمر عودة، وعودات عدّة أفلاك منها أقل من عودات فلكٍ واحدٍ، والتي من زمان الطوفان أكثر من التي من زماننا، ومـع ذلك فهي غير متناهية. وها هنا قومٌ يـرون المعدومات ذواتاً حاصلة متميزة بعضها عن بعض، والصنف الواحد منها كالســواد والبياض غير متناهي العدد. وإنْ لم يقل في هذه المعدومات التي في المستقبل إنَّ كل واحدٍ منها كذا بسبب أنّها معدومة، فلا يقال في المعدومـات التي في الماضي أن كل واحدٍ منها كذا، وإنْ قيل للمســتقبل كل واحدٍ، ولم يوجــب كل جملة، فكذلك ليقل في الماضي ولا يوجب جملة. وبالحري أنْ لا يقال جملة

Also, it is not appropriate to speak of a future or past set. [That] is because the set simply has no existence—[at least] not while it is past and future; neither is it more or less, finite or infinite, where [*infinite*] is not in the sense of *negation* but in the sense of *quantity,* so as not to have any endpoint. (Certainly, the past and future set is infinite [that is, indeterminate] in the sense of absolute negation, like the negation of what has no existence where it is like the negation of existence.) It is simply unforgivable of the apologist who says that the past entered into reality and, thus, it is impossible that it be infinite, while the future is not [similarly constrained]. In fact, we do not concede to him that the past is realized, but, rather, that each one of the past [events] has been realized, where the status of each one is not that of the whole of the past. Similarly, we might concede that each one of the future [events] can be realized, while the status of each one is not that of the whole of the future, such that the whole of the future is realized, possessing the whole of itself absolutely. The fact is that all the finite number of things that have been or will be realized are such that the second follows upon the first's ceasing to exist. Their set does not exist because what is understood by *set* is the collection. These are, however, simply not collected together in reality, even if each one of them exists individually at some moment during which the other does not exist. Certainly, they have been collected together in an intellectual depiction of them as existing, but the collection with respect to predication and intellectual depiction is different from the collection in existence. For there is the collection of *all humans* in that they are animal, which is decidedly not [the same as] some set of them [existing in reality].

مستقبلة ولا جملة ماضية، فإنَّ الجملة لا وجود لها البتة؛ لا فيما مضى ولا فيما يستقبل، ولا هي أكثر ولا هي أقل، ولا هي متناهية ولا غير متناهية ليس التي بمعنى السلب، بل بمعنى كمّ، فليس له نهاية. نعم، الجملة الماضية والمستقبلة غير متناهية بمعنى السلب المطلق؛ كما يسلب عمّا لا وجود له وكما يسلب الوجود. ولا عذر يقبل لمعتذرٍ يقول إنَّ الماضي دخل في الوجود؛ فلذلك يستحيل أنْ لا يتناهى، والمستقبل لم يدخل، فإنّه لا نُسلّم له أنَّ الماضي دخل في الوجود، بل كل واحدٍ من الماضي قد دخل في الوجود، وليس الحكم على كل واحدٍ حكماً على كلّية الماضي. كما أنّه قد نسلّم فيه أنَّ كل واحدٍ من المستقبل يجوز أنْ يدخل في الوجو وليس الحكم على كل واحدٍ حكماً على كلّيةٍ تكون للمستقبل حتى تكون كلّية المستقبل تدخل في الوجود، وتكون له كلّيته البتّة، بل والمتناهيات التي دخل في الوجود كل واحدٍ منها أو تدخل، على أنَّ الثاني يعقب عدم الأول، لا توجد لها جملة لأنَّ الجملة يفهم منها الاجتماع؛ وهذه لم تجتمع في الوجود البتة، وإنْ كان كل واحدٍ موجوداً بانفراده وقتاً لا وجود للآخر فيه. نعم؛ قد اجتمعت في وصف العقل لها بأنّها كانت موجودة، والاجتماع في الحمْل وفي وصف العقل غير الاجتماع في الوجود؛ مثل اجتماع كل إنسانٍ في أنّه حيوان، ولا جملة لهم البتّة.

(7) As for the second objection [namely, that any present motion would depend upon the existence of an infinite number of motions, which is purportedly impossible], the notion of *being dependent* used in it must be taken in either one of two ways. One may mean by it that, at some moment, there are two nonexisting things, and the condition for the future existence of one of them is that the second non-existing thing exist before [the other one], such that [the other's] existence depends upon it. In that case, if something in the past is nonexistent, and a condition for its existing is that an infinite number of ordered things exist, all of which are nonexistent, and then [all of them] begin to exist at some stipulated moment, then the existence of something that depends upon an infinite number of nonexisting things would be impossible. Alternatively, one may mean by [*being dependent* that] it does not come to exist unless certain things before have already come to exist, one before another *ad infinitum,* without there being some moment at which all of them are nonexistent. If they intend this, it is the very thing [we] sought to show, and so it cannot be a premise of a syllogism refuting itself.

(8) In the next objection, they show their ignorance of the precise difference between *each one* and *whole,* for, when each one of the things has a certain description, the whole need not have that description, nor must it have it as some determinate whole. Were that the case, the *whole* would be a part, since *each one* is a part. They do not see that each one of the future things is something that exists possibly, whereas the whole does not. It is just not true what they said—namely, that when each one passes into actual existence as something determinate, the whole does so as well—and so [they erroneously conclude that the things] do not go

(٧) وأمّـا الاعتراض الثاني فلا يخلو إمّا أنْ يعنـي بالتوقّف المذكور فيه، أنْ يكون أمران معدومان في وقتٍ، وشـرط وجود أحدهما في المسـتقبل أنْ يوجد المعدوم الثاني قبله؛ حتى يكون موقوف الوجود عليه. فإنْ كان الأمر على هذا، وكان أمراً في الماضي معدوماً، ومن شرط وجوده أنْ يوجد أمور بغير نهايةٍ في ترتيبها وكلها معدومة، فيبتدئ في الوجود من وقت ما يشـترط، استحال أنْ يوجد أمرٌ موقوف الوجود على أمور غير متناهيـة لا موجود فيها. وإمّا أنْ يعني به أنّه ليس يوجد إلاّ وقد وجد قبله أمور، واحد قبـل آخـر، لا نهاية لها، من غير أنْ يكون وقتٌ كلها فيـه معدومة. فإنْ أرادوا هذا فهذا نفس المطلوب، فلا يجوز أن يكون مقدمه قياسٌ على إبطاله.

(٨) وأمّـا بعد هذا الاعتراض، فإنّما جهلوا فيه الفرق بين كل واحدٍ وبين الكل؛ فإنّه ليس إذا كان كل واحدٍ من الأشـياء بصفةٍ يجب أنْ يكون الكل بتلك الصفة، بل لا يجب أنْ يكـون له كل حاصل، ولو كان كذلك لكان الكل جزءاً، إذ كل واحد جزء. ولا يرون أنَّ الأمـور التي في المسـتقبل كل واحدٍ منها جائز الوجـود، والكلّ غير جائز الوجود، فليـس حقاً ما قالوه إنّه إذا خرج كل واحدٍ إلى الوجود بالفعل حاصلاً؛ فالكل قد خرج،

on infinitely. The fact is that the situation is as we stated it—namely, that if ten finite things were to come into existence one after another such that one [comes to exist only] after the other passes away, then, undoubtedly, each one of these ten would actually have existed at some moment, whereas the whole would never have existed as something actual, for something like this whole as a whole simply has no existence.

(9) Those who deny that that the Creator himself has infinite power are frequently forced into what I'll relate. They allow that, before the first motion, it was possible for the Creator to have made a number of finite motions—for instance, ten—each one of which has a certain state of occurring for some duration and not occurring for some other duration, one happening after another without interruption. So, on their view, simultaneous with the possibility to make the first of [those possible motions] exist, as well as the motions right up to the presently existing ones, there is either the possibility that twenty motions exist, one after another, in the aforementioned way, such that each one occurs and does not occur for some duration, like we posited for the earlier ten—or, on their view, that is not possible. On the one hand, if they allow that it is possible, it would not be impossible that these ten [motions] exist in certain bodies and those twenty in other bodies, such that during the duration of those ten [motions], these twenty would exist, but the state of each one with respect to occurring and not occurring for some duration is like that of the other; this is absurd. If, on the other hand, they do not allow that it is possible, it follows that, during the state of not existing, the possibility of the motions' occurring and being brought into existence

فليس في غير التناهي، بل الأمر على ما قلناه: إنّه لو كانت عشرة متناهية تتوالى في الوجود واحداً بعد بطلان الآخر، فلا شكّ أنَّ هذه العشرة يكون كل واحد منها موجوداً بالفعل وقتاً مّا، والكل غير موجودٍ بالفعل البتة، فإنّه لا يكون لمثل هذا الكّل، من حيث هو كلّ، وجودٌ البتة.

(٩) وقد يلزم - هؤلاء الذين يمنعون أنْ يكون لذات الخالق هذا الاقتدار غير المتناهي - ما أقوله وهو أنّهم يجوّزون لا محالة أنْ يكون قبل الحركة الأُولى عدّة حركات متناهية يوجدها الموجد، لكل واحدٍ منها حالٌ من البقاء وغير البقاء محصّلٌ تتوالى عليه من غير انقطاع، وعددها عشرة مثلاً فلا يخلو إمّا أنْ يكون عندهم جائزاً، مع جواز إيجاد أولها إلى إيجاد الحركة الموجودة الآن، أنْ توجد عشرون حركة على التوالي المذكور، على أنَّ بقاء كل واحدٍ منها أو لا بقاءَه على نحو ما فرضناه لهذه العشرة؛ أو لا يكون ذلك جائزاً عندهم. فإنْ جوَّزوا لم يمتنع أنْ توجد تلك العشرة في أجسامٍ وهذه العشرون في أجسام أخرى، فيكون في مدّة تلك العشرة وجدت هذه العشرون، وحال كل واحدٍ في البقاء وغير البقاء كحال الآخر؛ وهذا محال. وإنْ لم يجوّزوا لزم أنْ يكون في حال العدم عددٌ لجواز وقوع الحركات وإيجادها مرتّبٌ، ويلزم لا محالة أنْ يكون ذلك ممّا لا يتناهى،

has a certain ordered numbering. Now, it inevitably follows that that [numbered ordering] is infinite, since there is no state that is the first one of possibility. In that case, there will be certain existing things, according to their system of thought, that are infinite in the past; but they forbid this. Other situations will also be forced [on them], one of which is what we pressed in the section on time—namely, that changes are successive, [for,] otherwise, one existence would not follow on another.[11] Also, [changes] have an existing subject, since there is no change unless it is in a subject; but, on their view, the subject would be the True One, since there is nothing else, but this is heresy! Keep yourself pure and renounce what the heretics say.

---

11. See 2.12.1.

إذ لا حـــالٌ هو حال أول جواز ؛ فتكون موجودات بالفعـــل على طريقتهم ، ليس لها نهاية في الماضي ، وقد منعوا هذا . وتلزم أُمور أخرى ممّا ألزمناه في باب الزمان أنْ تكون هناك تغيّـــرات متتالية ، وإلاّ لما كان وجود بعد وجود ، وأنْ يكـــون الموضوع لها موجوداً إذ لا تغيّر إلاّ بموضوع ، وأنْ يكون الموضوع ذات الأحد الحق عندهم ، إذ لا شيء غيره ، وهذا إلحادٌ ، تقدّس وتنزه عمّا يقول الملحدون .

Chapter Twelve

Following up on the claim that there is a point of smallness at which natural bodies are divested of their forms and that, in fact, each one of them has a certain limiting point less than which its form is not preserved;[1] *likewise, following up on the claim that no motion is the least, slowest, or shortest*[2]

(1) It is proper to add to these chapters an investigation into the continuous preservation of the forms belonging to bodies and whether they retain them while being divided infinitely. In other words, just as bodies are infinitely divisible with respect to smallness and yet preserve the form of corporeality, do they likewise preserve the rest of the forms that they have—as, for example, [the substantial forms of] being water, air, and the like?

(2) Forms that belong to [bodies] on account of mixture seem to be of a kind that decomposition reduces them to their simplest constituents, whose forms ceased to exist once the mixture was produced. Although the estimative faculty frequently imagines another sort [of decomposition], which, necessarily, is not accompanied by a reduction to the simplest constituents—and that is because the divisions have already reached the simplest constituents—[such an imagined decomposition] is not, in fact, a decomposition of [those bodies]. Still, for the sake of discussion it is fitting that we grant the division of the simple forms.

1. For the earlier reference to this topic, see 3.6.4.
2. For the earlier reference to this topic, see 3.6.6.

<الفصل الثاني عشر>

في تعقّب ما يقال إنَّ الأجسام الطبيعية تُخْلَع عند التصغّر المفرط صورها ، بل لكلٍ منها حدٌّ لا تُحفظ صورته في أقل منه. وكذلك تعقّب ما قيل إنَّ من الحركات ما لا أخف وأبطأ وأقصر منه

(١) وممّا يليق إلحاقه بهذه الفصول النظر في حفظ الأجسام للصور خلال الاتصال، وأنّها هل تبقى لها مع انقسامها إلى غير النهاية، أي هل كما أنَّ الأجسام لا تتناهى في الصغر انقساماً ، وتحفظ صورة الجسمية؛ أكذلك تحفظ سائر الصور التي لها مثل المائية والهوائية، وغير ذلك؟

(٢) أمّا الصور التي لها بحسب المزاج، فيشبه أنْ يكون ضَرْبٌ من التحليل يردها إلى بسائطها العادمة للصور المستفادة بالمزاج، وإنْ كان قد يُتوهم ضَرْبٌ آخر لا يجب معه الرجوع إلى البسائط، وذلك بأنْ تكون القسمة تتناول البسائط أيضاً ، لا أنْ تُحل إليها . لكن الأولى أنْ نجعل كلامنا في انقسام الصور البسيطة.

(3) We say that it appears, from the positions ascribed to the early periods [before] the Peripatetics, that these bodies terminate at parts that, when divided further, their forms cease to exist.[3] So they believed, then, that a certain amount of water is the smallest amount of water [possible], and the same for air and the rest of the [elemental] components. If that is their view about the simple constituents, then it would have been wisest if they then affirmed that about the composites that are thought to be the homoeomerous parts, such as flesh and bone. Now, a group of them had said that, if the situation is not like that, then, no matter how small any one of [the simple constituents] is, there can always be one smaller. Now, if that is possible for water, air, fire, and earth, as well as flesh, bone, and the like, then we should be able to take parts of the simplest components at whatever limit [of smallness we want]. In that case, what occurs from mixing [these indefinitely small simple components] would be like things that are generated from water, air, fire, and earth, and what [occurs] through the composition [of these minute mixtures] would be like animals, which are generated from flesh and bone; and so plants and animals of any given size could be generated. Thus, it would be possible for an elephant actually to be the size of a gnat. They can also say that it does not necessarily follow from the opposite of this that the gnat will be the size of an elephant, since the mixture requires that the parts be small, not large. So, when the parts are large and, while they are large, are brought into contact, then the action resulting from their mixture will not be what small [parts] would have produced. On account of this, whenever there are electuaries that are

3. The following discussion represents what might be called Avicenna's contribution to the theory of the *minima naturalia*. While the theory was certainly inspired by comments in Aristotle's *Physics* (see, for example, 1.4, which also seems to be the source for Avicenna's knowledge of the pre-Socratics on this topic), it was thought that advancements in the theory had to await Latin scholasticism. Recently, however, Ruth Glasner has shown that Ibn Rushd had made significant advances in the theory; see Ruth Glasner, "Ibn Rushd's Theory of *Minima Naturalia*," *Arabic Sciences and Philosophy* 11 (2001): 9–26. It would seem that Avicenna's contribution to this theory is still to be written.

(٣) فنقول: إنَّ الظاهر من المذاهب المنسوبة إلى صدور المشائين أنَّ هذه الأجسام تنتهي إلى أجزاء، إذا جُزئت بعد ذلك لم تكن الصورة فيها بموجودةٍ حتى يكون عندهم أنَّ للماء شيئاً هو أصغر صغيرٍ للماء، وكذلك للهواء ولسائر العناصر. وإذا كان قولهم في البسائط كذلك، فقولهم في المركّبات التي تُرى متشابهة الأجزاء كاللحم والعظم بذلك أحكم. وقد قالت جماعة منهم إنّه إنْ لم يكن الأمر كذلك؛ فجائزٌ أنْ يكون من كل صغيرٍ منها ما هو أصغر دائماً، وإذا كان يجوز ذلك في الماء والهواء والنار والأرض، واللحم والعظم وغير ذلك، فسيجوز أنْ نأخذ أجزاء البسائط بأي حدّ كان؛ فيكون منها ما يكون بالمزاج كالأشياء التي تتكون عن الماء والهواء والنار والأرض، وما يكون بالتركيب كالحيوانات التي تتكون عن تركيب اللحم والعظم، فجائز أنْ تكون المتكونات الحيوانية والنباتية على أي قدرٍ شئنا؛ فيكون من الممكن أنْ يحصل فيل في قدر البعوضة. ولهم أنْ يقولوا: ولا يلزم من مقابل هذا أنْ تكون بعوضة في قدر الفيل، إذ الامتزاج يقتضي صغر الأجزاء لا كبرها، فإن الأجزاء إذا كبرت وتلاقت وهي كبيرة، لم تفعل من الامتزاج ما يفعله الصغر، ولهذا ما كانت المعاجين التي تمزج قد يُعين على تكوينها حدّ من الدِّق،

mixed, and the larger parts in them prevent the infusion of the powers of some of them into others, a certain degree of mashing aids in bringing about [the electuaries]. They can say (or, perhaps, one of them even did say) that, if this possibility concerning the generation of animals from their elements were a real one, not only would it be an absolute possibility, but also it would be a possibility that, for the most part, refers to what exists. That is because the mixture of the lesser part precedes the mixture of the greater part, for the greater part takes in the lesser part. The same holds for the account concerning composition: It is more fitting that what is before should exist than what is after, and so it is more fitting that mixtures from the smallest parts should exist. In that case, elephants the size of cats (to say nothing of the size of gnats!) would not have been so rare as to verge on the impossible, albeit we would only equivocally call what is the size of a gnat an *elephant* (for the actions of elephants do not arise from this size). So this is what they say and why they say it. As for judging this claim, we will have to take on this role.

(4) This account necessarily follows in opposition to Anaxagoras and his account of mixture and belief that an aggregation of bodies that are homoeomerous parts and their being separated out in some way requires one sort of mixing to underlie another through which one thing will underlie another. He cannot escape it, for he associates the whole of generation with mixing and separating out. This, however does not necessarily follow on the basis of the principles of the Peripatetics. That is because, on their principles, it is not true that the mixture of the lesser part precedes the mixture of the greater part. That is because,

وكان أكبر الأجزاء فيها يمنع أنْ تنفذ قوى بعضها في بعض. ولهم أنْ يقولوا، أو عسى قائل منهم قال: إنَّ هذا الإمكان لو كان صحيحاً في تكوّن الحيوانات عن أُسْطُقساتها، لم يكن إمكاناً مطلقاً، بل كان يجب أنْ يكون إمكاناً أكثرياً بالقياس إلى الموجود. وذلك لأنَّ امتزاج الأقل قبل امتزاج الأكثر، فإنَّ الأكثر يحصل على الأقل. وكذلك القول في التركيب ووجود ما هو قَبْل أولى من وجود ما هو بَعْد، فتكون الامتزاجات عن أصغر الأجزاء أولى بالوجود. فكان يجب أن يكون وجود فيلة على قدر السنانير، فضلاً عن قدر البعوض، أمراً لا يندر ندوراً يلحق بالممتنع، وعلى أنّا كيف نسمي ما يكون على قدر البعوض فيلاً إلاّ باشتراك الإسم؛ فإنَّ الأفعال الفيلية لا تصدر عن هذا القدر، فهذا ما يقولونه ووجه ما يقولونه. وأمّا الحكم على هذا القول فيجب أنْ يكون منّا على هذه الصفة.

(٤) أمّا في مناقضة انكساغوراس وقوله بالخليط ورأيه أنّه مؤلفٌ من الأجرام المتشابهة الأجزاء، وأنَّ تميّزها على نحو ما يقتضي ضرباً من الاختلاط دون ضرب؛ يكون به شيء دون شيء، فهذا القول لازم لا محيص لانكساغوراس عنه، فإنّه ينسب التكوّن كله إلى الاختلاط والتميّز. وأمّا على الأصول التي للمشائين فإنَّ هذا غير لازم، وذلك لأنّه لا يحق على أُصولهم أنَّ امتزاج الأقل قبل امتزاج الأكثر، وذلك لأنَّ الأقل إنْ عنى به الأقل

if by *the lesser part* one means *numerically less,* then it does turn out true but is of no use to them, since their discussion concerns *what is less in magnitude.* Now, it is not necessarily the case, when the mixture of what is numerically less precedes the mixture of what is numerically greater, that the mixture of what is less in magnitude precedes the mixture of what is greater in magnitude. [That] is because the lesser magnitude exists in the greater magnitude in absolute potentiality, whereas the lesser number exists in the greater number in actuality. When what is less in magnitude ceases after being actual, no mixture is required of it at all. The fact is that, with respect to magnitude, it would be more fitting for the mixture of what is greater to precede the mixture of what is less, since the greater is what is bounded in the magnitude actually, whereas the lesser part is neither bounded nor actual, for every lesser magnitude is such [only] potentially.

(5) Moreover, it is not necessary, on the Peripatetics' principles, that the mixture resulting from the smaller part (should it result) be enough to bring about the species form, in which case size could well be a condition along with the mixture. That is because it is only on account of [for example] the soul—which, by actually being joined to a given body, produces a given species—that the body is perfectly prepared thereafter, being rendered such that [the soul] uses it as a tool for its [own] actions and motions. An example, then, is the fact that a human will be incapable of doing those things characteristic of a human unless his body is such as to perform human actions adequately. Not the least of [these human actions] are that he have the power and tool [that is, body] by means of which he can seek out and make a home (assuming there is no impediment), and by which he can fashion clothes, and [do] everything else a

في العدد صحّ ولم ينفعهم؛ لأنَّ كلامهم في الأقل في المقدار، وليس يجب إذا كان الأقل في العدد ‹امتزاجه› قبل امتزاج الأكثر في العدد أنْ يكون الأقل في المقدار امتزاجه قبل امتزاج الأكثر في المقدار، فإنَّ وجود الأقل مقداراً في الأكثر مقداراً وجودٌ بالقوة المطلقة، ووجود الأقل عدداً في الأكثر عدداً وجودٌ بالفعل. وإذا كان الأقل في المقدار معدوماً بعد بالفعل لم يجب له امتزاجٌ بتّةً، بل الأولى في المقدار أنْ يكون الأكثر في المقدار امتزاجه قبل امتزاج الأقل، إذ الأكثر محصورٌ في المقدار محصّلٌ، وأمّا الأقل فغير محصورٍ ولا محصّل، فإنَّ كل أقل من المقدار أقل بالقوة.

(٥) وأيضاً ليس واجباً على أُصول المشائين أن يكون المزاج الحاصل على أجزاء صغار - إنْ حصل - كافياً في حصول الصورة النوعية، فعسى أنْ يكون العِظم شرطاً مع المزاج. وذلك لأنَّ النفس الفاعلة بحصولها مقارنة لجسم ما نوعاً إنّما يستعد لها الجسم تمام الاستعداد، بعد أنْ يكون بحيث يصلح استعمالها إياه آلة لأفاعيلها وحركاتها مثلاً. فإنَّ الإنسان لن يتخلّق إنساناً إلاّ أنْ يكون بدنه بحيث يفي بالأفعال الإنسانية، ولا أقل من أنْ تكون له قوة وآلة يتمكن بها - إنْ لم يكن عائق - من اتخاذ الكِنّ وإحداثه، ويتمكن بها من إعداد الملبوس وسائر ما لا بُدّ للإنسان من وجوده له، وأنْ لا يكون بحيث تسفيه

human must do to exist—as well as not being such that strong winds blow him about as so much dust and that the predominating lower qualities in him [namely, hot, cold, wet, and dry] do not change him. So it would seem that the human soul exists as a form only on account of a body whose like characteristically performs human motions (assuming that nothing impedes it). Consequently, the occurrence of the mixture itself is insufficient to bring about the human species; nonetheless, it is on account of the occurrence of the mixture so prepared for a certain species that [the species] occurs, and is engendered in the equivalent of a certain place and source of origin, and is engendered from the equivalent of a certain matter, as well as there being a certain psychological faculty that acts through a tool having the power to produce motion and rest. Now, were this matter, together with the preparedness that the mixture gives it, a negligible trifle, it would be affected all at once by the present quality, and it would not preserve the form that the mixture gives it until natural motions have brought it to its perfecting form, and, instead, the psychological power to which the mixture gave rise would not be dependent upon something like this matter. Clearly, this proof is useful in refuting Anaxagoras only.

(6) As for ourselves, we say that, on close investigation, the body is seen to be divisible in two respects (both of which you have already learned),[4] one of which is by way of discontinuity and fragmentation, whereas the second is not by way of discontinuity. The divisibility of that which is not by way of discontinuity, fragmentation, and the disconnection of the parts is, rather, due to some accident specific to part

4. See 3.9.3.

السوافي وتحيله أدنى الكيفيات التي تغلب عليه. فيشبه أنْ تكون النفس الإنسانية لا تحصل صورة إلاّ لبدنٍ من شأن مثله – إنْ لم يعقه عائق – أنْ ينهض بالحركات الإنسانية. وإذا كان كذلك، فالمزاج نفسه غير كافٍ حصوله في أنْ يحصل النوع الإنساني، على أَنَّ لحصول المزاج المستعد لنوع ما مكاناً ومعدناً، في مثله يحصل ويتولد، ومادة عن مثلها يتولد، وقوة نفسانية تفعل بآلةٍ قوية على التحريك والتسكين. ولو كانت هذه المادة، مع استعدادها المزاجي؛ نزرة يسيرة لانفعلت عن الكيفية الحاضرة دفعة، ولم تحفظ صورها المزاجية ريثما تبلغها الحركات الطبيعية إلى صورتها الكمالية، بل مثل هذه المادة لا تتعلق بها قوة نفسانية مازجة. مبيّنٌ أَنَّ هذا القياس إنما ينتفع به في الردّ على انكساغوراس لا غير.

(٦) وأمّا نحن فنقول إنَّ الجسم يمعن في الانقسام على وجهين: أحدهما على سبيل الإنفصال والانفكاك، والثاني لا على سبيل الإنفصال، وقد علمت كلا الوجهين. فالذي يكون انقسامه لا على سبيل الإنفصال والإنفكاك وتباين الأجزاء، بل لعرضٍ يختص ببعضه

of [the body] or some relation specific to it—as, for instance, being contiguous, parallel with, or the like. From that [sort of divisibility], it is not necessary that [when] the simple body is divided, it should reach some limiting point at which it loses its form. That is because the form is spread throughout the whole of [the body] with which it corresponds. If some parts of the body were not to have their fair share of its form on account of their smallness, there would be an interval[5] of parts alike in [that] status such that either the body would cease or parts smaller than they (and less likely to bear that form) would remain, in which case this body would be an ordered series of parts, none of which have this form. This form occurs only through the collection of [these parts], where the collection *qua* collection provides nothing but number and its properties, while *qua* a collection of bodies it provides nothing beyond what the collection provides absolutely, save magnitude and its concomitants of shape and position. Now, none of that is fiery or earthy, so that [being fiery and earthy] does not exist in the parts taken separately, but in the whole, owing to the collection.

(7) [A part below a certain threshold of size] will equally not be like the mixture, for that results[6] from the various kinds of natures. Moreover, the mixture also spreads throughout when it becomes settled in that in which it becomes settled, and what is judged of it will be the same as for the simple form. This is something whose explanation does not require much effort. When something has this description, then it is obviously clear that every part of water is according to the [form] of water and that being divided in this way does not make the small part different from the whole.

5. Reading *buᶜd* with one of the MSS consulted by **Y**, as well as with **Z** and **T** for **Y**'s *yuᶜidd* (it predisposes).

6. Reading *ᶜan* with three of the MSS consulted by **Y**, as well as with **Z** and **T**, for **Y**'s preferred *ghayr* (different from)

أو إضافة ما تختص به. مثل مماسّة أو موازاة أو غير ذلك، فليس يجب من ذلك أنْ يكون الجسم البسيط يبلغ به الإنقسام إلى حدّ يكون في ذلك الحدّ فاقداً للصورة، لأن تلك الصورة فاشية في جميعه مطابقة له. ولو كان من أجزاء الجسم ما لا قسط له من صورته لصغره؛ لكان بعد أمثال له في حُكْمه، يفنى الجسم أو يبقى أصغر منه وأبعد من احتمال تلك الصورة، وكان حينئذ هذا الجسم منتظماً من أجزاء ليس ولا واحد منها على هذه الصورة. وإنّما تحصل هذه الصورة باجتماعها. والاجتماع بما هو اجتماع لا يفيد إلاّ العدد وخواصّه، وبما هو اجتماع أجسام لا يفيد زيادة على ما يفيده الاجتماع مطلقاً، إلاّ المقدار ولواحقه من الشكل والوضْع، وليس شيءٌ من ذلك نارياً ولا أرضياً حتى يكون غير موجودٍ في الأفراد، وموجوداً في الجملة للاجتماع.

(٧) ولا هو أيضاً كالمزاج؛ فإنَّ ذلك عن مختلفات الطبائع، ومع ذلك فالمزاج أيضاً فاشٍ عندما يستقر فيما فيه يستقر، وحكمه حكم الصورة البسيطة، وهذا ممّا لا يحتاج في إيضاحه إلى كثير سعي. وإذا كان الأمر على هذه الصفة؛ فواضحٌ بيّنٌ أنَّ كل جزء من الماء ففي مائية، وأنَّ الانقسام، على الوجه، لا يجعل الجزء الصغير مخالفاً للكل.

(8) As for being divided in the other way—that is, by way of discontinuity and being disconnected—it would seem that to exceed a certain degree of smallness is a cause of the body's no longer preserving its form, for, whenever bodies become smaller, they are increasingly disposed to being more quickly acted upon by other [bodies]. (This is something that will be explained to you.)[7] So, apparently, when the body exceeds its degree of smallness and separates off from its collective kind,[8] it would be impossible for it to retain its form; but, rather, it will alter into the bodies surrounding it and become continuous with them. As such, it will not maintain its form until it is mixed. So, if the situation is like this, then what is said must not be true—namely, that the smallest body preserving the form of earth is larger than the smallest body preserving the form of fire.[9] That is because the smallest [amount] that can exist as air inevitably is susceptible to the sort of generation and corruption to which the nature of fire is susceptible (and perhaps it is even better suited to that). Consequently, it will be of a character that it undergoes alteration into earth; but when it is of the character that it alters into earth, the earth into which it alters will be of a smaller volume than the volume of the altered fire, since, when fire undergoes alteration into earth, it becomes smaller. This is one of the principles of the Peripatetics and it is true.

(9) That is unless it is said that that small amount of fire is not of a character to be altered into some independently existing bit of earth but, instead, is continuous in that it becomes, in that case, a part of earth that is not numerically discontinuous from it such that it would actually exist apart from it. Instead, it would be like a drop of water that is continuous with a sea of water such that its actual existence as

---

7. This issue is briefly treated throughout 4.12, but especially par. 2.

8. Avicenna's use of *kullīyah,* translated here as "collective kind," is his preferred locution when he speaks about the so-called spheres associated with the elements earth, water, air, and fire; see, for instance, 4.10.2 and following.

9. While Aristotle mentions the transformation of elements into one another in book 4 of *De caelo* and book 2 of *De generatione et corruptione,* as well as throughout the *Meteorology,* I have not been able to find mention of this exact ratio.

(٨) وأمّا الانقسام على النحو الآخر، وهو على سبيل الانفصال والتباين، فيشبه أنْ يكون الإفراط في الصغر يصير سبباً لأنْ لا يحفظ الجسم صورته، فإنَّ الأجسام كلما صغرت ازدادت استعداداً لأنْ يفعل فيها غيرها بسرعةٍ، وهذا شيء سيتضح لك. فيشبه أنَّ الجسم إذا أفرط صغره وباين كلّيته، استحال أنْ يبقى على صورته زماناً، بل يستحيل من الأجسام المحيط به إليها ويتصل بها، فلا يكون بحيث يثبت على صورته إلى أن يمتزج. فإنْ كان الأمر على هذا؛ فيجب أنْ لا يحقّ ما يقال من أنَّ أصغر جسم حافظ للصورة الأرضية، هو أكبر من أصغر جسم هو حافظ للصورة النارية، وذلك لأنَّ أصغر ما يمكن أنْ يوجد ناراً، لا محالة هو قابلٌ من الكون والفساد ما تقبله طبيعة النار، وعسى أنْ يكون هو أولى بذلك. وإذا كان كذلك فمن شأنه أنْ يستحيل أرضاً، وإذا كان من شأنه أنْ يستحيل أرضاً، كانت الأرض التي استحال إليها أصغر حجماً من حجم النار المستحيلة، إذ النار إذا استحالت أرضاً صارت أصغر، وهذا هو أصل المشائين، وهو الحق.

(٩) اللّهم إلاّ أنْ يقال إنَّ تلك النار الصغيرة ليس من شأنها أنْ تستحيل أرضاً مفردة، بل على نحو الاتصال بأن تصير حينئذ جزء أرضٍ لا منفصلاً بالعدد عنها، موجوداً بالفعل دونها، بل كما تتصل قطرةٌ من الماء بالماء الغَمْر؛ بحيث يذهب وجودها بالفعل

an independently existing drop ceases. It consists in being only an addition to the whole of the sea; but it is such that we can posit it as independently existing, while it is not like that by being discontinuous and independently existing. Should one say this, they have been swept away by the sea of their own fancy. The alteration of [the smallest body of fire] need not necessarily occur wherever it encounter earth in its entirety, for many parts of the elements undergo alteration into one another in that very same region that is proper to its whole, while being a large part of appreciable size.[10] So imagine just how much more the fast-moving small body will alter! Moreover, it is not inevitably necessary that it be continuous; rather, it might perhaps alter into that nature, while remaining something contiguous.

(10) Let us now investigate the claim that, among motions, there is one that is so small that none smaller than it can be produced, and, relative to it, there would then also be a distance than which none is smaller, as well as a time. It is obvious from what preceded that it is impossible for some motion to exist of which none is smaller inasmuch as it is a part of some continuous motion,[11] and the same holds with respect to distance and time. As for [whether] there is a discontinuous and independently existing [smallest motion, distance, and time], it would not be odd to suspect that these things do have some claim to being limited in smallness. Now, it is most fitting—and the best case can be made [for the idea]—that the status of motion is like that of magnitude in that the smallness does not go beyond the nature of the magnitude, just as, in their opinion, it did not, for instance, go beyond[12] the nature of fire. [That] is because, when we posit some smallest distance, we ourselves know that, in itself, it is such that a certain division

10. For example, large quantities of air, while in the region of air, might undergo alteration so as to become rain, or water; or, conversely, large quantities of water, while in the region of water, might be heated so as to become steam, or air.

11. The reference is probably to 3.6.6.

12. **Y** seems to have inadvertently omitted the following phrase as a result of homeoteleuton, which occurs in both **Z** and **T**: *ʿan ṭabīʿat al-miqdārīyah ka-mā yakhrujuhu ʿindahuma mithlan.* The phrase corresponds with "the nature of the magnitude, just as, in their opinion, it did not, for instance, go beyond" in the translation.

قطـرة منفردة، وإنما يكون منها زيادة فـي جملة الغمْر، وتكون هي بحيث لنا أنْ نفرضها منفردة؛ ولا تكون كذلك بالانفصال والانفراد. فإنْ قال هذا قائل فقد أجحف في التحكّم. وليـس يجب - لا محالة - أنْ تقع استـحالته حيث يصـادف كلّية الأرض، فإنّ كثيراً من أجزاء العناصر يسـتحيل إلى غيره لا في نفس ذلك الحيِّز الذي يخص كله، وهو جزء كبير محسوس القدر، فكيف الصغير السريع الاستحالة؟ ومع ذلك فلا يجب أن يتصل لا محالة، بل قد يجوز أنْ يستحيل إلى تلك الطبيعة ويبقى مماساً.

(١٠) فلننظـر الآن فيما يقال من أنَّ فـي الحركات حركة لا يمكن إيجاد الأقل منها، فتكون فيها مسـافة أيضاً لا أقل منها وزمان كذلك، وأيضاً متحرك لا أصغر منه؛ فنقول: أمّـا امتناع وجود حركـةٍ لا أقل منها، على أنّها جزء من حركـةٍ متصلة، فأمرٌ ظاهرٌ مما سـلف، وكذلك في المسافة والزمان. وأمّا على سـبيل الانفصال والانفراد فغير بعيد أنْ يظنّ بهذه الأشياء أنّها تستحق التناهي في الصغر، وأمّا الأوْلى والأحق فهو أنْ يكون حكم الحركـة حكم المقدار؛ في أنَّ الصغر لا يخرجه عن طبيعة المقدارية، كما يخرجه عندهم مثلا عن طبيعته النارية. فإنّا إذا فرضنا أصغر مسافة، فنحن نعلم في نفسه بحيث يمكن

can be posited of it without the area's being fragmented; and that a certain limiting point shared by two of its parts can be posited in it; and that, when some mobile begins to move at its starting point, it inevitably reaches that shared limiting point; and [finally] that it is not impossible that something just so happens to oppose it and to bring it to rest when it reaches that limiting point, since it is of a character that it rests. In that case, that [motion] would be smaller than the smallest of motions. Now, the possibility of this is greater than that the magnitudes should become fragmented. [That] is because it is quite likely that magnitudes reach some limiting point at which the agent becomes incapable of causing fragmentation owing to [the magnitude's] smallness and the agent's [limited] power to make a division of [the magnitude] that divides [it], even if [the magnitude] in itself is divisible. Still, it is not impossible that the previously mentioned division follows upon it (assuming that there is some spatial magnitude) and [yet], at the limiting point of division, [the motion] comes against some cause that brings it to rest, and so there will not be some instant, to the exclusion of another, at which that is impossible.

(11) One point of inquiry still remains for us—namely, whether, among natural motions there is one than which none is faster, and, likewise, [whether], among them, there is one actually existing that which none is slower, even if, in the estimative faculty, one can imagine one slower than it. We say, then, that if there is among natural motions something like this, [in the case of the fastest motion] it will be the motion of the smallest thing[13] that can preserve its form,[14] [while in the case of the slowest motion] it will be a motion from among the slowest bodies undergoing rectilinear motion.

---

13. Reading with three of the MSS consulted by **Y**, as well as **Z** and **T**, *ḥarakat aṣghar mā* for **Y**'s preferred *juzʾ min ḥarakat aṣghar mā* (the smallest part of a motion).

14. Given Avicenna's earlier discussion about fire, the fastest motion would, then, correspond with the motion of the smallest amount of fire capable of independent existence.

أنْ تفرض له قسمة بغير جهة التفكّك، وأنّه يفرض فيه حدٌّ مشترك لجزئيه، وأنَّ متحركاً إذا ابتدأ يتحرك من ابتدائه؛ فإنّه لا محالة يوافي ذلك الحدّ المشترك، وأنّه لا يمتنع أنْ يعرض له مانعٌ ومسكّنٌ عند موافاة ذلك الحدّ؛ إذ من شأنه السكون. فتكون تلك أصغر من أصغر الحركات، وهذا أشــدّ إمكاناً من تفكّك المقاديــر. فإنَّ المقادير لا يبعد أنْ تبلغ حدّاً يعجز المفكّك عن تفكيكه لصغره وقوته أنْ يصيبه الفاصل بقسمته الفاصلة - وإنْ كان في نفسه منقســماً - لكنه لا يمتنع، إذا كان مســافة، أنْ تلحقه القسمة المذكورة، وأنْ تلحق عند حدّ القسمة عِلّة مسكّنة، فليس آنٌ يمتنع فيه ذلك دون آن.

(١١) وقــد بقي علينا من هذا الجنس بحثٌ وهو أنّه هل كما في الحركات الطبيعية حركة لا أسرع منها، فكذلك فيها حركة لا أبطأ منها في الوجود؟ وإنْ كان يمكن أنْ يكون في التوهم أبطأ منها؛ فنقول: إنّه إنْ كان في الحركات الطبيعية مثل هذا، فهو حركة أصغر ما يمكن أنْ تحفظ صورته، من أبطأ الأجرام المستقيمة الحركة حركة.

Chapter Thirteen

On the directions of bodies

(1) Now that we have learned how things stand for natural bodies and their finite and infinite powers, [both] with respect to increase and decrease, we should discuss the directions of bodies and the directions of their motions, since directions are among the set of necessary concomitants due to quantity.

(2) We say that, when we posit a certain interval, we posit it as either being rectilinear or in some other way [such as being circular or curved]. On the one hand, if we posit it as being rectilinear and it cannot proceed infinitely, then two extremities have been posited for it as well as two directions between them, one direction toward each extremity. If, on the other hand, it is circular or curved and then some division is posited in it, the common limiting point will formally[1] have a direction toward each one of the two divisions. By *interval* I mean any extension, whether some other extension can or cannot be posited in it. That [in which another interval] cannot [be posited] is the line, while that [in which] it can is a surface or body, for the surface has a single extension with respect to its being spread out, and the body has a single extension with respect to its being a solid. Now, the line is a single extension potentially and actually, whereas the surface can exist as this one [extension] alone as well as being considered as having two extensions. For example, if it is a square,

1. Reading *ʿalá hayʾah* with **Z** and **T** for **Y**'s *ʿalá hayʾatihi* (in the shape of it).

<الفصل الثالث عشر>

في جهات الأجسام

(١) وإذ عرفنا حال ما يعرض للأجسام الطبيعية وقواها من التناهي وغير التناهي، في الزيادة وفي النقصان، فحريٌّ بنا أنْ نتكلم في جهات الأجسام وجهات حركاتها، إذ كانت الجهات من جملة اللواحق بسبب الكميّة،

(٢) فنقول: إنّا إذا فرضنا بُعْداً، فأمّا أن نفرضه على الاستقامة، أو على جهةٍ أخرى، فإنْ فرضناه على الاستقامة واستحال ذهابه إلى غير التناهي، افترضت له نهايتان، وافترض ما بينهما جهتان إلى كل نهايةٍ جهة، وإنْ كان مستديراً أو منحنياً ففُرض له قطعٌ، كان للحدّ المشترك إلى كل واحدٍ من القسمين جهة على هيئة. وأعني بالبُعْد كل امتدادٍ، سواء كان يمكن أنْ يُفرض فيه امتدادٌ آخر أو لا يمكن، أمّا الذي لا يمكن فهو الخط، وأمّا الذي يمكن فالسطح والجسم؛ فإنَّ السطح له في انبساطه امتدادٌ واحدٌ، والجسم له في ثخنه امتدادٌ واحدٌ، والخط هو امتدادٌ واحدٌ بالقوة والفعل. وأمّا السطح فإنّه يجوز أنْ يوجد هو بعينه، ويعتبر له امتدادان؛ مثلاً إنْ كان مربعاً، كان له امتدادٌ من

it will have one extension from one side to the opposite [side] and another extension from the third side to its opposite [side]. The subject is one and the same, but the relation of the starting point from which it extends to the endpoint is different from the relation of that other starting point from which it is led to that other endpoint.[2]

(3) In general, whenever an extension is posited, the attending result is that exactly two directions belong to it insofar as it is such. Now, the common view among the masses—or even the apparent one among the speculative theologians[3]—is that the line has only two directions; the surface, four; and the body, six. Their belief about the line corresponds truly with what exists, whereas [what they say] about the rest calls for a closer examination. [The number of directions] that belongs to the surface *qua* surface is [derived] from the extremities. So, in fact, if the surface is a square and its primary limits, which are lines, are considered to the exclusion of points, then the situation is as supposed. If it is not a square, however, or that [namely, its being a square] is not taken into account, then its directions are more than that. So, if it is, for instance, a hexagon, then no limiting point is more suited than another with respect to being a direction, and so the surrounded surface, inasmuch as it is such, happens to have six directions; and if it has more [limits] than that, it will have more [directions] than that. Also, even if it is a square, but its being limited at straight lines is not the only thing taken into account and, instead, all of

2. So, for example, although the square below, whether considered as ABCD or as ADCB, is the same surface, it can also be considered as extending from AB to DC, which would be one direction, as well as from AD to BC, which would be another direction.

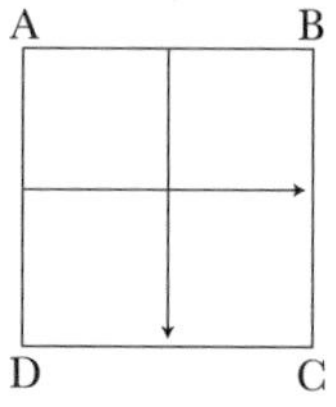

3. The phrase *ahl al-naẓar* commonly refers to the Islamic speculative theologians, and Avicenna is almost certainly referring here to their accounts of *jihah,* the term being translated "direction" but which can also mean "space," "spatial location," or even "side"; see al-Ashʿarī, *Maqālāt al-Islāmīn wa-ikhtilāf al-muṣallīn,* 316 were we see that mutakallimūn did ascribe six directions (*jihāt*) to the atom (*juzʾ*) as well as finding much of the same vocabulary that Avicenna himself will use to describe directions. (For a study of the early Muslim theologian's discussion of *jihah,* see Alnoor Dhanani, *The Physical Theory of Kalām,* chapters 3 and 4.)

ضلع إلى مقابله، وامتدادٌ آخر من الضلع الثالث إلى مقابله، والموضوع واحدٌ بعينه، لكنه بحسب الإضافة إلى مبدأ عنه يمتد إلى منتهى هو غيره، بحسب الإضافة إلى مبدأ غير ذلك المبدأ، يأخذ عنه إلى منتهى غير ذلك المنتهى.

(٣) وبالجملة كلّما افترض إمتدادٌ عرض عنه أنْ تصاب له – من حيث هو كذلك – جهتان لا غير. والمشهور عند الجمهور، أو عند أهل الظاهر من أهل النظر، أنَّ للخط جهتين لا غير، وللسطح أربع جهات، وللجسم ستّ جهات، أما رأيهم في الخط فصحيحٌ مطابقٌ للموجود، وفي سائر ذلك نظر. فالذي للسطح، بما هو سطح، من النهايات؛ فإنّه إنْ كان السطح مربعاً، واعتبرت نهاياته الأُولى التي هي الخطوط دون النقط، فالأمر على ما ظنّ. وإنْ لم يكن مربعاً، أو كان مربعاً ولم يعتبر ذلك، فإنَّ جهاته أكثر من ذلك، فإنَّه إنْ كان مثلاً مسدساً، فلا حدّ أولى من غيره بأنْ يكون جهة، فيعرض للسطح المحاط به، من حيث هو كذلك، أنْ تكون له ست جهات، وإنْ كان أكثر من ذلك عرض أكثر من ذلك. وإنْ كان أيضاً مربعاً، ولم يعتبر تناهيه إلى الخط المستقيم فقط، بل اعتبر له

the types of termination belonging to it are considered so as to include angles, it will have eight directions: four toward the lines, and four toward the angles. In actuality, the circle has only a single direction, whereas, in potentiality, it happens to have a potentially infinite [number of] directions. So no part of the circumference or some point in it, inasmuch at there is merely a circle, is more worthy of being assigned the direction than [any] other. Once you know this about surface, you also know it about body, as well as knowing how there are and how there are not six directions in the cube, parallelogram, and what is analogous. You also know how the directions of the four-sided pyramid are less than the directions of the cube, and how things stand in the case of the sphere.

(4) The popularity of this premise (namely, that every body has six directions) is due to two things: one is a common belief, the other is a specific consideration. The common belief explaining it is the common man's nonreflective impressions that animals, and specifically humans, are bounded by two sides (corresponding with the two hands), a back, a belly, a head, and feet. [In other words,] he has a right-hand side and a left-hand side, where the right-hand side is the direction with which he is most strongly inclined to begin a motion, whereas the left-hand side is what is opposite it. He also has an upward side and downward side, where the human's upward side is the direction closest to the head, while his downward side is the direction closest to his feet. In the case of quadrupeds, their upward side is the direction closest to their back, while their downward side is the direction closest to the belly and feet. Next, they have a front and rear: The front is the direction toward which they naturally move, as well as where the sensory faculty of vision is, while the rear is what is opposite it. Now, since, according to the [common man's]

جميع أنواع التناهي حتى إلى الزاوية، كانت له جهات ثمان؛ أربعٌ إلى الخطوط، وأربعٌ إلى الزوايا. والدائرة فلا جهة لها بالفعل إلاّ واحدة، وأمّا بالقوة فتعرض لها جهات لا نهاية لها بالقوة، فلا جزء من المحيط ولا نقطة فيه - من حيث هو دائرة فقط - هو أولى بأنْ يلي الجهة دون غيرها. وإذ عرفتَ هذا في السطح، فقد عرفت في الجسم - وعلمتَ أنَّ الجهات الست كيف تكون في المكعب والمستطيل الشبيه بالمكعب، وما يجري مجراهما، وعرفتَ كيف لا تكون. وأنّه كيف تنقص جهات المخروط الذي تحيط به أربعة سطوح مثلثات عن جهات المكعب، وكيف حال في الكرة.

(٤) وأمّا السبب في اشتهار هذه المقدمة؛ وهو أنَّ لكل جسم ست جهات فأمران: أحدهما رأي عامي، والآخر اعتبار خاصيّ، فالذي سببه رأي عامي هو أنّه لما سبق إلى أوهام العامة أنَّ الحيوان، وخصوصاً الإنسان، يحيط به جنبان عليهما اليدان وظهر وبطن ورأس وقدم، وكان له يمين ويسار؛ أمّا اليمين فالجهة القوية منه في ابتداء الحركة، واليسار ما يقابله. وكان له فوق وأسفل، أمّا الفوق للإنسان فالجهة التي تلي رأسه، والسفل منه فالجهة التي تلي قدمه. وأمّا في سائر الحيوان ذوات الأربع فالفوق منه الجهة التي تلي الظهر، والأسفل منه الجهة التي تلي البطن والقدم، فكان له قدّام وخلف؛ فالقدّام هو الجهة التي إليها يتحرك بالطبع - وهناك حاسّة الإبصار - والخلف ما يقابله، ولم يكن عندهم

belief, he has no direction other than these, he designates from his head to his toes his *length,* from his right side to his left side his *breadth,* and from his front to his rear his *depth.* So it is as if, once these extremities were initially posited there, these intervals were then posited corresponding with them, since the reality is that the intervals are posited only by positing the limits from which and toward which [the intervals] extend. Since this is so, [the man in the street] was under the impression that the directions are six—not being aware of any other, since there are no [other] names save for these—and so he was under the impression that this number was the maximum.

(5) He was reinforced in that [belief] on the strength of a specific sort of consideration as well—namely, that within bodies there can exist only three intersecting lines at right angles, and each intersecting line is terminated at two limits. So there are six limits and, thus, six directions. Be that as it may, these intersecting lines are only three precisely when one extension is originally assumed (being supposed, it should be added, without nature [in fact] requiring it) and the intersecting lines are, furthermore, arranged at right angles. If, in place of that one extension, however, another one were assumed that is oblique to the first, there would be three other lines intersecting at right angles numerically different from the former ones, as well as directions that are numerically different from the former ones.[4] In addition to that, it is not necessary that the specificity of the directions in a body differ such that, in every body *qua* body, there is one direction that, in itself, is right and another that, in itself, is left. That—by which I mean distinguishing potentially,

4. Avicenna seems to have something like the following two figures in mind:

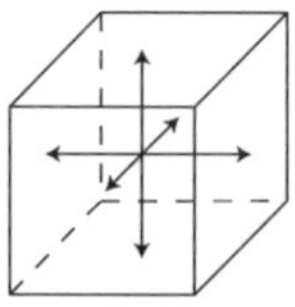

Figure 1

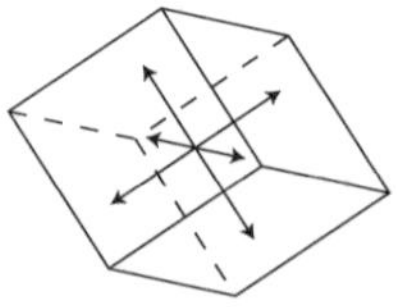

Figure 2

له جهة غير هذه، جعلوا طوله من رأسه إلى قدمه، وعرضه من يمينه إلى يساره، وعمقه من قدّامه إلى خلفه؛ فكأنه لما افترضت ها هنا هذه النهايات أولاً، افترض بعدها بحسبها هذه الأبعاد، إذ الأبعاد بالحقيقة لا تفترض إلاّ بافتراض النهايات التي عنها وإليها تمتد. فلّما كان هكذا، وقع في الأوهام إنَّ الجهات ستٌّ ولم يُشعر بغيرها، إذ لم تكن الأسماء إلاّ لهذه، فوقعت الأوهام على مبلغ هذا العدد.

(٥) وأعانَ على ذلك نوعٌ من الاعتبار خاصي، وهو أنَّ الأجسام يوجد فيها إمكان وقوع مقاطعات ثلاث على قوائم ولا يجوز غيرها، وتنتهي كل مقاطعةٍ إلى طرفي الخط الذي عليه المقاطعة، فتكون ستة أطراف، فتكون ستّ جهات. لكن إنّما تكون هذه المقاطعات ثلاثاً لا غير؛ إذا فُرض امتدادٌ واحدٌ أصلاً، ووضع وضعاً من غير أنْ يكون لاطبع يوجبه ثم تُرتّب عليه المقاطعات بقوائم. ولو فرض مكان ذلك الامتداد الواحد غيره، ممّا ليس موازياً له، لوقعت ثلاث مقاطعات أُخرى على قوائم غير تلك بالعدد، ووقعت جهات غير تلك بالعدد. ثم مع ذلك فلا يجب أنْ تختلف نوعية الجهات في كل جسم، حتى يكون في كل جسم - من حيث هو جسم - جهة هي بعينها يمين، وجهة هي بعينها يسار، إنّما يجب ذلك في الحيوان؛ أعني بذلك تميّز الجهات الست بعضها عن

naturally, and specifically the six directions from one another—is necessary only in animals. Certainly, every body that is near us seems to have an upward and downward side, either accidentally or by nature. As for what is accidental, it is according to whatever chance position [the body] has, and so whatever [part] of it is closest to Earth is the downward direction, while whatever is closest to the celestial sphere is upward (or, if there is no celestial sphere above that body, that part that is opposite what is closest to Earth [is called *upward*]). Still, it could be that this is not found with respect to [the element] earth, that is to say, the natural location of [earth]. [That] is because it appears to have no direction but upward, if by *direction* one means what is nearest to a thing's extremity, and Earth's extremity is a surface, and its surface is closest to the Heavens. Perhaps, however, a consideration of the directions does not require a relation to the surface but, instead, to each limit of some interval posited along with the body. When that is the case, the interval posited with respect to the Earth will have a direction vis-à-vis the center of its sphere, which is the center of the universe and around which there is rotation, as well as a certain direction vis-à-vis its surface, both effectively being extremities of the interval. In that case, the Earth also has a downward and upward direction, but the existence of the Earth's downward direction is not due to that to which it is related in the way the upward direction is. That is because the upward direction is an actually existing surface, whereas the downward direction is a point imagined by the estimative faculty.

بعض تميّزاً بالقوة والطبع والنوع. نعم، يشـبه أنْ يكون لكل جسـمٍ من ‹الذي› يلينا علوٌّ وسـفلٌ، إمّا عارض وإمّا بالطبع. أمّا العارض فعلى مـا يتفق من وضعه، فيكون ما يلي الأرض منه هو الجهة السافلة، وما يلي الفلك، أو ما يقابل ما يلي الأرض - إنْ لم يكن فوق ذلك الجسـم فلك - هو الفوق. لكن، عسى هذا أن لا يوجد في الأرض وهي موضعها الطبيعي، فيشـبه أنْ لا يكون لها جهة إلاّ الفوق، وإنْ عنى بالجهة ما يلي نهاية الشـيء، ونهاية الأرض سطح، وسطحها يلي السماء، فعسى أنْ يكون الاعتبار للجهات لا يقتضي النسبة إلى السطح، بل إلى كل طرفٍ لبُعْدٍ يُفترض في الجسم. وإذا كان كذلك، كان للبُعْد المفـروض في الأرض جهة عند مركز كرته الذي هو مركز الكل وعليه الدور، وجهة عند سطحه، وهما نهايتا البُعْد النافذ فيه. فيكون للأرض أيضاً جهة سفْل وجهة علو، وتكون جهة السفْل للأرض ليس وجوده لما يقاس عليه؛ كوجود جهة العلو. وذلك لأنَّ جهة العلو سطحٌ موجودٌ بالفعل، وجهة السفّل نقطة موهومة.

(6) Alternatively, it is not like that either, and, instead, the upward direction is also the limit of the interval that is continuous with the center at the surface and between a given point. Now, if that is the case, [one might ask,] how could it have two actual directions rather than two potential ones? [The answer is that] we have made instances of *projecting toward* and *standing exactly opposite,* one of the causes of actually dividing what is continuous, since what is contiguous and is projected toward by what is exactly opposite becomes determinate by becoming contiguous and being exactly opposite, just as in pointing. Therefore, the center and other extreme are what determine the existence of the projection of the posited interval. The issue, however, concerning this posited interval is how is it to be posited, to which we say: Earth never ceases to have a horizon, owing to the existence of that which stands above it. Now, all of that [which stands above it] are among the causes for positing intervals that pass through it, as if it would also have no actual up and down in this way, but, rather, [that] there would only be up, in the direction of its extremity toward its surface, were the Earth to exist independently and have no relation to external bodies. The truth, in fact, is this: were there no Heaven, [the Earth] would not have an *up* at all, in any way.

(7) It now remains that we resolve a certain doubt that is raised against this. It is said, then, that, were the estimative faculty to imagine that there is no Earth, but there is the Heavens, then how would it be up when there is no *up* except relative to *down*? How will [the Heavens] have a *down* when you have already posited that downward becomes something

(٦) أو لا تكون أيضاً كذلك، بل تكون جهة الفوق أيضاً طرف البُعْد المتصل بالمركز في السطح وبين نقطةٍ مّا. فإنْ كان كذلك، فكيف تكون له جهتان بالفعل؛ بل تكونان بالقوة. لكنا قد جعلنا أحد أسباب انقسام المتصل المسامتات والمحاذيات، وهو انقسامٌ بالفعل، إذ يتعيّن المماسّ والمسامت المحاذي بالمماسّة والمحاذاة، كما في الإشارة. فيكون إذن المركز والطرف الآخر ممّا يصير معيّن الوجود لمسامتة البُعْد المفروض. لكن الشأن في هذا البُعْد المفروض أنّه كيف يفترض؟ فنقول: إنّه لا تعدم الأرض وجود أُفق لها لوجود قائم عليها، وجميع ذلك من أسباب فرض الأبعاد الذاهبة فيه، فكأن الأرض لو انفردت أيضاً، ولم تكن لها نسبة إلى أجسامٍ خارجية، لم يكن لها بالفعل فوقٌ وأسفلٌ بهذا الوجه، بل فوقٌ فقط من جهة انتهائه إلى سطحه، بل هذا حقٌ؛ فإنّه لولا السماء لم يكن لها علوٌ البتة بوجهٍ من الوجوه.

(٧) فبقي الآن أنْ نحل ما يُتشكّك به على هذا فيقال: لو توهمنا أنَّ الأرض ليس لها إلاّ السماء، أفكان يكون لها علوٌ، والعلو لا يكون علواً إلاّ بالقياس إلى السفل؟ أفكان يكون لها سفل، وقد فرضتم أنَّ السفل ليس بمتعيّن إلاّ بتعيّن بُعْد؟ فإنَّ البُعْد لا يتعيّن لوجود

determinate only through an interval's becoming determinate? [That] is because the interval did not become determinate solely on account of the existence of the Heavens, but by a consideration of what stands above, which fixes Earth's horizon (or some other cause analogous to it). From this it necessarily follows that *up* becomes determinate through the existence of the Heavens, while [at the same time] not being determinate [because there is no *down*]; but this is a contradiction. The response is that two things are meant by *up* (one of which is the opposite of *down* and the second [of which] is the direction closest to the Heavens), just as two things are meant by *lightweight* (one of which is relative to the *heavy* and the other [of which] is that which tends to move so as to encounter the surface of the [lunar] sphere). So one of the two [senses of] *up* is said relative to *down* in just the way that one of the two [senses of] *light* is said relative to *heavy*, whereas the second is intelligible in itself, where its being intelligible does not require a consideration of the existence of its opposite. So it is not a necessary conclusion for whoever posits an actual direction closest to the Heavens that he understood that, because there is some direction that is not near the Heavens. Similarly, from our supposing that something is moved so as to encounter the surface of the [lunar] sphere, it does not necessarily follow that we judge that there is something that is moved toward the center. So the Earth, relative to the Heavens alone[5] without any other consideration, has a direction that is closest to the Heavens. So, if you use this sense to designate *up*, then [the Earth] has an *up*, while, if you do not and [instead] mean by *up* what is relative to *down*, then the Earth, without being considered insofar as it is related to the Heavens, has no *up*.

---

5. Reading *waḥdahā* with **Z** (in close parallel with **T**'s *waḥdahū*) for **Y**'s *fī ḥaddihā* (at its limiting point).

السماء وحده، بل باعتبارٍ قائم يجعل للأرض أُفقاً، أو سبباً آخر يجري مجراه، فيلزم من هذا أنّه يتعيّن العلو بوجود السماء ولا يتعيّن، وهذا خُلْف. فالجواب، إنَّ العلو يُعنى به شيئان؛ أحدهما المقابل للسفل، والثاني الجهة التي تلي السماء، كما أنَّ الخفيف يُعنى به أمران؛ أحدهما الذي بالقياس إلى الثقيل، والآخر الذي يريد في حركته ملاقاة سطح الفلك. فأحد العلوين مقولٌ بالقياس إلى السفل، وكذلك أحد الخفيفين مقولٌ بالقياس إلى الثقيل، والثاني معقولٌ بنفسه، لا يحوج تعقله إلى اعتبار وجود مقابله، فإنّه ليس يلزم لمن فرض جهة بالفعل تلي السماء، أن يكون تعقّل ذلك لأجل جهةٍ لا تلي السماء. وكذلك لا يلزم من فرضنا شيئاً يتحرك إلى ملاقاة سطح الفلك؛ أن نحكم أنَّ شيئاً آخر يتحرك إلى المركز؛ فللأرض، بالقياس إلى السماء وحدها من غير اعتبارٍ آخر، جهةٌ تلي السماء فإنْ سميتَ هذا المعنى علواً فلها علوٌ، وإنْ لم تسمّه علواً، وعنيتَ بالعلو ما يقال بالقياس إلى السُفْل؛ فليس للأرض – من حيث هي مقيسة بالسماء بلا اعتبارٍ آخر – علو.

(8) Starting from the beginning, up and down frequently belong to plants and animals by nature. So [for example] plants have one direction for [their] branches and another for [their] roots, one of which is naturally up and the other naturally down. Up, however, may accidentally become down, and down, up. Despite that, the upward [direction] preserves the sense that it is naturally up and, likewise, the downward [direction] preserves the sense that it is naturally down, just as when water is heated, it preserves the sense that is naturally cool. As for front and rear, the only things to which they belong, whether [those things] are moving or at rest, are animals, whereas they belong to moving bodies other than animals when [those bodies] are moving (for the direction toward which they are moving is their front, while the direction being left behind is their rear). If the motion of [things that are not animals] changes, however, so does their front and rear, whereas this is not the case for animals. [This is] because the front of animals does not correspond with every motion, but with the voluntary motion that is in the direction proper to their limbs, as long as it proceeds naturally and is not like walking backward, since that is not natural but is, in a sense, forced. So the up and down of inanimate bodies sometimes overlap with their front and rear (namely, when they move up and down) and at other times their up and down differ from their front and rear—namely, when their motion is not upward (that is, in the direction of the celestial sphere) or downward (I mean in the direction of the Earth). If they move horizontally, it will be a direction not subsumed under a direction [such as up and down].

(٨) ونبتدىء من رأسٍ ونقول: إنَّ الفوق والسُفْل بالطبع قد يوجدان للنبات والحيوان. فانَّ للنبات جهة أغصانٍ وجهة أُصول، وأحدهما بالطبع فوق، والآخر بالطبع أسفل. ولكن يعرض أنْ يصير الفوق أسفل والأسفل فوق، ويكون الفوق مع ذلك حافظاً لمعنى أنه بالطبع فوق، وكذلك يكون السُفْل حافظا لمعنى أنه بالطبع سُفْل، كما أنَّ الماء وإن سخن فهو حافظٌ لمعنى أنّه بالطبع بارد. وأمّا القدّام والخلف فليس إلاّ للحيوان، كان ساكناً أو متحركاً، وللأجسام المتحركة، غير الحيوان، حين تكون متحركة، فإنَّ الجهة التي إليها تتحرك هي قدّامها، والجهة المتروكة هي خلفها. لكنها إنْ تغيّرت حركتها تغيّر قدّامها وخلفها، ولا كذلك للحيوان؛ لأنَّ القدّام الذي للحيوان ليس بحسب كل حركةٍ، بل بحسب الحركة الإرادية التي إلى جهة أعضاء مخصوصة له، ما دام على النهج الطبيعي، لا كالقهقرى فإنَّ ذلك غير طبيعي، بل متكلف. فالأجسام غير الحيّة تارة يوافق فوقها وسُفْلها قدّامها أو خلفها، وذلك إذا تحركت إلى فوق أو إلى أسفل، وتارة يخالف فوقها وسُفْلها قدّامها وخلفها، وذلك إذا لم تكن حركاتها إلى فوق، أي نحو جهة الفلك، أو أسفل أعني نحو جهة الأرض، وإنْ تحركت عرضاً لم تدخل جهةٌ في جهة.

(9) We should now examine the states of these directions with respect to spheres that are moved upon themselves—in fact, the celestial sphere—and whether the claim that the celestial sphere has an up, down, right, left, front, and rear is meant in the way [that] it is said of other animals or [whether it] is said equivocally, as well as how those directions occur there. Before that, however, we will investigate the natural directions of things undergoing natural, rectilinear motion and how they occur.

(٩) فحـــريٌ بنا الآن أنْ نبحث عن أحوال هـــذه الجهات في الكرات المتحركة على أنفسها، بل في الفلك، وهل ما قيل إنَّ للفلك فوقاً وسُفْلاً ويميناً ويساراً، وقدّاماً وخلفاً، هو المعنى المقول للحيوانات الأخرى، أو باشـــتراك الاســـم؟ وإنَّ هذه الجهات كيف تكون هناك؟ وقبل ذلك ننظر في الجهات الطبيعية للمتحركات الطبيعية على الاســـتقامة، وإنّها كيف تكون.

Chapter Fourteen

The natural directions of rectilinear motions

(1) It is necessary for us now to undertake an independent verification of the account concerning the directions of natural motions and how they are delimited, beginning with the directions of rectilinear motions. From what we said earlier, direction must be delimited with respect to the interval,[1] and its being delimited must be vis-à-vis either a body or a nonbody. As we have explained, however, it is impossible to delimit a given direction in the void,[2] and so the delimitation must be vis-à-vis a body. Now, because what undergoes rectilinear motion leaves behind one direction and tends toward another, either each one of the directions is delimited by a body unique to it or a single body delimits both directions. Being delimited by a single body in the case where [the directions] are diametrically opposed to each other occurs only when one of the limiting points is at the maximal degree of proximity and the other at the maximal degree of remoteness. Now, the maximal degrees of remoteness and proximity pertaining to the body are delimited only in that there is something that surrounds and some central part as a direction in order that the single body get both of the two limiting points that it needs. The delimiting body, however, would have to be what surrounds, not an underlying body such as the central part. That is because, if it is what underlies, such as the central part, the proximity but not the remoteness would be delimited, whereas, in contrast, what surrounds delimits proximity and remoteness.[3]

1. See 3.13.2.
2. See 2.8, especially par. 10.
3. So, for example, imagine some circular body ABCD with center E. While E can delimit the maximal degree of proximity (namely, by being at E), it cannot, when taken alone, delimit the maximal degree of remoteness. In contrast, if one considers the circular line ABCD, it can delimit the maximal degree of proximity (namely, by being somewhere on the line ABCD), but it can also delimit the maximal degree of remoteness (namely, by being at the opposite extreme of some radii projecting inward from ABCD, which in effect is at point E).

<فصل الرابع عشر>

في الجهات الطبيعية للحركات المستقيمة

(١) ومـمّا يجب علينا أنْ نحقّق القول فيه أمر جهات الحركات الطبيعية، وأنّها كيف تتحدّد، ونبدأ بجهات الحركات المسـتقيمة، فنقول: قد سـلف من قولنا أنَّ الجهة - لا محالة - متحدّدة في البُعْد، وتحدّدها لا يخلو إمّا أنْ يكون عند جسمٍ أو عند لا جسم، ومحال، كما بيّنا، أن يكون في الخلاء تحدّد لجهة، فيجب أن يكون التّحدّد عند جسم. ولأنَّ المتحرك على الاستقامة يخلف جهة ويقصد جهة، فلا يخلو إمّا أن يكون كل واحدٍ من الجهتين يتحدّد بجسم على حدةٍ، أو تكون الجهتان تتحدّدان بجسم واحد، والتحدّد إنّما يكون تحدّداً متقابلاً بجسـم واحدٍ، إذا كان أحد الحدّين في غايةٍ القرب منه والآخر في غاية البُعْد منه. ولا تتحدّد غايةً البُعْد من الجسم كما تتحدّد غاية القرب منه؛ إلاّ بأنْ يكون على جهة إحاطة ومركز، حتى يكون الجسم الواحد يوجب الحدّين جميعاً، ويجب أنْ يكون الجسـم المحدّد محيطاً لا جسماً موضوعاً كالمركز؛ وذلك لأنّه إنْ كان موضوعاً كالمركز تحدد القرب منه ولم يتحدّد البُعْد، بل المحيط هو الذي يحدْد القرب منه والبُعْد عنه.

(2) As for when the delimitation [of direction] is by means of two bodies, then either it must be that one of them is like what surrounds, while the other is like the central part, or it is not like that. On the one hand, if one of them is like what surrounds and the other is like the central part, then what surrounds would be enough to make the interval have two limiting points even if that which is in the central part did not exist, in which case the delimitation by that which is in the central part is accidental. On the other hand, when the delimitation is by means of two bodies, but it is not the case that one of them surrounds and the other is something in the center, we say initially, in that case, that it cannot be that some part of one body's simple surface has, by its nature, a claim to be that toward which [the body undergoing rectilinear motion] tends and, with which it comes into proximity, while some other part of it is not such, when, in itself, it is one homogeneous surface belonging to one homogeneous body, whose relation to what is outside it is one homogeneous relation. Instead, its state toward what is outside it is the same in all directions. Now, that surface must have by nature, something outside [it] in every one of its directions, not just [in] some specific direction but not [in] another, such that [the surface] would have one direction lying next to certain places toward which bodies are moved and another direction [at which there is] a termination that has nothing outside of it, whether a void or plenum. On the contrary, it must either have nothing at all outside it, or, if there is something surrounding [it], there is something outside [it], whether a plenum or a void. In other words, [the surface] is such that, with respect to every place outside that which belongs to it, the estimative faculty can imagine some body that is naturally moved toward [that place] so as to become proximate with it. This, however (assuming [that the motion] is not in the direction of what surrounds and the central part), requires a homogeneous periphery. In that case, when the motion toward each one of those two bodies seeks the directional side[4] that is proximate to it, then, were we to imagine the movement

4. "Directional side" is a slight overtranslation of the Arabic *jihah,* which I have been translating "direction" but which can also mean "side." In the present context, Avicenna is using *jihah* in the sense of a given side of some body that functions as a *point of orientation* by which some direction is determined—hence the translation "directional side."

(٢) وأمّا إذا كان التحدّد بجسمين؛ فلا يخلو أنْ يكون أحدهما كالمحيط والآخر كالمركز، وإمّا أنْ لا يكون كذلك. فإنْ كان أحدهما كالمحيط والآخر كالمركز؛ كان المحيط كافياً في أنْ يجعل للبُعْد حدّين وإنْ لم يكن الذي في المركز، فيكون التحدّد بالذي في المركز بالعرض. فأمّا إذا كان التحدّد بجسمين، ولا يكون أحدهما محيطاً والآخر مركزاً، فنقول أولاً: إنّه لا يجب حينئذ أنْ يكون بعض سطح الجسم الواحد البسيط يستحق بطبعه أنْ يكون المتوجّه إليه وإلى القرب منه، وبعضه الآخر ليس كذلك، وهو في نفسه سطحٌ واحدٌ متشابهٌ من جسم واحدٍ متشابهٍ، نسبته إلى ما هو خارج عنه نسبة واحدة متشابهة، بل يجب أنْ تكون حاله إلى ما هو خارج عنه، من جميع الجهات، سواء. ويجب أنْ يكون له بالطبع خارجٌ من كل جهات ذلك السطح، ليس في جهةٍ بعينها دون جهة، حتى تكون جهة منه تلي أمكنةً وأجساماً تتحرك فيها إليه، وجهة نهاية ليس لها خارج، لا خلاء ولا ملاء، بل يجب إمّا أنْ يكون لا خارج له البتة، أو يكون الخارج المملوء أو الخالي - إنْ كان محيطاً به - أي يكون بحيث يجوز أنْ يتوهم في كل مكانٍ من الخارج الذي له جسم يتحرك إليه بالطبع، الحركة المقرّبة منه، وهذا يوجب إحاطة متشابهة، إذا كانت لا على جهة المحيط والمركز. فإذا كانت الحركة إلى كل واحدٍ من هذين الجسمين تطلب الجهة التي هي قربه، وجب أنْ يكون لو توهمنا التحرّك واقعاً من أحد

taking place from one of the two bodies toward the side of the other body whose direction is not the closest one, it would be moved toward the proximity of the first body, necessarily being moved only to that specific directional side rather than the one opposite it. [That follows] since it would have reached the other body—namely, that [side of the other body] that delimits the opposite directional side of the first body—and it would be impossible for the motion to be toward some directional side except the one opposite it. So it has been proven that our assumption concerning the delimitation of two directions by two bodies is absurd.[5]

(3) It cannot be said that one direction is delimited on one side, and the other on the other side, and that the two directions are naturally contrary, for the discussion at hand concerns the principle of a direction that is one and the same in species and what delimits it. So when the factor delimiting the direction that is one and the same in species does

5. Avicenna's argument seems to be this: In paragraph (1) he had asserted that direction is determined by either a single body or two bodies. On the one hand, if a single body determines the direction, it is the periphery of that body that, in fact, does the work, not the central part (see n. 4). On the other hand, if two bodies are required to determine direction, then they are either nested—in which case the body at the periphery is again what does the work—or they are not nested, and so they might be related thus:

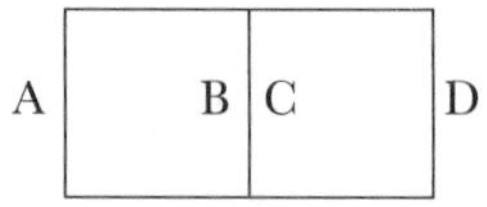

where A and B are the directional sides of one body and C and D are the directional sides of the other body. Given Avicenna's previous argument that the periphery of any given body must be homogeneous throughout, then, when there is rectilinear motion, no one part of the body is more worthy of being sought than another. Consequently, any body tending toward such a homogeneous body would stop once it reaches the closest side of the body toward which it is tending. Thus, it is impossible for an object moving from, for instance, side A of the first body to reach side D of the second body, for, as soon as it reaches C, it will have reached the second body and so arrived at that toward which it tended. Avicenna finds the conclusion absurd—probably because he thinks that it is a physical fact that something can move rectilinearly so as to cover the maximal degree of separation, which, in the present example, would be given by the distance between A and D—and so he rejects the initial assumption that gave rise to the absurdity: that directions are determined by two bodies (or, perhaps more exactly, two bodies that are not nested within one another).

الجسمين إلى الجهة التي لا تلي الجسم الآخر ، فيتحرك إلى قرب الجسم الأول أنْ يكون إنّما يتحرك إلى تلك الجهة بعينها لا من مقابلها ، لأنّها تؤدي إلى الجسم الآخر وهو محدّد الجهة المقابلة للجسم الأول ، ويستحيل أنْ تكون الحركة إلى جهةٍ إلاّ من مقابلها . فقد بانَ أنَّ ما فرضناه من تحدّد الجهتين بجسمين محال .

(٣) وليس يجوز أنْ يقال إنّه من جانب يحدّد جهة ، ومن جانب يحدّد أُخرى ، وأنَّ الجهتين متضادتين بالطبع ؛ فإنَّ كلامنا في الشيء من حيث هو مبدأً جهة واحدة بالنوع ومحدِّدها ، فإذا كان المحدِّد يحدّد الجهة الواحدة بالنوع لكونها قرباً منه ؛ فيجب أنْ يكون

so on account of [that direction's] proximity, then everything at the [same] proximity must also be a direction that is one and the same in species. In that case, its contrary is everything that is at the [same] remoteness from it. So, again, it comes down to its contrary being what surrounds. [That] is because the designated remoteness of the first body's surface must either be delimited by the nature of that other body or not. Now, on the one hand, if it must be delimited by the nature of that body, then [that nature] no more requires that one part of [the body's] surface [be delimited] than another; and, instead, it requires that it [be delimited] in every direction. So the remoteness will be delimited on every side by a body having that nature. If, on the other hand, it is not like that, then the delimitation of each one of [the portions of that sphere] will require some other direction, and what is proximate will delimit a single direction, while what is remote would delimit several directions, and that which is the counterpart of what is numerically one would be specifically many, all of which is absurd. If the bodies that are assumed to be around it are at that [same degree of] remoteness and they are assumed to be scattered in [different] directions, every one of them exchanging places with one of its co-mates, one [body] would [still] delimit the direction that another would (were it in its place) by delimiting a limit of remoteness reaching between it and the first body, since they are alike in that they naturally delimit that which is remote. [That] is because they have a certain position that is at the maximal degree of remoteness, and there is no difference between them in this respect, and [yet] it is in this respect that they delimit the remoteness. Also, the directions described by their positions on that first body would not differ specifically, but only numerically. Those bodies would also be like a single body surrounding the first body, in which

كل قــربٍ منــه هو جهة واحدة بالنوع. فيجب أنْ يكون ضدّهــا كل بُعْدٍ منه، فيعود إلى أنْ يكون ضدها محيطاً، لأنَّ البُعْد المقدّر من ســطح الجسم الأول، إمّا أنْ يقتضي تحدّداً بطبيعة ذلك الجسم الآخر أو لا يقتضي، فإن اقتضى تحدّداً بطبيعة ذلك الجسم، فليس أنْ تقتضيه قطعة من سطحه، أولى أنْ تقتضيه قطعة أخرى منه، بل يجب أنْ تقتضيه من كل جهة، فيكون البُعْد متحدّداً من كل جانب بجسم من ذلك الطبع. وإنْ لم يكن كذلك، كان التحديــد لكل واحدٍ منها يقتضي جهةً أخرى، ويكون القرب يحدّد جهة واحدة، والبُعْد يحدّد جهــات، ويكون مقابل الواحد بالعدد كثيراً بالنــوع، وهذا كله محال. فإنْ كانت الأجســام التي تُفرض حواليه بذلك البُعْد، وتُفرض جهات شتى، أيّها كان بدل صاحبه، حدّد الجهة التي يحدّدها الآخر لو كان مكانه؛ بتحديده طرف بُعْدٍ واصلٍ بينه وبين الجسم الأول. وتكون متشــابهة في أنّها بالطبع تحدّد البُعْد، لأن لها وضعاً مّا هو في غاية البُعْد، ولم يكن بينها - في هذه الجهة - خلافٌ، وكانت بهذه الجهة تحدّد البُعْد، وكانت الجهاتُ التي ترتسم بأوضاعها من الجسم الأول جهاتٌ لا تختلف بالنوع بل بالعدد، وكانت تكون تلك الأجسام كجسمٍ واحدٍ محيطٍ بالجسم الأول، فيكون حدوث الجهتين على سبيل مركزٍ

case two directions would come about, [delimited by] some central part and what surrounds. We have already noted, however, that, when it is like a central part and that which surrounds, the surrounding thing is sufficient to delimit both of the two directions, whereas the underlying body in the center [only] enters into the [whole] affair accidentally.[6]

(4) We maintain that not every body is suitable for delimiting direction. That is because the body whose character is to be moved rectilinearly is not suitable for delimiting direction, since either its nature requires that it be in that direction or not. So, on the one hand, if it does not require it, how can the direction be delimited by it when it might not be there? On the other hand, if its nature requires it to be in that direction, and yet it is possible that it accidentally not be in that direction while naturally seeking it, then, in the nature of that body, there would be a possibility that happens to belong to it to seek that direction; and so there would be[7] no part of that body except that, in its nature, there is the possibility to seek that direction. The possibility of seeking that direction, however, can be attributed to it only if that direction determinately exists. In that case, it is possible, with respect to its nature, that no part of that body happens to be in that direction, and yet that direction determinately exists in itself such that each part of [the body] seeks it. If this possibility does not exist, it is not because of something in the natures of each and every one of the parts right down to the very last single one, depending upon the number of parts marked off, but because of something external—namely, that there is nothing that carries it away from its natural location. If that is the case, then the direction is not itself delimited by this body owing to this body itself, but it is delimited by something else. It was assumed, however, [to be delimited] by this body, which is a contradiction.

---

6. See par. 2.
7. Reading *fa-kāna* with **Z** and **T** for **Y**'s *wa-ka-anna* (and as if).

ومحيط . وقد قلنا إنّه إذا كان على سبيل مركزٍ ومحيطٍ كفى المحيط في تحديد الجهتين جميعاً ، وكان الجسم الموضوع في المركز داخلاً في الأمر بالعرض .

(٤) ونقول ، إنّه ليس يصلح أنْ يكون كل جسم محدّداً للجهة ، وذلك لأنَّ الجسم الذي من شأنه أنْ يتحرك بالطبع على الاستقامة ، لا يصلح أنْ يحدّد الجهة ؛ لأنّه لا يخلو إمّا أنْ تقتضي طباعه الكون في تلك الجهة أو لا تقتضي ، فإنْ لم تقتض ، فكيف تتحدّد به الجهة؟ وجائز أنْ لا يكون هو عندها . وإنْ اقتضـ〈ـت〉 طباعه الكون في تلك الجهة ، وكان مع ذلك جائزاً أنْ يعرض له أنْ لا يكون في تلك الجهة ، وهو بالطبع يطلبها كان في طبيعة ذلك الجسم إمكان أنْ يعرض له طلب تلك الجهة ، فكان لا جزء لذلك الجسم إلاّ وفي طبيعته إمكان طلب تلك الجهة . ولكنه من المستحيل أنْ يوصف بأنَّ فيه إمكان طلب تلك الجهة ؛ إلاّ وتلك الجهة حاصلة . فيكون لا جزء لذلك الجسم إلاّ ويمكن في طباعه أنْ يعرض له أنْ لا يكون في تلك الجهة ، وتكون تلك الجهة حاصلة في نفسها يطلبها كل كل جزء منه . فإنْ لم يوجد هذا الممكن ، فإنّما لا يوجد لا لأمرٍ في طباع جزء جزء من الجسم ، إلى آخر أجزائه المعدودة بحسب عدد تلك التجزئة ، بل بسببٍ من خارج ، وهو فقدان ناقلٍ عن موضعه الطبيعي ، وإذا كان كذلك ، فالجهة غير متحدّدة الذات بهذا الجسم ، لذات هذا الجسم ، بل متحدّدة بشيءٍ آخر ، وقد فُرض بهذا الجسم ، هذا خُلْف .

(5) So it has been proven that not just any body, as chance would have it, can delimit a determinate direction. From that, it also becomes clear that a direction that is one and the same in species can naturally be delimited by a single body that is of a character that it not move at all rectilinearly, for that which delimits at the periphery cannot be some given ordering of disparate bodies. Indeed, some of those intervals cannot be such that they require that there exist, relative to them, some certain body upon which they necessarily follow, while other ones require some other body naturally different from [the first] upon which they necessarily follow. It is also impossible for that surrounding body to be divided into bodies whose species vary as chance would have it, without any necessity, while it remains the same. You also cannot maintain anything like this when that which delimits at the periphery is a single body. [That] is because the single body has no actual parts, even if, through varying external factors, parts can be marked off in it accidentally. As for the ordered position of bodies differing in species at the periphery of the interval that is farthest away from the surrounded body, [that order] is not something that just occurs unexpectedly and passes away. Otherwise, those bodies would [both] come to be at that periphery and depart from it, and the delimitation of that direction would exist determinately prior to them [but, again, it is the bodies that determine the direction].

(6) From this, we next know that what delimits at the periphery must be a single body that moves only circularly. In that case, it does not naturally include any directions that do not lead to it from the center and from it to the center. Now, the extremities of those [directions]

(٥) فقد بانَ أنّه لا يجوز أنْ يكون أي جسم اتفق محدّداً للجهة المعيّنة، وتبيّن من ذلك أيضاً أنَّ الجهة الواحدة بالنوع تتحدّد بجسم واحدٍ بالطبع، ليس من شأنه الزوال على الاستقامة البتة. فإنَّ المحدّد بالإحاطة لا يصلح أنْ يكون منتظماً من أجسام شتى، فإنّه ليس يجب أنْ تكون بعض تلك الأبعاد تستحق أنْ يوجد فيه جسمٌ بعينه يلزمه، وبعضٌ آخر يستحق جسماً آخر مخالفاً له بالطبع يلزمه. ولا يجوز أنْ يكون انقسام تلك الجهة المحيطة إلى أجسام، مختلفة الأنواع، اتفاقاً من غير وجوب وبقيَ كذلك. وليس لك أنْ تقول مثل هذا؛ إذا كان المحدّد بالإحاطة جسماً واحداً؛ فإنَّ الجسم الواحد لا أجزاء له بالفعل، وإنْ عرضت له تجزئة ما فبأشياء من خارج غير ثابتة. وأمّا ترتّب الأجسام المختلفة النوع في إحاطة أبعد البُعْد عن الجسم المحاط به، فليس ممّا يطرأ ويزول؛ وإلاّ كانت تلك الأجسام تحصل في تلك الإحاطة وتخرج عنها، ويكون تحدّد تلك الجهة حاصلاً قبلها.

(٦) فنعلم من هذا؛ أنَّ المحدّد بالإحاطة يجب أنْ يكون جسماً واحداً لا يزول، اللّهم إلاّ بالاستدارة. وإذا كان كذلك، لم يكن في ضمنه جهات بالطبع إلاّ التي تأخذ نحوه من

that stand opposite one another do not vary by nature, since they terminate at some definite individual bodies. Also their limits—some of which are at the maximal degree of proximity, while others are at the maximal degree of remoteness—are not delimited at limiting points that vary. This is pretty much what we have to hold.[8]

(7) We also maintain that the maximal degree of proximity to the delimited body whose proximity is sought by motion must not be the maximal degree of proximity of each part of it. [That] is because it is impossible that a single thing that is moving along a single interval (as, for example, a single line) should have reached every part of that which it is approaching. As for the maximal degree of remoteness, it is possible that there is a maximal degree of remoteness from all the parts when [the delimited body] is at the center. When a line originating from the surrounding [body] reaches the center and then passes through it, however, the limit from which it begins is at the maximal degree of proximity, but the [line's] other limit is not at the maximal degree of remoteness, since it is closer to what surrounds, even if the whole of it is not. Now, we have already said that it is not a condition of being proximate to what surrounds that it be proximate to the whole, but only to some part of it, even if it is at the maximal degree of remoteness from some other part of it. That is because it is not at the maximal degree of proximity to some part of it unless it is at the maximal degree of remoteness from what lies opposite it. [*What lies opposite* here] is by supposition rather than by nature, since the parts of what is circular have no opposite except by positing

8. Reading *an naqūla bihi* with **Z** and **T** for **Y**'s *an yaqūlu bihi haʾulāʾ* (they have to hold).

المركــز ، والتي تأخذ عنه نحو المركــز . واللواتي تعارضها فإنَّ نهاياتها لا تختلف بالطباع ، فإنّها تنتهي إلى أجســام واحدة بأعيانها ولا تتحدّد أطرافها بحدود مختلفة ، يكون بعضها غاية قربٍ وبعضها غاية بُعْدٍ ، على نحو ما وجب أنْ نقول به هذا .

(٧) ونقول إنَّ غاية القرب من الجســم المحدَّد المطلوب قربه بالحركة ، ليس يجب أنْ تكون غاية قربٍ من كل جزء منه ، فإنّه يستحيل أنْ يكون لمتحركٍ واحدٍ على بُعْدٍ واحدٍ ، كخــطٍ واحدٍ ، وصولٌ إلى كل جزء مــن المقترب إليه . وأمّا غاية البُعْد ؛ فيجوز أنْ تكون غاية بُعْدٍ من جميع الأجزاء ، إذا حصل عند المركز ، وإذا انتهى خطٌ من المحيط إلى المركز ثــم عدّاه . فإنَّ الطرف الــذي ابتدأ منه هو في غاية القرب ، والطرف الآخر ليس في غاية البُعْد ، فإنّه يلي المحيط ، وإن كان لا يلي كله . فقد قلنا إنّه ليس شرط القرب من المحيط أنْ يكون قريباً من كله ، بل من شــيءٍ منه ، وإنْ كان <في> غاية البُعْد من شــيءٍ آخر منه . وذلك لأنّه لا يقرب من شــيءٍ منه غاية القرب إلاّ صار على غاية البُعْد من مقابله بالوضع ليس بالطبع ، فإنَّ أجزاء المســتدير لا مقابلة لها إلاّ بالفرض الوضعي الإضافي المســافي .

some position, relation, and distance. So, even if there is a maximal degree of remoteness with respect to the distance, it won't be the maximal degree of remoteness with respect to the nature and the proximity and remoteness that are in the nature, since here it is not remote from this direction. In fact, here it happens to be where it is closer *qua* a single nature and single body.[9] So, by means of this, you know the form of direction toward which natural bodies move.

(8) Let us now discuss the direction of bodies that undergo circular motion. What is moved circularly is of two classes. One of them involves [a body] that does not move around its own center, but some center outside [of it]. This can, then, be assigned one direction toward which and another from which it moves, where one of them is like its front and the other its rear. As for the right- and left-hand directions, the direction that, if this were an animal, would be its right-side one is, by analogy, more aptly labeled *right* than its opposite, even though there is nothing in the nature of that body that requires that the two sides differ in the way that the two sides of an animal require that in the animal. As for this assumed mobile's up and down, what lies closest to the vicinity of the Earth seems to be its lower directional side, while the opposite of it is its higher directional side, that being assigned to it not because of the particular thing itself (as in the case of animals) nor because of its particular motion (as with heavy and light things that are moved), but relative to other bodies.

---

9. Avicenna's point in this discussion is that it is only between points or places on the periphery of what surrounds and the center point of what is surrounded that maximal separation exists between the directions. Put in geometrical terms, he is arguing that only in the case of the radius is there the maximal degree of separation between some point on a circle's circumference (which defines one direction, such as up) and the circle's center (which defines another direction, such as down). Assume circle ABCD with center E, and also assume a line AF (shorter than the diameter AC), which extends from A through E to F, where AF is longer than the radius AE by a distance x. In this case, then, at F, the line AF is closer by x to point C on the circumference than point E of the radius AE is to C. Consequently, the line AF, while longer than AE, does not have limits that are at maximal proximity and remoteness with respect to the circumference, for, while A is at maximal proximity to the circumference (since it is on it), F is not at maximal remoteness from the circumference (since it is closer to the point C on the circumference than E is).

فإنّها ، وإنْ كانت من حيث المسافة غاية البُعْد ، فليس من حيث الطبع ، ومن حيث القُرب والبُعْـد الذي في الطبـع بغاية البُعْد ، إذ لا بُعْد هناك من هـذه الجهة ، بل هناك اتفاق من حيث أنّها تلي طبيعة واحدة وجسماً واحداً . فبهذا نعلم صورة الجهات التي تتحرّك إليها الأجسام الطبيعية .

(٨) فلنتكلـم الآن في جهات الأجسـام المتحركة على الاسـتدارة : وأمّا المتحرك بالاسـتدارة فهو على قسمين : أحدهما المتحرّك لا على مركز نفسه بل على مركز خارج ؛ فهـذا يمكن أنْ تعيّن له جهة إليها يتحرك ، وجهة عنها يتحرك ، ويشـبه أنْ يكون أحدهما قدّاماً له والآخر خلفاً . وأمّا جهة اليمين واليسـار فيشـبه أنْ تكون الجهة التي لو كان هذا حيواناً كانت تلك يميناً له ، أولى أنْ تسمّى يميناً من مقابلها على التشبيه ، وإنْ كان لا شيء في طبيعة ذلك الجسـم يوجب أنْ تختلف به الجهتان ، كما يوجب جانبا الحيوان ذلك في الحيوان . وأمّا فوق هذا المتحرّك المفروض وأسـفله ؛ فيشبه أنْ يكون ما يلي ناحية الأرض جهته السافلة ، وما يقابلها جهته العالية ، يتعيّن ذلك له ، لا من ذاته بعينه كما للحيوان ، ولا من حركته بعينه كما للمتحركات الثقيلة والخفيفة ، بل بالقياس إلى أجسامٍ أخرى .

(9) As for [the second class, (that is,] what undergoes circular motion around a center internal to it and contained within it), what has been said about it—namely, that six directions can be determined for it, just like in animals—seems to have lost some of its direction. The fact is that, first, two poles and a great circle[10] are essentially to be determined with respect to it, where determining the poles and the great circle requires only [the moving body's] corporeality and its moving in the way just described [that is, rotating around an internal center]. In the case where [the body] is contained by some other body, there is one direction lying closest to that which contains it and another one opposite it that are determined for [the body] in such a way that it does not need to undergo the motion that belongs to it in order to be determined and, instead, has that even if it is at rest. Still, when its motion is considered along with that which contains it—where either certain relations between its parts are assigned, or certain points are posited in it, as well as between [other points] like them in what is contained and around which something is moved—then other directions can be determined for it. That is because, when three points are posited along the length of the motion (not its breadth, which is between [the rotating body's] two poles) and the middle one moves toward one of them and away from the other, and the direction toward which the middle one [moves]—relative to the horizon over which[11] this point rises—is the direction from which the natural motion begins (and its opposite is the opposite direction), then an eastward and westward direction are delimited. Likewise, there will be

10. While the Arabic *minṭaqah* literally means "belt," the over translation "great circle" seems appropriate, given the subsequent discussion.

11. Reading *ᶜalayhu* (over which [m]) with two of the MSS consulted by **Y** and **T**, which would have the pronoun refer to "horizon," for **Y** and **Z**'s *ᶜalayhā* (over which [f]), which would make the pronoun refer back to "this point," or perhaps all the way back to "direction."

(٩) وأمّا المتحرّك بالاستدارة على مركزٍ في داخله ويشتمل هو عليه، فيشبه أنْ لا يكون ما قيل فيه من أنّه قد تتحدّد له جهاتٌ ستٌ كما للحيوان أمراً على الجهة التي قيل، بل أول ما يتحدّد فيه، وعن ذاته، قطبان ومنطقة، ولا يحتاج في تحدّد القطبين والمنطقة إلى شيء غير جسميته وحركته التي هي على الصفة المذكورة. فإنْ كان محتوياً على جسم آخر تحدّدت له جهةٌ تلي ما يشتمل عليه؛ وجهة أخرى بخلافها؛ تحدّداً ليس يحتاج في ذلك إلى أنْ يكون متحركاً الحركة التي له، بل وإنْ كان ساكناً، كان له ذلك لكن إذا اعتبرت حركته على ما يشتمل عليه منها، ونوسب بين أجزائه، أو نقطٌ تفرض فيه وبين أمثالها من المشتمل عليه المتحرك حوله، فقد تتحدّد له جهاتٌ أخرى. وذلك لأنّه إذا فرضت - في طول حركته لا في عرضها الذي هو بين قطبيه - ثلاث نقطٍ، وكانت الوسطى تنحو إحداهما وتتباعد عن الأخرى، وتكون الجهة التي كانت فيها الوسطى بالقياس إلى الأفق الذي هذه النقطة طالعة عليهِ؛ هي جهة عنها ابتداء الحركة بالطبع، ومقابلها مقابل هذه الجهة فتتحدّد هناك جهة مشرقٍ وجهة مغرب. وكذلك تتحدّد هناك

delimited one direction toward the meridian and another toward what is below the Earth. In that case, the direction that is toward the meridian is that toward which the rising motion [tends] (where that is one of its extremes, because what is ascending is at its greatest proximity there), and then, gradually departing from it, [becomes] remote from it at the point where it sets. Now the extremity toward which the mobile tends is the front and its opposite is the rear, and so the meridian—relative to the ascending, eastward motion—is the front and what is opposite it the rear. Also, since the eastward direction is the direction from which the motion begins, it most deserves to be likened to the right-hand side of the animal, in which case westward is the left-hand side. That leaves the two poles, which delimit neither the interval delimited by front and rear (which most deserves to be depth) nor the interval delimited by the right- and left-hand sides (which most deserves to be breadth). So there is only the dimension of length [for them to delimit]. Of the two poles, the one deserving most to be compared with up is the South Pole, relative to the primary motion belonging to the celestial spheres, while relative to the secondary motion [belonging to the celestial spheres], the North Pole [deserves most to be compared with up].[12] [That] is because, were we to imagine some human rotating about himself, where his motion originates from his right-hand side, then his front is toward his face (that is, what is between his right- and left-hand side), where that faces the meridian, while his rear is what is toward his back. Now, when we make the eastward direction correspond with his right-hand side and the westward direction with the left-hand side, while his face [corresponds] with

12. The primary celestial motion refers to the apparent eastward, diurnal motion of the celestial bodies around the Earth (that is, the apparent motion of the stars, Sun, Moon, and other planets around the Earth approximately once every twenty-four hours), while the secondary celestial motion refers to the westward, sidereal motion of the planets as observed against the fixed stars. Thus, for example, while the Sun has an apparent eastward motion that makes it appear to circle the Earth from east to west once every twenty-four hours, it also appears to move from west to east approximately one degree every day, as measured against the fixed stars, and so completes this apparent westward motion approximately once every year.

جهــة تلي خط الزوال وجهة تلي ما تحت الأرض، فتكون الجهة التي تلي خط الزوال هي التي إليها الحركة الآخذة بالارتفاع وتلك غايتها، لأنّها تكون هناك أقرب ما تكون من المطلوع عليه، ثم تأخذ في مفارقته قليلاً قليلاً، والبُعْد عنه إلى أنْ تغرب عنه. والغاية التي إليها يتوجه المتحرك هو القدّام وما يقابله هو الخَلْف، فخط الزوال بالقياس إلى الحركة الشــارقة الطالعة قدّام، وما يقابله خلْف. ولما كانت جهة المشــرق، الجهة التي عنها مبدأ الحركة، فأولى ما يُشــبّه به من جهة الحيوان اليمين؛ فيكون المغرب هو اليســار. وبقي القطبان، يحدّدان البُعْد الذي هو غير البُعْد المحدّد بالقدّام والخَلْف؛ الذي هو أولى بأنْ يكون عمقاً، وغير البُعْد المحدّد باليمين واليســار الذي هو أولى أنْ يكون عرضاً، فليس له إلاّ أنْ يكون بُعْــد الطول. وأُولَيْ القطبين بأنْ يكون على جهة المقايســة علواً هــو الجنوبي في الحركة الفلكية الأُولى، والشــمالي في الحركة الثانية. فإنّا لو توهمنا إنســاناً يتحرك على نفســه مســتديراً، وتنبعث حركته من يمينه؛ لكان يكون قدّامه ما يلي وجهه، وهو ما بين يمينه ويساره. وذلك عند خط الزوال وخلفه ما يلي ظهره وإذا أطبقنا بين يمينه وجهة المشرق، وبين يساره وجهة المغرب، وبين وجْهِه وجهة خط الزوال؛ انطبق رأسه مع القطب الجنوبي

the direction of the meridian, the [top of] his head will correspond precisely with the South Pole. So, were he to rotate about himself like the Heavens, the head must be the South Pole, the face is the Midheaven,[13] while the right-hand side must be eastward.[14] Other than the fact that one of the poles is up and the other is down, there is absolutely no difference in them. [Whatever difference there may be] it is purely by comparison with an animal after directions are delimited owing to other factors. So, in that case, differences in the state of the poles are relative to the directions. That eastward is the right-hand side, however, is on account of something in the motion as it is related to the horizon, whether it [is or] is not compared to an animal, for the motion essentially originates from the eastward direction; and the same holds for the direction of the Midheaven, since the motion is toward it. So, when motion distinguishes eastward, westward, and the Midheaven relative to the horizon—and, moreover, when it distinguishes the limiting points—there is entailed a certain accidental distinction with respect to the poles. [This distinction, however,] is not owing to something primarily associated with the poles, but, rather, owing to a certain relation concomitant with it because of the distinction that happened to belong to something other than them. If you take some part of the celestial sphere

13. That is, the *Medium Coeli* of Ptolemaic astronomy, which, in the present context, seems to correspond with the direction facing the point where the meridian intersects the ecliptic.

14. It is difficult trying to picture the image Avicenna has in mind. The diagram below is one way of visualizing it.

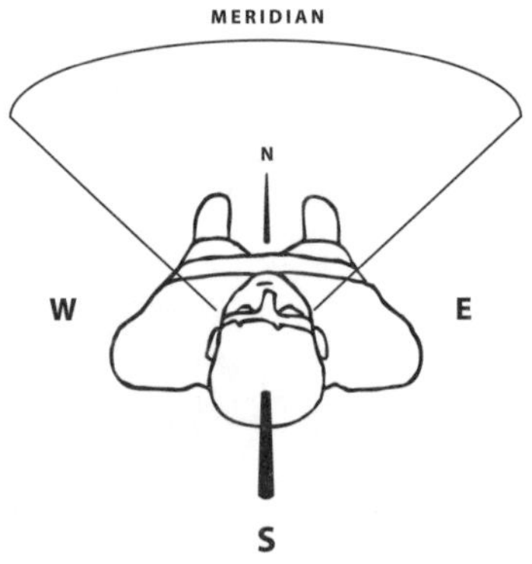

لا غير . ولو دار على نفسه مثل دَوْر السماء ، لكان الرأس يلزم الجنوبي ، والوجه يلزم وسط السماء ، وحيث اليمين يلزم المشرق ، إلاّ أنْ يكون أحد القطبين علواً والآخر سفلاً ، ليس لاختلافٍ البتة في أمر القطبين ، بل بالمقايسة الصرفة إلى الحيوان ، بعد أنْ تتحدّد جهات لأمورٍ أخرى ، فتختلف حال القطبين حينئذ بالقياس إلى تلك الجهات . وأمّا كون المشرق يميناً فهو لأمرٍ في الحركة مقيسة إلى الأُفق ، وإنْ لم يكن حيوان يقايس به ، فإن جهة المشرق لذاتها عنها تنبعث الحركة ، وكذلك حال جهة وسط السماء من كونها إليها الحركة . فإذا كانت حركة تميّز المشرق والمغرب ووسط السماء بالقياس إلى الأُفق ، ثم إذا تميّرت هذه الحدود ؛ لزم في القطبين أنْ يعرض لهما تميّز ما ، لا لأمر متعلق بالقطبين تعلّقاً أولياً ، بل لنسبةٍ تلحقه بسبب ما عرض لغيرهما من التميّز . هذا ، وأمّا إنْ أخذتَ جزءاً من الفلك

as undergoing motion and consider it itself, you will find that what is between the eastward and westward [sides] is the length of the distance. Now, [in this case, should] it strike you that what is between the two poles is the breadth of that length, then you consider how the state of directions will vary.

(10) While the two poles delimit two directions of that body itself and its motion, they do not essentially delimit up and down, nor is there any contrariety with respect to them, since there is no contrariety in the natures of what is in them. Instead, they delimit up and down by analogy and by a certain relation to animals. As for eastward and westward (and the same will hold for the Midheaven), the two delimit two directions for the body itself neither alone nor when taken together with its motion, but only relative to the horizon. In addition to the [body's] relation [to the horizon], one part of the motion itself must be distinguished from another part relative to the horizon. [That follows] since it has to differ such that part of it is that *from which* and part of it is that *toward which* [the motion tends]—one being the motion's place of origination, the other its destination. [So] each one has an opposite for which we do not at all have to take into account a certain relation or parallel with animals. Together with that, there occurs between them a certain species of contrariety or opposition. Now, given all of this, and the fact that right and left are assigned to the celestial sphere's motion and to animals [only] equivocally or by analogy, up and down are most suited to be that [namely, the species of contrariety or opposition belonging to the poles]. As for front and rear, it would seem that front belongs to the ascending part of the celestial sphere in a sense common to both it and others. That is because, if by *front* we mean some ultimate endpoint toward which the

متحركاً واعتبرته بنفسه، وجدتَ ما بين المشرق والمغرب طول المسافة، وحصل لك ما بين القطبين عُرْضاً لذلك الطول، فانظر إلى حال هذه الجهات كيف تختلف.

(١٠) أمّا القطبان فيحدّدان جهتين لذات الجسم وحركته، ولا يحدّدان بذاتيهما فوقاً أو سُفْلاً، ولا يكون فيهما تضادّ؛ إذ لا تضادّ في طباع ما هي فيه، بل يحدّدان فوقاً وسُفْلاً بمقايسةٍ ونسبةٍ إلى حيوان. وأمّا المشرق والمغرب، وكذلك وسط السماء، فليسا يحدّدان جهتين لذات الجسم وحده ولا لذاته مأخوذة مع حركته، بل بمقايسته إلى الأفق. ثم بعد المقايسة فإنَّ نفس الحركة توجب تميز بعضها من بعض بالقياس إلى الأُفق، إذ توجب أنْ تكون متخالفة؛ فيكون بعضها عنه، وبعضها إليه، وبعضها منبعث الحركة، وبعضها متجه الحركة، ولكل واحدٍ مقابل. ولا نحتاج في ذلك إلى أنْ نراعي مقايسة ومحاذاة مع حيوانٍ البتة، ومع ذلك فقد تقع بينها - نوع ما - مضادّة أو مقابلة. ومع هذا كله، فإنَّ اليمين واليسار تقع على جهات الحركة التي للفلك والتي للحيوان باشتراك الإسم أو باشتباهه والفوق والسُفْل أولى بذلك. وأمّا القدّام والخَلْف فيشبه أنْ يكون الجزء الطالع من الفلك قد يوجد له قدّام؛ بمعنىً يعمّه وغيره، وذلك لأنّا إن عنينا بالقدّام نهاية ما يتحرك إليه

ascending part is moving, the celestial sphere would have no front, for its motion has no [ultimate] endpoint toward which it tends. If we mean some endpoint toward which the ascending thing is moving—that is, the thing over which it is ascending—then that endpoint is projecting toward the thing that defines the horizon, and, by defining the horizon, it defines the ascending. [That] is because, when [some body] ascends over [the horizon], continuing to move along until it is projecting at [the zenith of] the meridian, and thereafter moves away from it until it sets at that same horizon, then, if nothing defined the horizon, there would be no horizon, and so there would be neither an ascending nor a meridian, since the delimitation of these directions is relative to [the horizon]. This is the way [in which] we should conceptualize these directions, understanding that these six directions are delimited for the celestial sphere inasmuch as it is undergoing circular motion. As for the direction of the surface that is closest to the Earth and its opposite, it has that inasmuch as it is a body having shape and position, not inasmuch as it is moving.

الجزء الطالع مطلقاً ، لم يكن للفلك قدّام فإنّه ليس لحركته نهاية إليها تقصد . وإنْ عنينا نهاية ما يتحرّك إليه الجزء الطالع - وهو طالعٌ على شيء - فتلك النهاية هي مسامتة الشيء الذي حدّد الأفق فحدّد الطلوع بتحديد الأفق ، فإنّه إذا طلع عليه لا يزال ينحو نحوه إلى أنْ يسامته في خط الزوال ، ثم يعرض عنه إلى أنْ يغرب عنه في ذلك الأفق بعينه . فإنْ لم يكن محدّداً للأفق لم يكن أفق ، فلم يكن طلوعٌ عليه ولا كان خط زوالٍ فلمّا كان محدّداً تحدّدت هذه الجهات بالقياس إليه . فهكذا يجب أن نتصور أمر هذه الجهات ، ونعلم أنَّ هذه الجهات الست تتحدّد للفلك من حيث هو متحرك على الاستدارة . وأمّا جهة السطح التي تلي الأرض والتي تقابلها فذلك له من حيث هو جسمٌ على شكله ووضعه ، لا من حيث هو متحرّك .

FOURTH BOOK: ON THE ACCIDENTS OF THESE NATURAL THINGS AND THEIR INTERRELATIONS, AS WELL AS THE THINGS THAT ARE NECESSARY CONCOMITANTS OF THEIR INTERRELATIONS

Chapter One

Of the subjects contained in this book

(1) In this book, we must independently investigate how motion is one and many; how two motions coinciding in concert are and are not to be compared in relation to their speed; how one motion is and is not contrary to another; how motion is natural; whether place is natural and how it is so; whether every body has a natural place; and how motions are not natural and how many kinds are non-natural. We shall also bring together all of motion's specific differences and explain the interrelations between[1] the motive powers and motions.

1. Reading *mā bayna* with **Z**, **T**, and two MSS consulted by **Y** for **Y**'s preferred *ta'thīr* (influence).

<المقالة الرابعة>

في عوارض هذه الأمور الطبيعية
ومناسبات بعضها من بعض
والأمور التي تلحق مناسباتها

<الفصل الأول>

في الأغراض التي تشتمل عليها هذه المقالة

(١) يجب أنْ نحقّق في هذه المقالة أنَّ الحركة كيف تكون واحدة، وكيف تكون كثيرة، وأنَّ الحركة كيف تكون مضامّة مطابقة لحركةٍ أخرى تقايسها في السرعة والبُطْء؛ وكيف لا تكون، وكيف تكون الحركة مضادّة لحركةٍ أخرى، وكيف لا تكون. وأنَّ الحركة كيف تكون طبيعية، وأنَّ المكان هل يكون طبيعياً، وكيف يكون طبيعياً وهل لكل جسم مكانٌ طبيعي، وأنَّ الحركات كيف تكون غير طبيعية وكم أقسام غير الطبيعية؟ وأنْ نجمع جميع فصول الحركة، وأنْ نعرف مناسبات ما بين القوى المحرّكة والحركات.

Chapter Two

On the numerical unity of motion

(1) There are a number of ways in which motion is one: either one in number, one in species, or one in genus (whether the proximate or remote genus). Let us, then, investigate what is one in number before the others. We say that a group of Parmenides' followers, and some Platonists who stood with them, wholly denied that being one or even being itself is to be attributed to motion.[1] They asked: How can being be attributed to motion when none of [the motion] ever determinately exists? They also mentioned other puzzles, which we have already presented in the preceding chapters on motion[2] and time.[3] For example, they asked: How could being one be attributed to motion when there is no motion that cannot be divided into past and future, nor is there any motion that does not have two periods of time belonging to it, whereas those who affirm that motion is one make, as a condition for it, that its time be one? Also, how could motion be one when everything that is one is something complete with respect to that in which it is one, and everything that is complete has a fixed existence, the parts being present (if it has them), whereas motion has no fixed existence despite the fact that it does have parts? In what preceded, we have already explained the case involving motion's existence so as not to need to worry about these puzzles.[4] At present, we need to explain the case involving motion's being one and [show] how to resolve the sophism that they have raised.

1. Cf. Aristotle, *Physics* 1.2.
2. The reference may be to 2.1.5.
3. See 2.10.2–4.
4. See 2.1.1–11.

<الفصل الثاني>

في الحركة الواحدة بالعدد

(١) الحركـة تكون واحدة على وجوه. فإنّهـا إمّا أنْ تكون واحدة بالعدد، وإمّا أنْ تكون واحـدة بالنوع، وإمّا أنْ تكون واحدة بالجنس؛ إمّـا بالجنس الأقرب وإمّا بالجنس الأبعد. فلنحقّق الواحد بالعدد قبل غيره فنقول: إنَّ قوماً من آل برمانيدس ومَنْ شـايعهم مـن أصحاب أفلاطن منعوا كل المنع أنْ تكون الحركة توصف بالوحدة؛ بل بالهوية، وقالوا كيف توصف الحركة بالهوية، ولا يحصل شيء منها موجوداً حاصلاً؟ وقالوا – سائر ما قد فرغنا منه من الشكوك في باب الحركة والزمان فيما سلف – مثل قولهم كيف توصف الحركة بالوحدة، ولا حركة إلاّ منقسـمة إلى ماضٍ ومستقبل، ولا حركة إلاّ ولها زمانان، ومثبتو وحدة الحركة يشترطون أنْ يكون زمانها واحدا. وكيف تكون الحركة واحدة، وكل واحدٍ فإنّه تامّ فيما هو فيه واحد، وكل تامّ فهو قارّ الوجود حاضر الأجزاء إنْ كانت له، والحركة لا وجود قارّ لها، مع أنَّ لها أجزاء؟ . ونحن – فيما سلف – قد بيّنا الحال في وجود الحركة بياناً لا نلتفت معه إلى هذه الشكوك. والآن، فيحق علينا أنْ نبيّن الحال في وحدة الحركة، ونبيّن أنَّ الشبهة التي أوردوها منحلّة.

(2) We say: We ourselves have already explained that motion is said both of the first perfection (which we have described) and of traversing a distance.[5] The unity of the first perfection is through the unity of its subject, together with the unity of time during which it exists, which is continuous, and is like the rest of the attributes [in that] their subject's being one, [taken] alone, is not enough that they are individually one. [This] is because, when a single subject has the accident of white in it, and then [that accident] ceases, but thereafter it happens to have another instance of white in it, this [latter] white is not the same individual one that the first one was. So the motion is one (in the sense that we have designated) when the subject is one and the same during one and the same time, where the unity of the time is its continuity. So every motion having this description is individually one. Also, [motion] is inevitably in the thing undergoing the motion with respect to some thing that is one. Examples would be one continuous distance; or some white toward which the mobile is directed without temporally stopping at some [other] degree [of whiteness] during the alteration; or, again, one quantity or the like. This account is no less deserving of being included as a condition of motion's unity than time is. If mentioning time is necessary and even sufficient, that is not because it is a distributive attribute of all the conditions by which motion is one, but, rather, because it necessitates the remaining condition, and, from [time], the mind is moved to [that condition] and entails it. (You know the difference between what is distributed, what necessitates, and what is entailed.) As for motion that is in the sense of a traversal, this account [that is, time] is most deserving of being a condition for it. So the things that must be one in order that the motion be one are the mobile, the distance (and what is analogous to it), and the time. So, necessarily, what is undergoing motion, the distance (or that with respect to which there is motion), and the time are one—that is, one in number in its entirety.

---

5. See 2.1.5–6.

(٢) فنقــول: قد بيّنا نحن أنَّ الحركة تقال للكمــال الأول الذي وصفناه وتقال لقطع المســافة. فالكمال الأول وحدته بوحدة الموضوع له، مــع وحدة زمانٍ موجودٍ فيه؛ التي هي اتصال، وكســائر الصفات التي لا تكفي في كونها واحدة بالشخص، كون موضوعها واحــداً فقط، فانَّ الموضوع الواحد إذا عــرض فيه بياضٌ، ثم عُدم، ثم عرض فيه بياضٌ آخر، لم يكن هذا البياض هو بعينه الأول بالشخص. فتكون الحركة - بالمعنى الذي أشرنا إليــه واحدة، إذا كان الموضوع واحــداً بعينه، في زمانٍ واحدٍ بعينه، ووحدة الزمان هي اتصاله. فكل حركةٍ بهذه الصفة فهي واحدة بالشــخص، وتكون لا محالة في متحركٍ فيه واحد، مثل مســافةٍ واحدةٍ بالاتصال، ومثل بياضٍ يتوجه إليه المتحرك بالاستحالة اتجاهاً لا يقــف فيه عند حدّ زماناً، ومثل كمّ واحدٍ، أو غير ذلك. وليس هذا المعنى بأولى في أنْ يدخل شرطاً لوحدة الحركة من معنى الزمان، وإنْ كان لا بُدّ من ذكر معنى الزمان. وإنْ كان معنى الزمان يكفي ذكره؛ فذلك ليس لأنّه متضمنٌ جميع الشروط التي بها تكون الحركة واحدة، بل لأنّه يقتضي الشرط الباقي، وينتقل الذهن منه وإليه ويلتزمه، وأنت تعلم الفرق بــين المتضمــن والمقتضي والملتزم. وأمّا الحركة، التي هي بمعنــى القطع؛ فهذا المعنى أولى بأنْ يكون شــرطاً فيها، فالأمور التي يجب أنْ تكون واحدة، حتى تكون الحركة واحدة، هــي المتحرك والمســافة وما يجري مجراها والزمان، فيجــب أنْ يكون المتحرك واحداً، والمسافة - أو ما فيه الحركة - واحداً، والزمان واحداً، أي واحداً بالعدد في جميعه.

(3) Motion's being many follows when the things that provide the motion with a certain quality and type of divisibility are many. These things are three: the mobile, that with respect to which [there is the motion], and the time. So, if there are many things undergoing motion while the distance is one and the same, there are many motions. Also, when there are many things undergoing motion and the time is one and the same, the distances and that with respect to which there is motion must be numerically many. Again, when there are many things undergoing motion and the distance is one, the periods of time are many. [That] is because, [when] the things undergoing motion are many, while the distance is one [they can do so] only when the things undergoing the motion follow one another sequentially over that distance, since two bodies do not simultaneously traverse one and the same distance any more than they are simultaneously in one place. Also, there simply cannot be many things undergoing motion during many periods of time when that with respect to which [there is motion] is numerically one, unless it is with respect to many distances, for [distance] can remain one and the same even after the traversal. As for the [categories of] quantity, quality, and the like, none is numerically one and the same when several mobiles undergo motion with respect to [these categories] in temporal succession. [That is] because the quality that belongs to *this* thing undergoing the motion insofar as it is numerically one is in no way shared in common with some other thing undergoing motion, nor is it like the distance.

(٣) فإنَّ كثرة الحركة تتبع كثرة الأشياء التي تفيد الحركة كمّاً مّا ونمطاً من الانقسام، وهذه الأشـياء هي هذه الثلاثة: المتحرك، وما فيـه والزمان. فإنْ تكثّر المتحرّك، وكانت المسافة واحدة بعينها، تكثّرت الحركات وإذا تكثّر المتحرّك - والزمان واحدٌ بعينه - لزم تكثّر المسـافات وما فيه الحركة بالعدد. وإذا تكثّر المتحرّك، والمسـافة واحدة، لزم تكثّر الزمان. فإنه لا يتكثر المتحرك والمسـافة واحـدة إلاّ وتكون المتحركات تتعاقب على تلك المسافة، إذ لا يقطع جسمان معاً مسافةً واحدةً بعينها، كما لا يكونان في مكانٍ واحدٍ معاً ولا يجوز أنْ يتكثّر المتحرّك في أزمنةٍ كثيرة وما فيه واحد بالعدد البتة، إلاّ في المسافات، فإنّهـا يجـوز أن تبقى بعد القطع واحدة بعينها. وأمّا الكمّ والكيف وغير ذلك، فلا يكون كيـفٌ واحدٌ بعينه، ولا كمٌّ واحدٌ بعينه بالعدد، يتحرّك فيه متحركون عدّة في زمانٍ بعد زمـان، لأنَّ الكيفية التـي لهذا المتحرّك، من حيث هي واحدة بالعدد، لا يشـاركه فيها متحرك آخر بوجه لا كالمسافة.

(4) We suppose that all of this entails that the mover is numerically one and that, when several things move something together, it is only as one thing, since the collection [of movers] becomes a single mover, since one of them alone would not produce the motion. Still, if it is possible that something produce a motion, and, either just before or simultaneous with its ceasing to produce the motion, there comes to exist a certain relation to that moved body together with some other mover—in that case, then, this thing that is undergoing the motion is aptly one through one motion. An example would be a piece of iron that is no longer being influenced by some magnet but, that we were then to imagine [that the iron] is all at once passed on to some other [magnetic] nature, and [that] the iron happens to be where it is attracted by the other magnet (and there is no time between the ceasing of the first and the beginning of the second's influence, the time and distance being continuous). Similarly, were some water heated by one fire and, then, without delay it immediately and successively came upon other fires until it reached a certain degree of hotness, this motion would not be many, but one, except by comparison. [That] is because the thing undergoing the continuous motion just happens to be many in the way in which we say *a lot,* sometimes in the sense of actually distinct and separate [things] and other times by making comparisons.[6] In fact, time also is actually divided in this way—namely, when it is compared with the starting and ending points of things coming to be during it and then instants are marked off in it accordingly. The same thing happens in our case: With the arrival of each mover,[7] a first instant of its time is posited, being posited in the time by making a comparison [with the given arrival]. As a result of that, the time is accidentally many, and so the motion is accidentally many. In this case, because of the time,

6. For example, when we say that a child rides the merry-go-round *a lot,* we may mean that he rode it at a number of different times throughout the day, or we may mean that he rode it only once but remained on it much longer than the other children, or longer than one might have expected.

7. Reading *muḥarrik* with two of the MSS consulted by **Y**, as well as **Z** and **T**, for **Y**'s preferred *mutaḥarrik* (mobile).

(٤) ونظـنّ أنّه يلزم هذا كله أنْ يكون المحرّك واحداً بالعدد، وأنَّ العدّة إذا اجتمعت على تحريك شيءٍ فإنّما هي كشيءٍ واحد، إذ تصير الجملة محركاً واحداً، إذ ولا واحد منهـا يحرّك وحده. لكنه إنْ أمكن أنْ يكون شـيء يحرّك وقبـل أنْ ينقطع تحريكه أو مع انقطاع تحريكه - تقع هناك مناسـبةٌ للجسم المتحرك مع محرّكٍ آخر؛ كما يتخلّص حديدٌ مثـلاً من تأثير مغناطيس لو توهمناه قد اسـتحال إلى غير طبيعته دفعة، وحصل الحديد حيـث ينجذب إلى مغناطيس آخـر، ولم يكن بين تعطّل الأول وابتداء تأثير الثاني زمان، واتصل الزمان والمسـافة. فبالحري أنْ يكون هذا المتحرك واحداً بحركةٍ واحدة، وكذلك لو سُـخّن ماء بنار تلحقه عقيب نار، من غير وقوع فتور، حتى بلغ حدّاً من السـخونة. فبالحري أنْ لا تكون هذه الحركة متكثرة، بل تكون واحدة، إلاّ على جهة المقايسـة. فإنَّ الشـيء المتحد بالاتصال قد يعرض له التكثّر على مـا قلنا مراراً؛ تارة من جهة التفكيك والقطع بالفعل، وتارة من جهة المقايسات. فإنّ الزمان ينقسم أيضاً بالفعل على هذه الجهة، وذلـك إذا قيس بمبادىء أُمور كائنة فيه وغاياتها، فارتسـم<تْ> فيه، بحسـب ذلك، آناتٌ، فيكون في مسألتنا أيضاً يُفرض، عند ورود كل محرّكٍ، آنٌ أول من زمانه، يفترض في الزمان بالمقايسة. فيعرض من ذلك أنْ يتكثّر الزمان، فيعرض من ذلك أن تتكثر الحركة، ولا تكـون حينئذ الحركة واحدة للزمان من هذه الجهة، ومـن حيث أنَّ الزمان واحدٌ في

the motion is not one from this perspective [namely, accidentally not one], whereas, inasmuch as the time in itself is one, the motion in itself is one. This is like what happens to the motions of the celestial sphere relative to the [Sun's] rising and setting such that the time and motion are divided accordingly without the continuity being severed. Also falling under this sort would seem to be the audible sound of a string plucked a single time so as to continue [sounding] for a while, which is called a *musical note.* [That] is because, in the particular instances of natural things and observing their states, you will learn that this musical note does not result from the pick's being on the string; rather, the strumming sound of the string results only by the pick's causing a vibration in it that pushes the air[8] and so produces a sound. Thereafter, it continues to vibrate like that, one strumming sound occurring after another, until it stops vibrating. Those strumming sounds preserve the continuous sound that is heard (if there is, in fact, something continuous like what we hear, and [if] the strumming sounds are not so small as to be imperceptible). Know that the common action itself [occurring] at a single instant does not require[9] that the motions be a single thing. [That] is because a single instant might be an endpoint of some locomotion while being a starting point of some alteration, both of which belong to a single body, where the two motions are not one.

(5) Equally just taking the terminus *a quo* or *ad quem* alone as a condition is not enough for the unity of motion. [That] is because being away from the terminus *a quo* may be not only toward the terminus *ad quem* but also toward the privation without crossing an intermediary. Equally, reaching the terminus *ad quem* may be all at once without crossing an intermediary. The two motions are not even one in species, let

8. Literally, "by strumming the string so that the pick, when it pulls away, forces the string to depart from its position to another with a certain power and strength that hits the air so as to push it on."

9. Reading *yūjibu* (masc.) with **T**, which is in close parallel with **Z**'s *tūjibu* (fem.), for **Y**'s *yūjidu* ([it] exists).

ذاتـه ، تكون الحركـة واحدة في ذاتها . وهذا مثل ما يعرض لحـركات الفلك بالقياس إلى الشـروق والغروب ، فينقسـم الزمان ، وتنقسم الحركة ، بحسـب ذلك ، انقساماً لا يقطع الاتصال . ويشـبه أنْ يكون الصوت المسـموع من الوتر المنقور بنقرةٍ واحدة الباقي زماناً ، الذي يسـمّى نغمة ، هو من هذا القبيل ، فإنَّ هذه النغمة ، سـتعلم في جزئيات الطبيعيات ومشاهدة أحوالها ، أنّها ليست تحدث عن وقع المضراب على الوتر ، بل إنّما تحدث من قرع الوتـر المدفوع بالمضراب عن وضعه المنصرف ، عند مفارقة المضراب إلى وضعه ، انصرافاً بقـوةٍ وحميّةٍ يقرع ما زحمه من الهواء فيصوّت . ثم لا يزال مهتزاً كذلك ، فيحدث قرعٌ بعد قرع إلى أنْ يهدأ ، وتكون تلك القروع مسـتحفظة لصوتٍ مسـموعٍ على الاتصال ، إنْ كان بالحقيقة متصلاً كما يسـمع ، ولم تكن القطوع من الصغر بحيث لا تُحسّ . واعلم أنَّ نفس الاشـتراك في الآن الواحد لا يُوجب أنْ تكون الحركات متحدة ، فإن آناً واحداً قد يكون منتهى نقْلةٍ ومبتدأ استحالةٍ كلاهما لجسم واحد ، ولا تكون الحركتان واحدة .

(٥) وأيضاً فإنَّ اشـتراط ما منه ، أو مـا إليه وحده ، غير كافٍ في وحدة الحركة . فإنَّ ما منه قد يفارق لا إلى الذي إليه وحده بل إلى العدم ، من غير سـلوك واسـطة ، وما إليه يواصل دفْعه أيضاً من غير سـلوك واسـطة ، فلا تكون الحركتان واحدة بالنوع فضلاً

alone one in number. Also, taking as a condition that they be simultaneous is not enough for that, because sometimes [the mobile moves] away from the terminus *a quo* toward the terminus *ad quem* through disparate mediums. In the case of distance, sometimes [the mobile] tends in a straight line toward the terminus *ad quem* and, at other times, [it travels] twisting and turning. In that case, the two motions are not one in species, let alone one in number. Likewise, sometimes [the motion] is from black to white via shades of grey, while at other times it is from yellow, then red, then dark ochre, and at even other times it is from chartreuse and then green. Now, when the two are taken as a condition [of the motion's being one] together with the previously noted conditions, then making them conditions, is superfluous. [That] is because, when that through which [there is motion] is made one, it is precisely from some starting point that is one toward some endpoint that is one. The former implies this account.

(6) So the motion that is numerically one is something continuous with respect to its time and whose distance and subject are one as well. The best candidate is that one that is the same in species, in which there is no variation. Now, [such motion] is seldom found in locomotion, for the natural ones accelerate at the end, whereas the forced ones decelerate the closer they are to the end. The continuous motions most deserving of being one are the rectilinear (if the estimative faculty can picture something's being continuous along an angle)[10] and the circular. That is most appropriate of what is complete and not deficient; for being complete is one of the attributes of being one, while what is deficient is some portion of what is one. It is also most fitting that what is complete should be such that it cannot be increased without being repeated. Now, while it is the case that when circular motion makes a complete rotation,

10. For this conception of continuous, see 3.2.8.

عن العدد . وأيضاً فإنَّ اشتراطهما معاً غير كافٍ في ذلك ، لأنَّ ما منه قد يفارق إلى ما إليه من متوسطاتٍ شتى . أمّا في المسافة فقد يقصد ما إليه ممّا منه على الاستقامة ، وقد يقصد على تقويسٍ وتحنيةٍ ، فلا تكون الحركتان حركة واحدة بالنوع ، فضلاً عن العدد . وكذلك قد تؤخذ من السواد إلى البياض من تطريق الدُكْنة ، وقد تؤخذ من طريق الصفرة ثم الحمرة ثم القُتمة ، وقد تؤخذ من طريق الفستقية ثم الخُضرة ، وإن اشترطا مع الشرائط المذكورة كان اشتراطهما فضلاً . فإنَّ الطريق إذا جعل واحداً ، لم يكن إلاّ عن مبدأ واحدٍ إلى منتهى واحدٍ ، ويضمن ذلك هذا المعنى .

(٦) فالحركة الواحدة بالعدد هي المتصلة في زمانها ، ومسافتها واحدة ، وموضوعها واحد ، وأُولى ذلك المستوية بالنوع الذي لا اختلاف فيها ، وقلّ ما توجد في المكانية ، فإنَّ الطبيعية تشتد آخراً ، والقريبة القسرية تفتر آخراً . وأُولى الحركات المتصلة بالوحدة هي التي على الاستقامة أو الاستدارة ، إنْ توهم للمتصلة على الزاوية وجود ، وأُولى ذلك ما تمَّ ولم ينقص ، فإنَّ من صفات الواحد أنْ يكون تامّاً ، والناقص بعض الواحد . وأولى ما يكون تاماً ما ليس من شأنه أنْ يزاد عليه بلا تكرّر ؛ وهو الحركة المستديرة إذا تمّت الدورة ، فلا

it is not increased but only repeats, that is not so for rectilinear motion as such. [That] is because, when rectilinear motion is completed, its completion is not because it is rectilinear, but on account of the fact that no distance remains, like [at] the edge of the universe. From the truth of this claim, the claim that the rectilinear line has the best title to being complete comes up short, because [the rectilinear line] has a beginning, middle, and end, whereas the circle has none of that. Indeed, if the circle is complete, the motion along it does not need to be [that is, come to an end]. [That is] because rectilinear motion comes to an end and is complete, whereas circular motion [need not] either come to an end or be complete. As a preliminary, it is not the case that whatever is complete possesses a beginning, end, and[11] middle. Instead, what is one with respect to the totality is more complete than the many in which only this threefold division [of beginning, end, and middle] exists. This, however, is only one species of completeness and is considered only with respect to what possesses number, whereas the circle has unity of form. It is not susceptible to increase, precisely [because] and for no other reason than [that] it is a circular line, while, if what is rectilinear is not so susceptible, it is not because it is rectilinear, but for some other reason. Now, when circular motion completes a rotation, it starts again, and so each rotation is one; and our discussion concerns a rotation's being one.

(7) So this is what we have to say about motion that is one in number. Let us now discuss motion that is one in species and genus.

---

11. Secluding *ṭaraf* (limit) with **Z** and **T**, which apparently appears in only two MSS consulted by **Y**.

يزاد عليها بل تكرّر . ولا كذلك المستقيمة ، من حيث هي مستقيمة ، فإنَّ المستقيمة إذا تمت فليس تمامها لأنَّها مستقيمة ، بل لأجل أنَّ المسافة لم تبق ، كقطر العالم . ويسقط من تحقّق هذا قول من قال : إنَّ الخط المستقيم أولى بالتمام ؛ لأنَّ له ابتداءً ووسطاً وانتهاءً ، ولا شيء من ذلك للدائرة ، فإنّه إنْ كانت الدائرة تامّة ، فليس يجب أنْ تكون الحركة عليها تامّة ، لأنَّ الحركة على المستقيمة تتناهى وتتم ، وعلى المستديرة لا تتناهى ولا تتم . فأمّا أولاً ، فليس كل ما هو تام فهو ذو ابتداء وانتهاء ووسط ، بل الواحد في الجملة أتمّ من الكثرة التي لا يوجد هذا التثليث إلاّ فيها ، بل هذا نوعٌ من التمام . ولا يعتبر هذا التمام إلاّ في ذي عدد ، والدائرة وحدانية الصورة ، وإنّما لا تقبل الزيادة لا لشيء غير أنّها خط دائرة ، والمستقيم إنْ لم يقبل فليس لأنّه مستقيم ، بل لسببٍ آخر . وأمّا الحركة المستديرة فإنّها إذا تمت دورة ابتدأت منت رأس ، فتكون كل دورةٍ واحدة ، وكلامنا في دورةٍ واحدة .

(٧) فهذا ما نقوله في الحركة الواحدة بالعدد . ولنتكلم الآن في الحركة الواحدة بالجنس والنوع .

Chapter Three

On motion that is one in genus and species

(1) Since motion is like the rest of the accidents[1] with respect to the judgments that follow upon being an accident, it is many and one just as other accidents are many and one. So, just as white, for example, is numerically many only when its subject or the time [during which it exists] is many, so likewise is motion. Also, just as white is not many in species and genus simply because the subject is many, whether with respect to species or genus—rather, the whiteness of snow and the phoenix are one in species (since they do not differ by some admixture of another color), while the whiteness of snow and sands[2] [are one in genus][3]—so, likewise, the simple fact that the subject is many in either genus or species does not require that the motion be many in species or genus. That is because something's being many in species is a consequence of the specific differences' being many, whereas the relations of accidents to their subjects falls under the accidents' set of accidental judgments. You have learned, however, that the accidentality of the essences of accidents belongs only to the necessary[4] accidental accounts, not constitutive ones. Now, the accidental relations of the things themselves to

1. Reading *aʿrāḍ* with **Z** for **Y**'s and the **T**'s *aghrāḍ* ("goals" or "tendencies").

2. See Lane, *Lexicon*, s.v. *ḥajar* for this metonymical use of *ḥijārah*.

3. The text seems incomplete. Avicenna is providing examples of when white might exist in different subjects and yet be either specifically or generically alike. The example of snow and the phoenix present two things that are the same species of white: namely, pure white with no tinge of any other color. Thus, it would not be unreasonable to think that he is providing here an example of two things that, while generically white, are not the same species of white—hence the suggested addition.

4. Reading *al-lāzima* with **Z** and **T** for **Y**'s *li-azmina* (belonging to or on account of the periods of time). **Y** would seem to have some support for his reading, since *azmina* does appear two lines below; however, he has been misled, since Avicenna will there take up the second way in which motions differ numerically and why it need not imply a specific or generic difference in the motion. Moreover, at *Kitāb al-madkhal* 1.13, Avicenna discusses how necessary accidents (using the same vocabulary as here) can be specific differences while not being constitutive differences that produce a new species, which is the very point of the present context.

<الفصل الثالث>

في الحركة الواحدة بالجنس والنوع

(١) ولمّا كانت الحركة مشاركة لسائر الأعراض في الأحكام التي تتبع العرضية، كان تكثّرها وتوحدها يشاكل تكثّر الأعراض الأخرى وتوحدها. فكما أنَّ البياض، مثلاً، إنَّما يكون متكثّراً بالعدد، إذا تكثّر موضوعه أو زمانه، فكذلك الحركة. وكما أنَّ البياض لا يكون متكثّراً بالنوع أو متكثّراً بالجنس، لنفس تكثّر الموضوع بالنوع أو بالجنس، بل يكون بياض الثلج وبياض الققنس - إذا لم يختلفا بمخالطة لونٍ آخر - واحداً بالنوع، بل بياض الثلج والحجارة. فكذلك ليس نفس تكثّر الموضوع بالنوع أو بالجنس يوجب تكثّر الحركة بالنوع أو بالجنس، وذلك لأنّ تكثّر الشيء بالنوع يتبع تكثّر الفصول وإضافات الأعراض إلى موضوعاتها، من جملة الأحكام العرضية للأعراض. فقد علمتَ أنَّ العرضية لماهيات الأعراض إنّما هي من المعاني العارضة اللازمة دون المقوّمة، وإضافات الذوات العرضية

their differing subjects are accidental features belonging to them, not constitutive of them so as to constitute specific differences, whereas there being many individuals does not depend upon essential differences, but upon accidents. As for periods of time, they do not at all differ in species as periods of time, but individually (if they must), because they are divisions of one continuous thing. Being joined to that which differs individually but not specifically, however, in no way requires a differentiation that makes for a different species.

(2) Motion's species may differ by things that do constitute the motion's essence—namely, [the category] with respect to which [there is motion] and the termini *a quo* and *ad quem*. So, when a species of one of these differs, the motion is different in species. So, when that with respect to which [there is motion] differs, while the termini *a quo* and *ad quem* remain consistent, the species of motion is different. An example would be that one of two motions is a rectilinear motion from some starting point to some endpoint, while the other is a circular motion from the one to the other.[5] The same holds when that with respect to which [there is motion] remains consistent, whereas the termini *a quo* and *ad quem* differ, as in the case of ascending and descending. Necessarily, then, when there is a change of one of these in species (whether with respect to itself or with respect to certain conditions and states included among the ones upon which the motion depends), the motion is not one in species. So, if all of [the motions] are with respect to place, or quality, or quantity, then they are one with respect to the highest genus, while, if they agree with respect to some lower genus, such as being colored, they are one with respect to the lower genus.

5. In this case, the rectilinear motion is a motion with respect to the category of place, while the circular motion is with respect to the category of position.

إلــى موضوعاتها المختلفة أمــور عارضة لها ، لا مقوّمة إياها تقــويم الفصول . وأمّا تكثّر الأشــخاص فليس متعلقاً بالفصول الذاتية ، بل بالعــوارض . وأمّا الأزمنة فلا تختلف من حيث هي أزمنة بالنوع البتة ، بل بالشخص - إنْ كان لا بُدّ - لأنها أقسام متصل واحد ، ومقارنة ما يختلف بالشخص دون النوع لا توجب البتة مخالفة فصلية منوّعة .

(٢) فعســى الحركة يختلف نوعها باختلاف الأمور التي تقوّم ماهية الحركة ؛ وهي : مــا هي فيه ، وأيضــاً ما منه ، وما إليه ، فإذا اختلف نوعٌ واحدٌ من هذه ، اختلفت الحركة بالنوع . فإنَّه إذا اختلف ما فيه ، واتفق ما منه وما إليه ؛ اختلف نوع الحركة ، مثل أنْ تكون إحدى الحركتين من مبدأ إلى منتهى على الاســتقامة ، والأخرى منه إليه على الاستدارة . وكذلــك إذا اتفــق ما فيه واختلف ما منه وما إليه ؛ مثل الصاعد والهابط ، فيجب أنّه إذا اختلف شــيء من هذه في النوع في نفســه أو في شرائط وأحوال داخلة في تعلّق الحركة بهــا ، كانت الحركة غير واحدة في النوع . فإنْ كانت كلّها مكانية ، أو كلها كيفية أو كميّة ، كانــت واحدة في الجنس الأعلى ، وإنْ اتفقت في جنسٍ أســفل كمــا في اللونية ، كانت واحدة في الجنس الأسفل .

(3) Now, whether it is with respect to the species or accidentally that circular motion with respect to place[6] differs from rectilinear motion might present a problem, for it apparently seems that being straight and being curved are accidental features of the line, not specific differences, which makes it immediately seem that one line can be posited as [both] rectilinear and curved. In that case, how can the species of rectilinear lines differ from the species of curved lined? Otherwise, were we to order them so that one species accompanies what is rectilinear and the other what is curved, then every accident could constitute a species; but that is not the case. Thus, when the straight line does not differ from the circular in species, how could rectilinear motion differ from circular motion in species so that a difference in [the lines] is [reflected] in the two [motions] as well? (This consideration concerns the rectilinear and circular motions that involve [change of] place, not the circular motion that involves position, as you have learned.)[7] We also say: A similar problem arises for ascending and descending—namely, it would seem to appear that ascending does not differ in species from descending with respect to the starting point and endpoint inasmuch as they are the limits of a certain interval, but rather, inasmuch as they are directions, one of which lies upward and the other downward. Now, the motion depends upon the starting point and endpoint only inasmuch as they are limits[8] of the distance, not inasmuch as one of the distance's two limits happened to be in one direction, while the other was in a different direction. [That] is because the motion is complete as a motion when it starts along this interval from beginning to end, even if the starting point were not so as to be upward (that is, closest to the Heavens) and the endpoint so as to be downward (that is, closest to the Earth). So, when the situation is such, this falls under one of motion's necessary accidents,

6. For example, circumambulation, as opposed to rotating in place, which is circular motion with respect to position.

7. For the two types of circular motion, see 3.14.8–9.

8. Reading the dual with **Z** and **T** for **Y**'s singular "limit."

(٣) لكنه قد يشكل الحال في أنّه هل الحركة المكنية المستديرة تخالف المستقيمة في النوع أو تخالفها بعرض؟ فإنّه يشبه أنْ يظنّ أن الاستقامة والانحناء من الأمور التي تعرض للخط، لا من الأمور التي هي فصول. ويسبق إلى الظنّ أنَّ الخط الواحد يصلح أنْ يوضع للاستقامة والانحناء، وإذا كان كذلك، فكيف يكون نوع الخطوط المستقيمة مخالفاً لنوع الخطوط المنحنية؟ اللّهم إلاّ أنْ نجعل تركيبها مع الاستقامة نوعاً، ومع الانحناء نوعاً آخر، فيكون كل عرضٍ من شأنه أنْ يقوّم نوعاً، وليس الأمر كذلك. فإذا كان الخط المستقيم لا يخالف المستدير في النوع، فكيف تكون الحركة على المستقيم تخالف الحركة على المستدير بالنوع لأجل اختلافها فيهما؟ وهذا الاعتبار في المستقيمة والمستديرة التي تكون مكانية، لا المستديرة التي تكون وضعية، على ما علمت. ونقول كذلك يشكل الحال في أمر الصاعد والهابط، ويشبه أنْ يظنّ أنَّ الصاعد لا يخالف الهابط في النوع، في المبدأ والمنتهى، من حيث هما طرفان لبُعدٍ، بل من حيث هما جهتان: إحداهما تلي عُلّواً، والأخرى تلي سُفْلاً، والحركة لا تتعلق بالمبدأ والمنتهى إلاّ من حيث هما طرفا مسافة. وأمّا من حيث عَرض أنْ كان أحد طرفي المسافة في جهةٍ والآخر في جهةٍ أخرى، فذلك ليس ممّا تتعلّق به الحركة، فإنَّ الحركة تتمّ حركة إذا ابتدأت في هذا البُعْد من مبتدئه إلى منتهاه ولو لم يكن المبتدأ بحيث يكون عُلْواً وهو أنْ يلي السماء، والمنتهى بحيث يكون سُفْلاً وهو أنْ يلي الأرض. فإذا كان الأمر كذلك، كان هذا من الأعراض اللازمة للحركة، لا من

not one of the factors internal to its essence. So it does not produce some difference in species. The same holds for the difference between motions in that they are either natural or forced, for it, too, is a difference in factors outside of the motion's essence, even if necessary. So these are the puzzles that seem immediately to come to mind.

(4) As for ourselves, we say that these doubts arise only for locomotion, for they do not appear in instances of motions with respect to quality, quantity, and the like. Indeed, it is known by everyone that blackening differs from whitening on account of the difference of the termini *a quo* and *ad quem,* even if that through which [there is motion] is, as it were, one and is something passed through in each case, conversely with the other. Similarly, the process [starting at white and going through] yellow to red to black is different in species from the one going through green to indigo to black, even if the starting point and endpoint are one [and the same].

(5) This presents a problem only in the case of locomotion and requires that locomotion not be a genus, but only a species, where descent would differ from ascent by certain accidents falling under a single species, just as the literate man differs from the illiterate one. Just as *man* is taken in the definition of *literate man* and *illiterate man* and predicated of both, while not being their genus, but a subject, so, likewise, *locomotion* would be predicated in that way of descending and ascending. So it would be as if, with respect to descending, the underlying subject is rectilinear motion from a given starting point to an endpoint. So, as a result of that [alone], its being a motion would be made complete; however, it accidentally happens that this starting point was up and so the motion accidentally happened to be a descent. The same holds for the initial puzzle. For example, it is just accidental that locomotion is sometimes rectilinear and sometimes circular. [That] is

الأمور الداخلة في ماهيتها ، فلم يكن الاختلاف به اختلافاً في نوعه . وكذلك الاختلاف الذي بين الحركات ؛ في أنْ تكون طبيعية أو قسرية ، فإنّه أيضاً اختلاف في أمور خارجة عن ماهية الحركة وإنْ كان لازماً ؛ فهذه هي الشكوك التي يظنّ أنّها تسبق إلى الذهن .

(٤) أمّا نحن فنقول : إنَّ هذه الشكوك لا تعرض في غير النُقْلة ، فإنّه لا يعرض في مثل الحركات التي في الكيف ، والحركات التي في الكمّ ، وغير ذلك . فإنَّ التسوّد معلومٌ من حاله ، عند كل أحد ، أنّه مخالفٌ للتبيض بالنوع ، لأجل مخالفة ما إليه وما عنه ، وإنْ كان الطريق كأنه واحدٌ ومسلوكٌ في كلٍ بالعكس من الآخر . وكذلك التصفّر إلى التحمّر إلى التسوّد ، مخالفٌ للتخضر إلى التنيّلية إلى التسوّد في النوع ، وإنْ كان في حال المبدأ والمنتهى واحداً .

(٥) إنّما يشكل هذا في أمر النُقْلة ، ويقتضي أنْ لا تكون النُقْلة جنساً ، بل تكون نوعاً فقط ، ويكون النزول يخالف الصعود بأعراض تحت نوع واحد ، - كما يخالف الكاتبُ الأُميَ - وأنّه ، كما أنَّ الإنسان مأخوذٌ في حدّ الكاتبً والأُمي ومحمولٌ عليهما ، وليس جنساً لهما ، بل موضوعاً ، كذلك النُقْلة محمولة على ذلك الوجه ، على النزول والصعود ، فكأن أصل الموضوع في النزول هو حركة مستقيمة من مبدأ إلى منتهى ، فيتم بذلك كونه حركة ، لكن عرض لهذا المبدأ أنْ كان فوق ، فعرض للحركة أنْ صارت نزولاً . وكذلك الحال في التشكّل الأول مثلاً ؛ أنّه عرض أنْ كانت النُقْلة تارة في مستقيم ، وتارة في مستدير .

because the motion is not realized as a motion insofar as it happens to have some length through which there is the motion (like the circular distance) and what is shorter (like the rectilinear [distance]), such that the essences of the two motions, as a result, would differ specifically. So these are the lingering doubts that can be raised in this chapter. So we must resolve them, which requires that we first[9] show that locomotion is one genus and that the situation is not of this form.

(6) So we say that, in truth, neither the straight line nor the circular line undergoes alteration so as really to become the other. That is because the very being of the line is to be a limit of a surface, while the being of the surface is to be a limit of a body. So, as long as the body does not happen to lose its configuration, neither does the surface, and so nothing happens to the line. Also, if the body is dry, it is not bendable, whereas it is if it is wet inasmuch as the continuity of the convex curve is either broken up or extended (and conversely in the case of concavity). If the continuity of the convex curve is broken up, then the line is divided into a plurality of lines, whereas, if it is extended, that very line itself also ceases to be and another line comes to be, for one [and the same] line does not, by extension, become longer than what itself is. So, when it is impossible that either of these two lines [that is, the rectilinear and curved lines] be transferred to the nature of the other (not even in the estimative faculty), then, if the estimative faculty does that (namely, isolate the line from the surface), it makes the line have two directions as well as two sides not in the extension of [the line]. In that case, however, it has not been taken as a limit of a surface, for what possesses two directions <and two sides>[10] is a surface [and] not its limit, which is a line. So the estimative faculty took something other than a line. In fact, it took a thin body and then imagined that it was a line. So the belief that one and the same line is a subject of two things is groundless.

---

9. Adding *awwalan* with **Z** and **T**, which is omitted in **Y**.

10. All the texts agree in having simply *jihatayn* (two directions), but at 3.13.3, Avicenna was explicit that a line alone is what has two directions, not a surface, as the present text would suggest. Thus, if Avicenna is to be made consistent, something like the suggested emendation seems required. Alternatively, he may have intended *jānibayn*, which would have the sense of "two (flanking) sides".

فإنَّ الحركة ليست تتحقّق حركة بما يعرض لها من طولٍ ما تتحرك فيه كالمسافة المستدير، وقصره كالمستقيمة، حتى تختلف بذلك ماهيتا حركتين اختلافاً منوعاً. فهذه هي الظنون التي يمكن أنْ تُظنّ في هذا الباب، فيجب أنْ نحلّها، ويلزمنا أولا أن نبيّن أنَّ النُقْلة جنسٌ واحدٌ، وأنَّ الأمر ليس على هذه الصورة.

(٦) فنقول: إنَّ الخط المستقيم بالحقيقة والمستدير لا يصحّ أنْ يستحيل أحدهما إلى الآخر في الوجود، وذلك لأنَّ هوية الخط في الوجود أنْ يكون طرف السطح، وهوية السطح أنْ يكون طرف الجسم، فما لم يعرض للجسم زوالٌ عن هيئته لم يعرض للسطح، فلم يعرض للخطّ البتة. والجسم إذا كان يابساً لم يقبل التحنية، وإذا كان رطباً قبل التحنية، بأنْ يكون اتصال الحَدْبة يتفرق، أو يكون اتصال الخَدْبة يمتد، والتقعير بالعكس. فإنْ تفرّق اتصال الحَدْبة، فقد انقسم الخط خطوطاً، وإنْ امتد فقد بطل أيضاً ذلك الخط بعينه، وحدث خطٌ آخر، فإنَّ الخطّ الواحد لا يصير أطول ممّا هو بالمدّ. فإذا كان هذان الخطّان يستحيل انتقال أحدهما إلى طبيعة الآخر، ولا في الوهم أيضاً، فإنّ الوهم إنْ فعل ذلك مفرداً للخط عن السطح، جعل الخط ذا جهتين وجانبين لا في امتداده، فلم يأخذه طرف سطح، فإنَّ ذا الجهتين سطحٌ لا طرف الذي هو خطٌ، فيكون الوهم قد أخذ غير الخط، بل أخذ جسماً دقيقاً فتخيله خطّاً، فالذي ظنّ أنَّ الخط هو واحدٌ بعينه، موضوعٌ للأمرين، فقد ظنَّ باطلاً.

(7) The individuals of one species of accidents differ either through their subjects or through certain accidents joined to them, where this is of two kinds. That is because those accidents either are joined (for example, in the way that literate is with musical) [and so] do not necessarily follow in a primary way upon [their subjects], or they do follow upon their [subjects] in a primary way (as white is joined to a surface). Now, it is not merely owing to the multiplicity of the subject that the rectilinear line is set apart from the circular one, for rectilinear lines are really set apart from circular ones, and it is not because of [any] other two accidents, whatever they might by chance be. So rectilinear and circular apply to the nature of the line in a primary way, and thus they are either specific differences or primary accidents. On the one hand, if they are specific differences, they have produced [different] species. On the other hand, if they are primary accidents and if the primary accidents are the necessary ones belonging to the nature of what has the accident, then the individuals of the species are alike with respect to it. If they just happen to occur in a certain situation without being necessary, then they happen to have a certain passivity that is a concomitant of the matter. In that case, it would not be out of the question for you, through an act of the estimative faculty, to imagine [the passivity's] disappearing from what has the accident or [imagine] its not existing. So there would be no inconsistency in your imagining that the consequential accident disappears. In that case, what has the accident could exist while not differing from another, as a result of this primary accident that is a consequent of the passivity. The situation is not like that for the rectilinear and circular line. [That] is because, if the matter in each one of them does not have this description by which the line of [each]

(٧) وأشخاص النوع الواحد من الأعراض؛ تختلف بموضوعاتها أو بأعراض تقارنها، وهذا على قسمين؛ وذلك لأنّه إمّا أنْ لا تكون تلك الأعراض تلحقها لحوقاً أولياً، مثل كتابة تجتمع مع موسيقى، وإمّا أنْ تلحقها لحوقاً أولياً كالبياض يجتمع مع السطح. ومفارقة الخط المستقيم للمستدير ليس لأجل كثرة الموضوع فقط، فإنَّ هذه المفارقة موجودة بين مستقيمين وبين مستديرين، وليس لعرضين آخرين كيف اتفق، فإنَّ الاستقامة والاستدارة تنال طبيعة الخط نيلاً أولياً، فلذلك إنّما يمكن أنْ يكونا إمّا فصولاً وإمّا أعراضاً أولية. فإنْ كانت فصولاً فقد نوّعت، وإنْ كانت أعراضاً أولية؛ فالأعراض الأولية إنْ كانت لازمة لطبيعة المعروض له، استوت فيه أشخاص النوع. وإنْ كانت تعرض في حالٍ من غير لزوم؛ فتعرض لانفعالٍ يلحق المادة، فلا يبعد توهم زواله عن المعروض له، أوْ لا وجوده فلا يبعد توهم زوال العارض التابع، فيجوز أنْ يكون المعروض له يوجد، ولا يخالف الآخر بهذا العارض الأولي التابع للانفعال. وليس كذلك الحال في الخط المستقيم والمستدير، فإنّه - إنْ لم تكن المادة في كل واحدٍ منهما على هذه الصفة التي بها صار خطٌ مستقيماً

becomes either rectilinear or circular, then that very line itself would not exist. [That] is because we asserted in what went before[11] that, with [the cessation of] the dryness, the rectilinear ceases to exist and comes to be curved (or, to be exact, the line that was rectilinear ceases to exist, and another line that is curved comes to exist), whereas, if their change were accidental, the line would not cease to exist. Therefore, the difference between them is neither through some nonprimary accident nor through some primary accident that is not necessary. So, then, the rectilinear and circular are antipathetic to one another in just the way that specific differences and the concomitants of specific differences are antipathetic to one another, which indicates a difference in species among the things. Now, because motion in the genus of white is different from motion in the genus of black on account of that with respect to which there is motion, so likewise is the rectilinear and the circular.

(8) Also failing to show a proper understanding of this general rule is the erroneous claim that there is contrariety among the natures of things celestial, since there is convexity and concavity among them. [That] is because, if the primary subject of convexity and concavity is the very same body and they are joined together in one [and the same] sphere, then they are not contraries. If their subject is two separate surfaces, and it is impossible that the concave one be the subject of the convexity and the convex one of the concavity (as we have made clear),[12] then, again, they are not contraries, since those two subjects of theirs are not receiving one of them after another successively, and there simply is no other subject (as we have explained).[13] We'll explore the doubt raised concerning ascending and descending later.[14]

11. See par. 5.

12. See par. 5, where Avicenna argues that if the concavity is changed (or extended), then a new line is created with a different curve.

13. See par. 6, where Avicenna argues that rectilinear and circular are related to the line as their subject, either as specific differences or as necessary accidents, understood in the sense of propria—as, for example, risibility is related to human.

14. The reference seems to be to 4.7.3–4.

أو مستديراً – لم يكن نفس ذلك الخط موجوداً. لأنّه قد أثبت، ممّا تقدّم، أنّه مع اليبوسة تُعْدم الاستقامة ويحدث الإنحناء، بل يعدم الخط الذي كان مستقيماً، ويوجد خط آخر منحني، ولو كان تغيرهما بعرض لكان الخط لا يُعدم، فليس إذن الخلاف بينهما بعارضٍ غير أولي، أو بعارضٍ أولي غير لازم. فإذ الاستقامة والاستدارة متعاندان تعاند الفصول أو لواحق الفصول اللازمة؛ التي يدل تعاندها على اختلاف الأشياء في النوع، ولأنّ الحركة في نوع السواد غير الحركة في نوع البياض – لاختلاف ما فيه الحركة – فكذلك المستقيمة والمستديرة.

(٨) ويسقط من تصوّر هذا القانون قول مَنْ ظنّ أنّ في طبائع الأمور السماوية تضادّاً؛ لأنّ فيها تقبيباً وتقعيراً، فإنّه إنْ كان الموضوع الأول للتقبيب والتقعير هو الجسم نفسه، واجتمعا في كرةٍ واحدة، فليسا بمتضادين. وإنْ كان موضوعهما سطحين متفرقين يمتنع أنّ المقعّر منهما يقبل التقبيب، والمقبّب التقعير، على ما أوضحناه. فليْسا بمتضادين، إذ ليس موضوعاهما ذانك يقبلان تعاقبهما، ولا موضوع آخر البتة على ما بيّناه. وأمّا التشكّك المورد من حال الصاعد والهابط فسنحقّقه بعد.

(9) As for [differences in] speed, motions do not differ at all in species owing to them. How could it be [so] when they are both accidents of every kind of motion that is subject to increase and decrease, whereas the specific difference is not subject to either, and, in fact, one continuous motion progresses gradually from fast to slow? So they fall under the factors that belong to the motion relative to [another] motion, not the factors that belong to [the motion] in itself. Now, it has been supposed that, when *fast* is predicated of rectilinear and circular [motion], it is done so equivocally, but that is not the case. Even if, on reflection, it seems necessary that there can be no interrelation or correspondence between the two (just as there cannot be one between a line and a surface), nevertheless, magnitude is predicated of both univocally. It is not predicated equivocally because the definition of fast and slow is one [and the same] in both cases—namely, in the case of each, *fast* is that which traverses a longer magnitude in the same period of time. Now, just as the rectilinear is a magnitude, so likewise is the circular; and, again, just as being longer with respect to what is rectilinear is that with respect to which there is a potential inclination and increase, so likewise is being longer with respect to what is circular, when the period of time does not differ. Therefore, this is not an equivocal term, and, instead, the [same] definition is taken in both. Now that we have discussed the unity of motions, it seems fitting that we resolve the puzzles mentioned about it.

(٩) وأمّا السـرعة والبُطْء فلا تختلف بهما الحـركات البتة اختلافاً بالنوع، وكيف وهما يعرضان لكل صنف من الحركات ممّا يقبل الأشـدّ والأضعف، والفصل لا يقبلهما، بـل تكون الحركة الواحدة بالاتصال، تتدرج من سـرعةٍ إلى بُـطْء، فهما من الأمور التي تكـون للحركة بالإضافة إلى حركة، لا من الأمـور التي تكون لها في ذاتها. وقد ظنّ أنَّ السـرعة، إذا قيل على المستقيمة والمستديرة، كانت باشتراك الإسم، وليس كذلك، وإنْ كان النظر ربّما أوجب أنّه لا تصحّ المقايسـة بينهما، ولا المناسـبة فيهما، كما لا تصح بين الخط والسطح، مع قول المقدار عليهما بالتواطؤ. أمّا أنّه ليس يقال باشتراك الاسم؛ فلأنَّ حدّ السـرعة والبُطْء فيهما واحد، وهو أنّ السـريع في كل واحدٍ منهما هو الذي يقطع مقداراً أطول في الزمان الواحد، وكما أنَّ المسـتقيم مقدارٌ، فكذلك المسـتدير. وكما أنَّ الأطول في المسـتقيم ما فيه الميل بالقوة والزيادة، كذلك الأطول في المستدير، والزمان غير مختلف، فليس إذن هذا باشتراك الإسم، بل الحدّ يتناولهما معاً. وإذ قد تكلمنا في وحدة الحركات، فحريٌ بنا أنْ نحلّ الشكوك المقولة فيها.

Chapter Four

Resolving the doubts raised against motion's being one

(1) The claim of those who say that all motion can be divided into past and future is untrue.[1] [That] is because, as you know, motion in the way that we ourselves have independently explained it[2] is not divisible into past and future but is always between the past and future. As for motion in the sense of *traversing*, the motion is really a traversal only in a past period of time. Moreover, if it is divisible into past and future, it is so [only] potentially; for, when, in the period of time corresponding with [the motion], one posits a certain instant that accidentally divides [the motion], the instant is not some actually determinate thing. In general, when [the motion] is divided, it is so only accidentally owing to divisions of the time or distance. Now, a condition for the unity of motion is precisely that its time and distance not *actually* be divided, not that [the time and distance] be such that they cannot be divided or are not potentially [divisible]. This is [simply] not a condition for the unity of motions[3] and many other things.

(2) The first thing to say to them about their claim concerning how [motion] could be one when it is not complete is that *one* in the sense of *complete* is different from *one* in the sense of *continuous*. Now, when something is not one in one sense, it does not necessarily follow that it is not one in the other. Moreover, on the one hand, the motion on whose definition

1. For the objections presented here and in the next paragraph, see 4.2.1.
2. See 2.1.5–6, esp. 6.
3. **Z**, **T**, and two of the MSS consulted by **Y** read *kammīyāt* (qualities).

<الفصل الرابع>

في حل الشكوك الموردة على كون الحركة واحدة

(١) أمّا قول أولئك أنْ لا حركة إلاّ وهي منقسمة إلى ماضٍ ومستقبل، فهو قولٌ غير صحيح. فإنك تعلم أنَّ الحركة على النحو الذي نحقّقها نحن، ليست ممّا تنقسم إلى ماضٍ ومستقبل، بل هي دائماً بين ماضٍ ومستقبل. وأمّا الحركة التي بمعنى القطع، فإنَّها لا تحصل حركة قطعاً إلاّ في زمانٍ ماضٍ، ومع ذلك، إنْ كانت الحركة تنقسم إلى ماضٍ ومستقبل، فإنّها تنقسم بالقوة. فإنّه إذا فرض في الزمان الذي يطابقها آنٌ عرض لها أنْ تنقسم، لا أنَّ الآن يكون حاصلاً بالفعل. وبالجملة فإنّها إذا انقسمت، فإنّما تنقسم بالعرض، ولأجل انقسام الزمان، أو لانقسام المسافة. وإنّما الشرط في وحدة الحركة هو أنْ لا يكون زمانها ومسافتها منقسمين بالفعل، لا أنْ يكونا بحيث لا ينقسمان ولا بالقوة، ولا هذا شرطٌ في وحدة الحركات، وكثير من الأشياء.

(٢) وأمّا قولهم: إنّها كيف تكون واحدة ولا تكون تامّة، فأول ما يجابون به عن ذلك أنَّ الواحد، بمعنى التام، غير الواحد الذي بمعنى الاتصال، فلا يجب أنْ لا يكون الشيء واحداً بمعنى، إذا لم يكن واحداً بمعنى آخر. وأيضاً، فأنَّ الحركة التي شرحنا

we are commenting is indivisible and preserved in the thing undergoing motion as something complete and remaining the same until [the motion] ends. On the other hand, if motion in the sense of traversal exhausts the rectilinear interval, it is complete. Also, if it completes a rotation, it is something complete to which nothing more is added. Since something is complete when no part of it lies outside of it, and [since] motion in the sense of traversal exists on the condition that the traversal occurred, [then], when there is no part of it that has not already occurred and there remains no anticipated part [of it] that lies outside, it is complete. In this case, it is one in [both] ways [of being one].

(3) Now, some responded to this [objection that motion cannot be one when it is not complete] by saying that, inasmuch as the parts of motion might cease to exist, while the form is preserved despite the loss of those parts, it is like the form of a house that remains one and the same even when one brick after another gradually crumbles but is [in turn] replaced.[4] In this case, the form is numerically one, even though it is preserved by the succession of [new] materials. The same holds in the case of each individual plant and animal, as well as psychological habits that remain one and the same despite the decomposition, exchange, and change of the [humoral] mixture (only the states of being affected cease and are renewed). Similarly, the form of the shadow in the stream whose flowing waters[5] are [ever] changing remains one and the same. They said that, because the emanative principle (that is, the Creator most high) is one, the form (in other words, the emanation proceeding [from Him]) is also one in relation to its procession from Him. So, as long as the material is at the limit of receptivity (even should it be by replacement), that very same form remains.

---

4. Cf. Philoponus, *In Phys. ad* 185b11ff., 45.27–47.24.
5. Literally, "matter."

حدّها لا تنقسم، وهي محفوظة في المتحرّك، تامّة ثابتة بعينها إلى أنْ تنتهي. وأمّا الحركة بمعنى القطع، إنْ استوفت البُعْد المستقيم، فهي تامّة، وإنْ أتمت دائرة فهي تامّة لا تزيد عليها، إذ كان التامّ ما ليس منه شيء خارجاً عنه، وكان وجود الحركة بمعنى القطع، وهو على أنَّ القطع حصل. فإذا كان ليس شيء منه إلاّ وقد حصل ولم يبق خارجاً منتظراً؛ فهو تام، وهو حينئذ واحد من وجهين.

(٣) وقد أجاب بعضهم عن هذا بأنْ قال: إنَّ مَثَلَ الحركة في أنّها قد تعدم منها أشياء، وتكون الصورة مع عدم تلك الأشياء محفوظة، هو مثل صورة البيت التي تستحفظ واحدة بعينها، مع نقص لبنة لبنة، وسد الخلل الواقع عند النقص بما يقوم مقامها، فتكون الصورة واحدة بالعدد، وإنْ استحفظت بمواد متعاقبة، وكذلك صورة كل شخص من النبات والحيوان. وكذلك تبقى الملكات النفسانية محفوظة واحدة بعينها، مع التحلّل والاستبدال وتغيّر المزاج، وإنّما تبطل الانفعالات وتتجدّد، وكذلك صورة الظل تبقى واحدة بعينها في النهر الجاري المتغيّر المادة. قال: لأن مبدأ الفيض، وهو الباري تعالى، واحدٌ والصورة - وهي الفيض الصادر - واحدٌ بالقياس إلى صدوره عنه. فما دامت المادة في حدّ القبول - ولو بالتعاقب - كانت تلك الصورة بعينها مستحفظة.

(4) I am not particularly impressed by answers like these. It does not seem to be true that the things subject to generation and corruption have a permanent form that does not undergo [any] alteration—that is, unless we grant that the parts that existed in the generated things are permanent, being preserved until the moment of corruption, neither leaving nor ceasing to be. They are joined to a single form or single power, that form or power being preserved [during] the decomposition of the rest of those parts, where that which comes to take the place of one balances out that which it replaces.

(5) Our position is that the emanative principle's being one is insufficient to maintain that the emanation is one. [That] is because, when the one emanative principle emanates many things, the emanation is many as a result of the many [emanated things], regardless of whether they are many things occurring at a single time or [whether] they occur successively over many [times]. Indeed, it is known with certainty that the form subsisting in the second brick, owing to the act of composition and the form that that one has relative to the existing brinks, is not the same one by which the first brick that was removed subsists, which accidentally belongs to that one owing to the relation, given that these states are not transferred from their materials. Instead, the individual [forms] are corrupted through the corruption of their individual bearers. Given that, the form of the present brick is not the same one that was before, but it is only similar to that one whose place it is taking. So, likewise, were the damage not constantly being repaired such that there is a replacement, the form would pass away. Moreover, if, in the gradual

(٤) وليس يعجبني أمثال هذه الأجوبة، ولا يصحّ عندي أنْ يكون للكائنات الفاسدة صورة ثابتة لا تستحيل، اللّهم إلاّ أنْ نقضي بثبات أجزاء وجدت في الكائنات من أول الكون محفوظة إلى وقت الفساد، لا تفارق ولا تبطل، وتكون مقارنة لصورةٍ أو قوةٍ واحدةٍ، تلك الصورة أو القوة تستحفظ التحلّل الواقع في سائر تلك الأجزاء، وتسدّ مسدّه بما تورده من البدل.

(٥) ونقولّ إنّه ليس يكفي في ثبات الفيض واحداً، كون مبدئه المفيض واحداً. فإنَّ المبدأ المفيض الواحد، إذا أفاض على أشياء كثيرة، كان الفيض متكثّراً بتكثّرها، سواء كانت متكثّرة حاصلة في زمانٍ واحدٍ، أو كانت متعاقبة التكثّر. فإنّه يعلم يقيناً أنَّ الصورة القائمة في اللبنة الثانية من التركيب، والصورة الإضافية التي لها بعينها إلى اللبن الموجود، ليست هي بعينها ما كان يقوم باللبنة الأولى المنتزعة، ويعرض لها بعينها من الإضافة. إذ كانت هذه الأحوال لا تنتقل من موادها؛ بل تفسد أشخاصها بفساد أشخاص حواملها. فإذا كان كذلك، لم تكن صورة اللبنية الآن هي بعينها التي كانت قبل، بل تكون شبيهة بتلك، تسدّ مسدّها، فكما أنّه لو لم يتدارك النزْع بالالتئام حتى يتعوض، لكانت الصورة

replacement of the bricks, [the building] is deprived of that very organization that is that form, [the form that] would have come to be would be different in species [from the first form]. Also, if one were not to observe the damages during the entire period of time until the restoration is complete, the one observing the form that came to be would suppose that it is the first form, even though it is different. Similarly, when one is not negligent[6] in restoring the damages, continuously making the needed repairs, he supposes that the second [form] is the first without anything new coming about. So this account of theirs is altogether incorrect—that is, unless, in the set of accidents, there is one that is such as to be transferred from one subject to another, or one subject after another is transferred to it.

(6) The same might be supposed about light and darkness. So [for example], when the source of light or darkness moves, [the light or darkness] is outwardly seen to move with [the moving source]. Also when what receives them moves but the source of light and darkness remains at rest, [the light and dark] move in [that] recipient. Still, it seems that light and darkness (or the shadow in the flowing water) are one and the same individually, since the light that occurs is a certain attribute and state of some recipient resulting from some agent, and, when the recipient undergoes alteration, its attribute no longer remains. So, when the recipient undergoes alteration absolutely, the attribute and state no longer remain absolutely. When *this* recipient undergoes alteration, then *this* attribute and *this* state do no remain; and, when *this* attribute and *this* state do not remain, then what does remain is not some individually enduring thing. On the contrary, at every instant, there is another individual belonging to a whole of a species that remains

6. Reading *lam yuhmil* with **Z** and **T** for **Y**'s *lam yamhal* (does not tarry), which may actually be an inadvertent transposition of the *h* and *m*.

تبطل. ثم إنْ أخذ في إعادة لبنة لبنة، على ذلك النظم بعينه، تكون تلك الصورة قد حدثت، وتكون صورة أخرى بالنوع، حتى لو لم يشاهد الانتقاص المستمر زماناً إلى أنْ يردّ إلى العمارة، لكان مشاهد الصورة الحادثة يظنّ أنّها هي الصورة الأولى، وإنْ كانت أُخرى. وكذلك إذا لم يهمل العمارة إلى الانتقاص، بل لم يزل المسْتَرِمّ يرمّ؛ ظنَّ أنَّ الثانية هي الأُولى من غير حدوث أمر، فهذا القول منها غير صحيح البتة. اللّهم إلاّ أنْ يكون في جملة الأعراض عرضٌ من شأنه أنْ ينتقل من موضوع إلى موضوع، أو ينتقل إليه موضوع بعد موضوع.

(٦) كما عسى أنْ يظنّ من أمر الضوء والظلمة، فإنَّ المضيء والمظلم إذا انتقلا انتقلا معه في ظاهر الأمر، وإذا انتقل القابل، وسكن المضيء أو المظلم، انتقلا في القابل. لكن يشبه أنْ لا يكون الضوء والظلمة أو الظّل في الماء السائل واحداً بعينه بالشخص، إذا كان الضوء الواقع هو صفة أو حالٌ لقابلٍ عن فاعل، فإذا استحال القابل لم تبق صفته، فإنْ استحال القابل مطلقاً، لم تبق الصفة والحال مطلقة، فإذا استحال هذا القابل لم تبق هذه الصفة وهذه الحال، وإذا لم تبق هذه الصفة وهذه الحال، لم يكن الباقي ثابتاً بالشخص، بل يكون كل آن شخصاً آخر، من جملة نوع مستحفظ على الاتصال. وهذا كما يعرض

continuously. This is like what happens to the stream of liquid that remains stationary with respect to what is parallel to it or stands opposite of it; for it does not necessarily follow from the fact that, when one parallel or opposite part after another continues to exist in the liquid, the parallel things in the liquid are individually preserved. Likewise, illumination and darkening track [the cases of] being parallel and exactly opposites, except that when, at every moment, one sensibly observes some light just like the one that was, he reckons that that is one and the same lasting thing. It is like air moving in a dark house, for you know that, when the air in it moves, the darkness of [the air] moves in it. So the darkness moves about accidentally, but, when [the moving darkness] is succeeded by its like, [the motion] is not sensed. The same would hold if darkness were replaced with red and no motion is sensed as a result of some tactile sensation or the like. In this case, then, vision does not pick out any motion at all, and one would reckon that the red he is experiencing at every moment is the first. [Yet] it is different from it since it is in a different part. In fact, [if you imagine that] there chanced to be some river whose banks were of a uniform height and slope and whose bed is the same throughout (whether flat or curved) and [that] the water flows in it wholly undisturbed by winds or some difference in the parts at the bottom of the river or the like, [then] you yourself would reckon that this water is one and the same water, tranquil and at rest, since you cannot sense any differences between a part that has passed you and another that has arrived before you. Similarly, when you do not sense altering differences in the darkness and light connected with something, you reckon that the darkness or light is the former one itself.

للسيّال مع الساكن من أمر الموازاة والمحاذاة. فإنّه ليس إذا كان لا يزال يوجد في السائل جزء موازٍ بعد جزء أو محاذٍ، يلزم من ذلك أن الموازاة التي في السائل تكون محفوظة بالشخص، كذلك ما يتبع الموازاة والمحاذاة من إضاءةٍ وإظلام. إلاّ أنَّ الحسِ إذا شاهد في كل وقتٍ ضوءاً كالذي كان، حسب ذلك شيئاً واحداً بعينه راهناً؛ كالحال في بيتٍ مظلم متحرك الهواء. فإنّا نعلم أنَّ الهواء الذي فيه إذا تحرّك يحرّك فيه ظلمته، فتكون الظلمةً متحركة ومتنقلة بالعرض، لكن إذا كان إنّما يعقبها مثلها لم تحسّ به. وكذلك لو كان بدل الظلمة حُمرة، وكان لا يحسّ بالحركة من جهة اللمس أو غيره، فإنَّ البصر لا يدل حينئذ على حركةٍ البتة، ويحسب أنَّ كل ما يلقاه من الحمرة كل وقتٍ هي الأولى، ويكون غيرها؛ لأنّها في جزء غير. بل لو اتفق أنْ كان نهرٌ غير مختلف الشطوط، بارتفاعٍ وانحدار، وأسفله مستوٍ متشابه، مسطّح أو مقبّب، وفيه ماء يسيل من غير أنْ تكون هناك علّة تموّج؛ من ريحٍ أو اختلاف أجزاء قرارٍ أو غير ذلك، فإنك تحسب هذا الماء ماء واحداً بعينه، راكداً ساكناً، إذ لا يمكنك أنْ تحسّ بفصولٍ بين جزء عدّاك وجزء وصل إلى سمتك، وكذلك إذا لم تحسّ بفصول الاستحالة في الظلمة والضوء لاتصال الأمر، حسبتَ أنَّ الظلمة أو الضوء هو ذلك بعينه.

(7) There is a puzzle that is to be noted about this: namely, that if it is not one, it is thus many; but it cannot be infinitely many, and so it is finitely many, in which case there is one of two alternatives. On the one hand, each one belonging to that plurality might remain only for an instant; but it was thought to be some continuously existing thing, and so, from a finite number of instants, one continuous period of time would be composed, which is absurd. On the other hand, each one of them might remain for some period of time, along with the flow of the subject; but this is what they denied, in which case you must uncover its solution[7] from principles that you have independently verified.

(8) Next, a doubt comparable to the ones we have mentioned might be raised about heavenly motion, even if it differs from them a bit. So it is said that it must be either one or many. If it is one, then how can it be one and not complete? Indeed, we find some part of it that lies outside of that which has yet to occur; but whatever is one is complete. If it is many, how does one speak of its number and units? We say that *motion* is either in the sense that we speak[8] of it (and so is one and remains in [the mobile] as it is undergoing motion), or it is that which is in the sense of traversal (in which case each rotation is one motion, except that the rotations are renewed only by supposition).[9]

(9) Since we have completed the discussion about motion's being one, we should discuss the comparison between motions with respect to their speed—namely, the account designating when motions are in concert.

---

7. Reading *ḥallahu* with **Z** and **T** for **Y**'s *jumla* (whole).
8. Reading *naqūlu* with **Z** and **T** for **Y**'s *yaqūlu* (he says).
9. For Avicenna's analysis of rotation, see 2.1.20–23.

(٧) وأمّا التشــكّك الذي يقال في هذا ؛ وهو أنّه إنْ لم يكن واحداً فهو إذن كثير ، ولا يجــوز أنْ يكــون كثيراً غير متناهٍ ، فيكون كثيراً متناهياً ، فــلا يخلو إمّا أنْ يكون كل واحدٍ من ذلك الكثير لا يبقى إلاّ آناً ، وقد كان يرى موجوداً على الاتصال ، فتكون الآنات المتناهية يتألف منها زمانٌ متصلٌ واحدٌ ، وهذا محال . أو يكون كل واحدٍ منها يبقى زماناً ، مع ســيلان الموضوع ، وهذا ما ينكرونه . فيجب أنْ تعرف حلّه من الأصول التي تحقّقتها .

(٨) وبعد هذا ، فقد يشــكّك في أمر الحركة الســماوية بتشكّكٍ يناسب الشكوك التي ذكرناها ، وإن كان مغيّراً عنها يسيراً . فقيل إنها لا تخلو إمّا أنْ تكون واحدة أو تكون كثيرة ، فإن كانت واحدة فكيف تكون واحدة وليست بتامّة؟ فإنّا نجد منها شيئاً خارجاً عنهــا لم يحصل بعد ، وكل واحدٍ تام ، وإنْ كانت كثيرة فكيف يقول : عددها وآحادها ؟ فنقول ؛ أمّا الحركة بالمعنى الذي نقوله فهي واحدة باقية فيه أبداً ما تحرّك ، وأمّا الذي بمعنى القطــع فيشــبه أنْ يكون كل دورةٍ حركة واحدة ، إلاّ أنَّ الــدورات لا تتحدّد إلاّ بالوضع .

(٩) وإذ قد فرغنا من الكلام في وحدة الحركة ، فبالحري أنْ نتكلم في التقايس الذي بين الحركات ، في سرعتها وبطئها ، وهو المعنى الذي يسمّى تضامّ الحركات .

Chapter Five

On motions that are and are not in concert

(1) It is normal practice for people, in one turn, to say that any motion that is completed in a shorter time is faster and so to say that this alteration is faster than this locomotion, where the sense of *faster*, in this context, is that which is conveyed to the end in the shorter time. Alternatively, they refuse, in turn, to say that the tortoise's moving a hand's width in a quarter of an hour is faster than a horse's moving a parasang[1] in an hour. In fact, they count the tortoise's motion as slower, even though it reached its goal or came to rest in a shorter period of time, whereas they count the horse's motion as faster, even though getting to the end took a longer amount of time. Thus, this fastness and slowness must have some sense other than the first one—namely, that what is fast is that which traverses a longer distance (or what is analogous to distance) in an equal time or that which traverses equal [distances] in a shorter period of time. Thus, when we want to compare the speed of two motions, we must take into account that [category] with respect to which there is [the motion]. So, if there can be a comparison of the increase and decrease and the intensity and weakness between two things with respect to which there is motion, then the speed of the two motions, as well as the increase and decrease of the two things, can be compared.

1. That is, 3.5 miles or 5.6 kilometers.

<الفصل الخامس>

في مُضامَّة الحركات ولا مضامّتها

(١) من عادة الناس أنْ يقولوا مرّة، في كل حركةٍ تتمّ في زمانٍ أقصر، إنّها أسـرع، فيقولون إنَّ هذه الاستحالة كانت أسرع من هذه النُقْلة، فيكون معنى الأسرع في هذا الموضع هو الذي ينتقل إلى الغاية في زمانٍ أقصر. أو يمتنعوا مرة أخرى عن أنْ يقولوا إنَّ حركة السلحفاة، من مبدأ شبرٍ إلى منتهاه في ربع ساعة، هي أسرع من حركة الفرس فرسخاً في ساعة، بل يعدّون حركة السلحفاة بطيئة، وإن كانت تبلغ المقصد، أو تنتهي إلى السكون في زمانٍ أقصر، ويعدّون حركة الفرس سريعة، وإنْ كانت طويلة الزمان إلى المنتهى. فيجب أنْ يكون لهذه السرعة وهذا البُطء معنى آخر غير الأول؛ وهو أنَّ السريع هو الذي يقطع من المسافة - أو ممّا يجري مجرى المسافة - ما هو أطول في زمانٍ مِثْل، أو الذي يقطع المِثْل في زمانٍ أقصر. فيجب إذا أردنا أنْ نقايس بين حركتين في السرعة والبطء، أنْ يكون ما فيه الحركة مُراعىً. فإنْ أمكن بين الشيئين اللذين فيهما الحركة مقايسة بالزيادة والنقصان، والاشتداد والضعف؛ أمكنت المقايسة بين الحركتين في السرعة والبُطْء. والمقايسة بين الشيئين في الزيادة والنقصان.

(2) Quantity is equal in two ways: one, actually; the other, potentially. The one that is actually [equal] occurs when one of the two [quantities] can be made to coincide with the other such that the whole of one coincides with the whole of the other. In that case, either the two limits (assuming that they have limits) [of the one] may actually coincide with the two limits [of the other], or one of them may exceed that part that coincides with the other. In the first case there is equality, while in the second there is a difference in size.[2] The second way, which is [to be equal] potentially, is that the two magnitudes are such that there cannot be a coinciding or [measurement of] excess between them—as, for example, the rectilinear and circular line, and the triangle and square. Now, it is obvious that the triangle does not coincide with the square nor does the rectilinear coincide with the circular in the former way. Still, it might be supposed that there is this coincidence among them potentially. In the case of the triangle, it is such that it can be divided into sections that can be rearranged [to form] a square, in which case that [rearranged] triangle can be constructed on top of that [original] square, so as either to coincide with and actually be equal to it or to exceed it and so actually have a greater size. Before that, however, it was not, in fact, in any obvious way, actually equal or greater. So it is in this respect that the triangle is said to be equal to the square. The same holds for the circular line. If it could be so worked so as to be changed into a rectilinear line, then, by making [the reworked line] coincide with it, it would be such that it either exceeds or falls short of or equals it. Again, however, as long as it is circular, this coinciding does not actually work, save potentially (assuming that that is possible). When one thing and its limits do not

2. Literally, "a difference in increment and decrement."

(٢) والمساواة في الكميّة على وجهين: أحدهما بالفعل، والآخر بالقوة، أمّا الذي بالفعل فأنْ يكون انطباق أحدهما ممكناً بالآخر، حتى ينطبق كله على كله، فينطبق الطرفان - إنْ كان لهما طرفان - على الطرفين بالفعل أو يفضل أحدهما على مطابق الآخر، فيكون في الأول مساواة، وفي الثاني تفاوت بزيادةٍ ونقصان. والوجه الثاني الذي بالقوة؛ هو أنْ لا يكون المقداران بحيث يمكن أنْ يكون بينهما مطابقة وفضل، مثل خطٍ مستقيم ومستدير، ومثل مثلثٍ ومربع. فظاهر أنّه لا ينطبق المثلث على المربع هذا الانطباق، ولا المستقيم على المستدير، لكن قد يظنّ أنَّ هذا الانطباق فيهما بالقوة. أمّا المثلث فهو بحيث يمكن أنْ يقطع قطوعاً يردّ إلى نظامٍ يكون منه مربع، فحينئذ يمكن أنْ يركّب ذلك المثلث على ذلك المربع، فينطبق عليه ويساريه بالفعل، أو يفضل عليه، فيزيد عليه بالفعل، وقبل ذلك لم يكن مساوياً ولا زائداً بالحقيقة بالفعل الصريح. فمن هذا القبيل يقال إنَّ المثلث مساوٍ للمربع، وكذلك المستدير، لو أمكن أنْ يعمل به ما يغيره إلى الاستقامة، لكان يكون بحيث يزيد على المستقيم أو ينقص عنه أو يساويه بالانطباق عليه. فما دام مستديراً فليس يمكن أنْ يعمل به هذا الانطباق بالفعل اللّهم إلاّ بالقوة، إنْ

coincide with another and its limits, it is not actually equal to it. When there is nothing with respect to which it is equal in the way mentioned, however, and there is nothing that exceeds the equal part, then neither does the one actually exceed the other nor the other actually fall short of it. (From what has been explained to you before,[3] you can judge that it is not in the power of the rectilinear line to be changed to the point that it coincides with the circular line while existing as the same thing, so its status, in this [case]—when you are being precise—is not like that of the triangle and square.)

(3) Someone could say that we know with certainty that the arc is greater than the chord, and that the chord is smaller than it. So, when there exists some difference in smallness and largeness, it is just fitting that there be equality.[4] Some of those who are exact responded to this, saying: There sometimes is a comparison of greater and lesser between two things; nonetheless, a comparison of equality cannot occur between them. [That] is because you know with certainty that an acute angle [formed] of two straight lines is greater than the angle that comes to be from an arc and a straight line,[5] while being smaller than another. [Yet] it is impossible that there be an angle of the two-straight-lines sort that is equal to something of the other sort. We say that the acute [angle formed] of two straight lines is greater than an angle resulting from [a curved and straight line] only because the curved angle actually exists within that [smaller acute rectilinear angle] and another greater [rectilinear angle], where the other [angle] is greater than [the first one] of two straight lines only because the [first's] two straight lines exist in

3. See 4.3.6.

4. Avicenna had just argued that, speaking precisely, there cannot be a comparison between rectilinear and circular lines. The would-be-objector is observing that, since judgments of "greater" and "smaller" are known to apply to the chord and the curved line that delimits that chord, one should likewise be able to speak of "equality" between curved and straight lines. In the Arabic, the example is more striking, since *qaus* (translated here as "arc") also means "bow," while *watar* (translated as "chord") can also mean "the string of a bow." Of course, the string of a bow must be shorter than the bow itself, even though we think that, if the string were longer, it could equal the length of the bow.

5. That is the so-called horn angle, which is the angle between a circle and a line tangent to it.

أمكن ذلك. والشيء إذا لم يكن منطبقاً على غيره، ونهاياته على نهاياته، لم يكن مساوياً له بافعل، وإذا لم يكن فيه ما يساويه على الوجه الذي قيل، وزيادة على ما يساويه، لم يكن زائداً عليه بالفعل، ولا الآخر ناقصاً عنه بالفعل. ومما سلف بيانه لك، تحكم بأنَّ المستقيم ليس في قوته أنْ يتغيّر إلى أنْ ينطبق على المستدير وهو موجود بعينه، فليس حكمه في هذا - إذا رجعت إلى التحقيق - حكم المثلث والربع.

(٣) فإنْ قال قائل: إنّا نعلم يقيناً أنَّ القوس أعظم من الوتر، والوتر أصغر منه، فإذا وجد تفاوتٌ في الصغر والكبر، فبالحري أنْ تكون هناك مساواة. وقد أجاب عن هذا بعض المحصّلين فقال: يكون بين شيئين تناسب الزيادة والنقصان، مع استحالة أنْ تقع بينهما مناسبة المساواة. فإنّا نعلم يقيناً أنَّ زاوية مستقيمة الخطّين حادّة؛ هي أعظم من زاويةٍ حادثةٍ عن قوس ومستقيم، وأصغر من أخرى، ويستحيل أن تكون من قبيل مستقيمة الخطّين زاوية مساوية لشيءٍ من قبيل الأخرى. وإنّما قلنا إنَّ الحادّة، المستقيمة الخطّين أعظم من زاويةٍ منهما، لأنَّ الزاوية القوسية توجد بالفعل في تلك وزيادة أخرى، وإنّما كانت الأخرى أعظم من مستقيمة الخطّين، لأنَّ مستقيمة الخطين توجد فيها وزيادة، فهذا

it and it is greater.[6] So this is one response. Additionally, how could we concede that the arc is actually greater than the chord when there cannot exist in the arc that with which the rectilinear line coincides so that there is a coincidence with two generically similar things? How can there be an actual comparison at all between them? That might occur potentially or through an act of the estimative faculty such that, if the circular line could be made rectilinear, something like it and something greater would exist with respect to it. So, therefore, difference and equality are sometimes considered as actual; sometimes they are potential in the sense that there is some ultimate existential basis, as between the triangle and the square; and sometimes [they] are considered as remote—namely, that the thing is such that, *if* (counterfactually) it were susceptible to change, *then* it would be described as solely larger or solely smaller or solely equal; but, again, this is a remote consideration.

(4) So the local motions that are comparable are those that undergo motion with respect to that which is comparable. So, if the like is traversed in a like period of time, the speed is equal; if one traverses a longer [distance] in the same amount of time or the same [distance] in a longer time, the motions are unequal and, in fact, they differ in being greater and lesser. Next, if that with respect to which they undergo motion is not comparable either actually or potentially, then the motions are [likewise] not comparable either actually or potentially. So rectilinear and circular motion are not, in fact, comparable save in that very remote way just mentioned.

---

6. The present discussion draws on Euclid's description of the horn angle in *Elements* 3.16, as well as Euclid's definition of when two magnitudes can be compared: namely, when they have some ratio to one another, (book 4, defn. 4). Avicenna's point might be explained by reference to the following diagram:

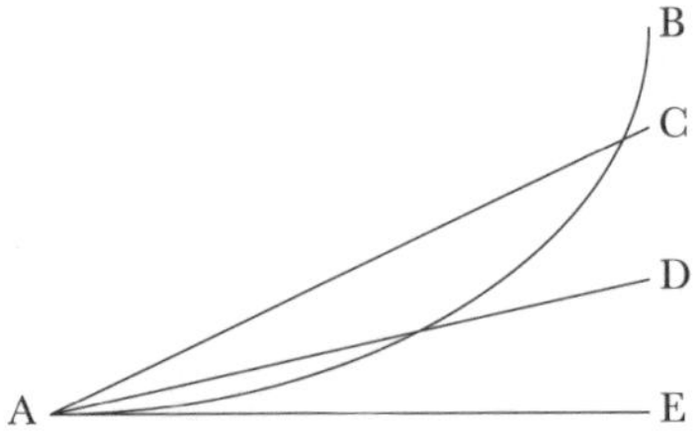

the rectilinear angle DAE is smaller than the rectilinear angle CAE, because DAE is contained in CAE. Similarly, the horn angle BAE is smaller than DAE because the horn angle is contained in DAE.

جواب. ومع ذلك فكيف نسلم أنَّ القوس أعظم بالفعل من الوتر، وليس يمكن أنْ يوجد في القوس ما ينطبق عليه المستقيم انطباقاً مع انطباق المتجانسين، وكيف تكون بينهما مقايسة البتة بالفعل؟ عسى أنْ يكون ذلك بالقوة، أو عسى أنْ يكون ذلك بالتوهم، بحيث أنَّ المستدير لو أمكن استقامته؛ لكان حينئذ يوجد فيه مِثْل وزيادة، فيكون إذن اعتبار التفاوت والمساواة مرّة بالفعل ومرّة بالقوة المستندة إلى الوجود، كالحال بين المثلث والمربع. ومرّة باعتبارٍ بعيدٍ وهو أنْ يكون الشيء بحيث لو كان يقبل التغيّر لصار إلى صفة الزيادة لا غير، أو النقصان لا غير، أو المساواة لا غير، وهذا اعتبار بعيد.

(٤) فالحركات المتقايسة المكانية هي التي تكون ما تتحرك فيه متقايساً، فإنْ كان المِثْل يقطع في زمانٍ مِثْلٍ، فالسرعة متساوية، وإنْ كان الأطول يقطع في زمانٍ مِثْلٍ، أو المِثْل يقطع في زمانٍ أطول؛ فالحركات غير متساوية، بل متفاوتة بالزيادة والنقصان. فإنْ لم يكن ما تتحرك فيه متقايساً بالفعل ولا بالقوة، فالحركات غير متقايسة بالفعل ولا بالقوة، فتكون المستقيمة والمستديرة لا تقايس بينهما بالتحقيق إلاّ المقايسة المذكورة البعيدة جداً.

(5) As for comparing qualitative motions, some are close and others farther removed. Close [comparisons] are those where that with respect to which there is motion admits comparing truly similar things, such as black and black and hot and hot. So, when something undergoing a [qualitative] motion has started from some quality similar to some other quality from which another moving thing starts, and then it ends at something like what the other ends at in the same period of time, and [when] they both coincide in reaching any resting point that the estimative faculty might imagine (were they to rest at it), then the one would be equal to the other in speed. If one has not yet reached [that ending point] (even if both were to rest during the intervening period of time), then its quality is weaker and it takes some additional time, in which case it is slower than the other, and the other is faster than it. So that with respect to which there is motion must be one, as well as the starting and ending points—that is, [one] in species. [The comparison of qualitative motions] that are farther removed considers contraries, such that, if, in the contrariety, one of the termini *a quo* and *ad quem* is a limit and the other is that other limit exactly opposite it (or it falls short of the limit and is closer to the middle), then the one on one side is like the other (or is alike with respect to closeness to the center). So the consideration would be, for example, [that] this instance of whitening is faster than or equal to this instance of blackening such that a certain relation of some termini *a quo* and *ad quem* and that with respect to which there is whitening is like the relation of its exact opposite on that side that is blackening. (This way, however, is not established in a principled way.)

(٥) وأمّا المقايسة المعتبرة في الحركات الكيفية، فمنها وجهٌ قريبٌ ومنها وجهٌ بعيد، فالوجه القريب هو أنْ يكون ما تتحرك فيه قابلاً لقياس المشابهة الحقيقية، مثل سوادٍ وسواد، وحرارةٍ وحرارة. فإذا كان متحركٌ ما قد ابتدأ من كيفيةٍ شبيهةٍ بكيفيةٍ أخرى ابتدأ منها متحرك آخر، ثم انتهى إلى شبيه ما انتهى إليه الآخر في زمانٍ واحدٍ، وكان كل موقفٍ متوهم يتوافيان فيه متشابهين - لو وقفا عليه - فهو مساوٍ له في السرعة وإن كان لم ينته إليه بعد. ولو وقفا جميعاً في وسط الزمن، كانت كيفيته أضعف وبقي زمان، فهو أبطأ منه، فيكون الآخر أسرع منه، فيجب أن يكون المتحرك فيه واحداً، والمنتهى والمبتدأ واحداً؛ أي في النوع. وأمّا الوجه البعيد، فأنْ يكون الاعتبار بالضّد، حتى إنْ كان أحد المنتهى إليهما والمبتدأ منهما طرفاً في التضاد، والآخر ذلك الطرف الآخر لنظيره، أو إنْ كان دون الطرف وأقرب إلى الوسط، كان الآخر من ذلك الجانب كذلك، أو على مِثْل ذلك القرب من الوسط. فيكون الاعتبار مثلاً هو أن هذا يبيضّ أسرع من هذا وهو يسودّ، أو مساوٍ له، حتى تكون نسبة ما منه ابتدأ وما إليه انتهى، وما كان فيه إلى البياض، كنسبة نظرائه من ذلك الجانب إلى السواد، وهذا وجهٌ غير محقّق بحسب الأصول.

(6) Sometimes it happens that two things are comparable absolutely, while not being comparable relative to a given thing. So [for example] large and small with respect to water *qua* water is different from the large and small with respect to air *qua* air, for the maximally large amount of water is unlike that of air; and the same holds for smallness. [An example would be that] when water undergoes rarefaction so as to become a large quantity of air, the motion has a certain limiting point that falls below the limiting point for air when it undergoes rarefaction so as to become a large quantity of fire.[7] So, when these motions are taken absolutely with respect to being large and small, they are comparable, whereas comparing what it is to be a large quantity of fire with a large quantity of air is not possible. So the rarefaction of air (that is, the motion that makes it larger) is not comparable to the rarefaction of water. Also, the condensation of one cannot [be compared] with that of the other, for the large or small quantity of the latter is not of the [same] species as the large or small quantity of the former. The comparison takes place, rather, between two instances of the rarefaction of air or two instances of [that of] water. The same holds for flying and walking: They might turn out to be comparable *qua* moving through some rectilinear distance, but inasmuch as this one is the flight of an eagle and this one the flight of a sparrow (to say nothing of walking), the [long] flight of an eagle and the [long] flight of a sparrow are not comparable. Instead, an eagle's flight is to be compared only with that of an eagle's, and a sparrow's with that of a sparrow's. The same holds for [comparing] cases of grogginess that follow upon honey wine and those following upon grape wine.

7. In his discussion concerning the void (see 2.9.17 & 20–21), Avicenna observed that natural substances can undergo rarefaction and condensation such that their natural volume becomes larger or smaller without the need of positing a void. Here, he seems to suggest that there is some natural limit of rarefaction for each substance beyond which there is a substantial change. Thus, if one begins with an equal volume of water and air and then rarefies the two, the size that the rarefied water can reach before turning into air is less than the size that the rarefied air can reach before turning into fire. See also 3.12.8.

(٦) وقد يعرض أن يكون شيئان متقايسين على الإطلاق، ولا يكونان متقايسين بالنسبة إلى شيء، فإنّ الكبير والصغير في الماء - من حيث هو ماء - غير الكبير والصغير في الهواء من حيث هو هواء، لأن غاية الكبر في الماء ليست مثل غاية الكبر في الهواء، وكذلك في الصغر. وإذا تخلّخل الماء إلى كبر الهواء، كان للحركة حدٌّ دون حدّ تخلّخل الهواء إلى كبر النار. فإذا أُخذت هذه الحركات في الكبر مطلقاً وفي الصغر مطلقاً، كان ذلك متقايساً، وأمّا مقايسة الكبر الناري إلى الكبر الهوائي فليس بجائز، فالتخلخل الهوائي وهو الحركة إلى الكبر - لا يقايس بالتخلْخل المائي ولا تكاثفه بتكاثفه، فإنَّ كبر هذا ليس من نوع كبر ذلك، ولا صغره من نوع صغره، بل المقايسة تجري بين تخلخلي هوائين أو تخلخلي مائين، وكذلك حال الطيران والمشي. وأمّا من حيث الحركة في مسافةٍ مستقيمةٍ فقد يصح التقايس، وأمّا من حيث هذا طيران النسْر، وهذا طيران العصفور - فضلاً عن المشي - فلا يتقايس طيرانٌ نسْريٌّ وطيرانٌ عصفوريٌّ، بل الطيران النسري يقاس بالطيران النسري، والعصفوري بالعصفوري، وكذلك النحلي العسلي بالنحلي العسلي،

In this context, then, one must take into account the thing with respect to which there is motion, whether it is being taken absolutely or conditionally, and then consider the time. So, if that thing does not differ in species, one can be compared with the other (and sometimes the comparison is not with respect to the nature of the species, but the nature of the species together with some accident). What is undergoing the motion is not taken as a condition in this context. [That follows] since a difference in it does not bring about a difference in the motion—that is, unless it is taken as a condition in the configuration of the motion and that with respect to which there is motion, like the sparrow for the flight of a sparrow, for [whether] the distance of the sparrow's motions [is long or short] relative to its flight is different from the distance of the motion of what is not a sparrow.

(7) Also, in this context, equivocal terms and things like them might lead to error. For example, it might be supposed that this knife becomes sharp faster or slower than this tone does, but *sharp* means something different in the two cases. Similarly, one might suppose that this inflamed eye has healed faster than this paralyzed hand. [That] is because, just as the humoral mixture of the eye and its activity is different in species from [those of] the hand, so, likewise, the soundness or lack thereof of its activity is different in species from those of the hand, and so the motion with respect to them is not of a single species. That is [so] unless we consider *health* absolutely, in which case the two motions are not one with respect to species, but genus; but you have already learned that generic comparisons are not true.[8]

---

8. See, for instance, *Kitāb al-jadal* 1.5.

والنحلـي العنبي بالنحلي العنبي. فيجب أنْ يراعـى في هذا الباب معنى ما فيه الحركة، ويراعى أخذه مطلقاً أو بشرط، ثم ينظر إلى الزمان، فإنْ لم يختلف ذلك في النوع، صحّ التقايس به، فربما كانت المقايسة لا لطبيعة النوع، بل لطبيعة النوع مع عرض. وأمّا المتحرك فلا نأخذه شـرطاً في هذا الباب، إذ لا يغيّر اختلافه اختلاف الحركة، اللهم إلاّ أنْ يكون مأخوذاً شـرطاً في هيئة الحركة، وفيما فيه الحركة؛ كالعصفور للطيران العصفوري، فإنَّ مسافة حركات العصفور في طيرانه غيرمسافة حركات ما ليس بعصفور.

(٧) وقد يغلط في هذا الباب اشتراك الاسم أو أشباهه، مثل أنْ يظن أنَّ هذا السكين تحدّ أسرع أو أبطأ ممّا يحدّ هذا الصوت، لكن الحدّة فيهما بمعنى مختلف. وكذلك قد يظنّ أنَّ هذه العين الرمدة قد صحّت أسرع مما صحّت هذه اليد المفلوجة. فإنّه كما أنَّ مزاج العين وفعله غير فعل اليد في النوع، فكذلك سلامة فعله أو فساد فعله غير الذي منهما لليد في النوع، فلا تكون الحركة فيهما من نوع واحدٍ؛ اللّهم إلاّ أنْ نعتبر الصحة مطلقاً، فلا تكون الحركتين في النوع، بل في الجنس، وقد علمنا أنَّ ذلك التقايس الجنسـي ليس بالحقيقي.

(8) Here there arises a question that one might ask: [Imagine] that some moving thing [*x*] traverses a certain distance, but that distance begins to undergo alteration at the very same time that [*x*] begins to move such that the alteration quits at the point at which [*x*'s motion] stops and is completed, and so the locomotion stops at the same time [that the alteration] does. In that case, can this instance of alteration be said to be equal to this motion? The answer is that this is a mistake and cannot be maintained. That is because, while the distance is equal to that which has undergone the alteration, the motion is not equal to the alteration, save with respect to time only. Also, the locomotion is not traversing something that the alteration is traversing. That is because the motion traverses a given distance, since it was a change from the beginning of [the distance] to its end, whereas the alteration traversed what is between two qualities, since it was a change, not from some limiting point of a given distance to another, but from one quality to another. [That] is because that which underwent the alteration as such did not proceed from some limiting point of a given distance to another limiting point, but from one quality to another, except that one quality after another was continually being renewed in it in such a way that the thing's substrate was undergoing uninterrupted renewal.

(٨) وها هنا مسـألة ربّما سـأل عنها سـائل وقال: متحركٌ قطع مسـافة وكانت تلك المسـافة تبتدىء تسـتحيل مع ابتداء حركته، حتى انتهت الاستحالة إلى الحد الذي تقـف عنده وتتم لديـه، فوقفت النُقْلة معها، فهل من الممكن أنْ يقال إنَّ هذه الاسـتحالة مساوية لهذه الحركة؟ فالجواب أنَّ ذلك خطأ لا يجوز أنْ يقال، وذلك لأنَّ المسافة مساوية للمسـتحيل، وأمّا الحركة فليست مسـاوية للاسـتحالة، إلاّ في الزمان فقط. ولا النقلة قطعت شيئاً ممّا قطعته الاستحالة، . وذلك لأنَّ الحركة قطعت مسافة إذ كانت تغيّراً من مبدئها إلى منتهاها، والاسـتحالة قطعت ما بين كيفيتين، إذ كانت تغيّراً لا من حدّ مسافةٍ إلى أخرى، بل من كيفيةٍ إلى أُخرى. إذ المستحيل - من حيث هو - لم يخرج من حدّ مسـافةٍ إلى حدٍ آخر، بل خرج من كيف إلـى كيف، إلاّ أنه لم يزل يتجدّد فيه كيفٌ بعد كيف؛ لا على استقرار تجدّد الشيء في محله.

Chapter Six

On the contrariety of motions and their opposites

(1) Since we have mentioned the similarity and difference among motions, we should now discuss the contrariety of motions. So we say: Generically different motions, such as locomotion, alteration, and augmentation, might be joined together. So, if, at some moment, some of them are prevented from joining together with others, that is not because their natures *qua* locomotion, alteration, and augmentation make that necessary, but because of some additional, extrinsic causal factor. Motions that are included under a single genus—as, for example, blackening and whitening (which fall under the genus of quality)—occur in the way previously mentioned.[1] So blackening and whitening agree in genus and share a common subject, yet they are opposites that cannot simultaneously be joined together. In other words, [blackening] (as is whitening) is a fact about existence, not something that is said relative to something else. Also, the difference between them is greater than that which is between one of them and yellow or the like—namely, it is the maximal degree of difference. These are the situations whereby one thing becomes a contrary of another. So whitening is the contrary of blackening, just as white is the contrary of black. The same likewise holds for the category

1. See 4.5.5.

<الفصل السادس>

في تضادّ الحركات وتقابلها

(١) وإذ قد قلنا في تساوي الحركات وتفاوتها، فأولى ما نتكلم فيه هو حال تضادّ الحركات فنقول: أمّا أولاً؛ فإنَّ الحركات المختلفة الأجناس؛ مثل النُقْلة والاستحالة والنمو، فقد تجتمع معاً. فإن امتنع بعضها عن الاجتماع مع بعض في وقتٍ ما، فليس ذلك لأن طباعها - من حيث هي نُقلة واستحالة ونمو - توجب ذلك، بل لأمرٍ زائدٍ وسببٍ من خارج. وأمّا الحركات الداخلة تحت جنسٍ واحدٍ؛ مثل التسوّد والتبيّض الواقعين في جنس الكيفية، على النحو من الوقوع المذكور، فإنَّها قد تكون متضادّة. فإنَّ التسوّد موافق للتبيّض في الجنس، ويشاركه في الموضوع، ولكنه مقابلٌ له يستحيل اجتماعه معه، وهو معنى وجودي، كما أنَّ التبيّض معنى وجودي، وليس مقولاً بالقياس إلى الآخر. وبينهما من الخلاف أكثر ممّا بين أحدهما وبين التصفّر وغيره، وهو غاية الخلاف. وهذه هي الأمور التي بها يصير الشيء ضدّاً للشيء؛ فالتبيّض ضدّ التسوّد، كما أنَّ البياض

of quantity, for augmentation is the contrary of diminution. Indeed, even if one says that small is not the contrary of big, but, rather, its correlative, this can be undermined by the fact that the small and big that are commensurate in species are said absolutely, but not relatively. [That] is because there is some other consideration in augmentation and diminution that does away with saying that, since increasing is only an increase relative to decrease, and yet the increase and decrease that are the termini *ad quem* that exist in nature are not relative. So you will find that the situation concerning augmentation and diminution is just like that of whitening and blackening, and the same holds for the case of rarefaction and condensation. As for motions that are with respect to position, it would seem that there is no contrariety, just as there is no contrariety in circular motions, which you will learn shortly.[2]

(2) As for motion with respect to place, the genus of circular [motion] falling under [the category of place] is in no way whatsoever the contrary of the genus of rectilinear [motion]. That is because the specific differences of contrary motions, despite agreeing in genus, must be opposites that are inevitably antipathetic to one another, while being related to something or other upon which the motion is dependent. It is not [enough] that the two things undergoing motion should be contraries [to make] the motions contraries, for contraries might accidentally be undergoing some motion that agrees in species. So, when, by force, something hot has an accidental motion downward (and so it is like the stone), then there are two species of two motions that do not differ in themselves, but differ only by force and nature. Force and nature, however, do not make something to be different, for the heat that forcibly comes to be in some body and that naturally rises is a heat whose action is the same.

2. See par. 6, specifically, and also pars. 7–8.

ضدّ السـواد ، وكذلك في مقولة الكم أيضاً ؛ فإنَّ النمو ضدّ الذبول. فإنّه، وإنْ كان لقائل أنْ يقـول إنَّ الصِغر ليس بمضـادّ للكِبر ، بل هو مضايف له ، وكان يجوز أنْ يبطل هذا بأنَّ الصغير والكبير ، اللذين بحسـب النوع ، يقالان علـى الإطلاق لا بالقياس ، فإنَّ في النمو والذبـول اعتباراً آخر يُغني عن أنْ يقال ذلك ، لأنَّ الحركة إلى الزيادة إنّما هي زيادة بالقياس إلـى النقصان ، وعلى أنَّ الزيادة والنقصان اللذين يتوجهان إليه موجودان في الطبع ليسـا بالقياس. وسـتجد الحال في النمو والذبول ، كما في التبيّض والتسوّد ، وكذلك الحال في التخلّخل والتكاثف. وأمّا الحركات التي في الوضع ، فيشبه أنْ لا يكون فيها تضادّ ، على نحو ما لا تضادّ في الحركات المستديرة ، وستعلم هذا عن قريب.

(٢) وأمّا الحركة المكانية ؛ فإنَّ الجنس المسـتدير منها غير مضادّ للجنس المسـتقيم بوجهٍ من الوجوه ، وذلك لأنَّ فصول الحركات المتضادّة - مع الاتفاق في الجنس - يجب أنْ تكون متقابلة متعاندة لا محالة ، وتكون منسوبة لا محالة إلى أمرٍ من الأمور التي تتعلق بهـا الحركة. والحركات ليس كونها متضادة هـي أنَّ متحركها متضادان ؛ فإنَّ الأضداد قد يعـرض لها أنْ تتحـرك حركة متفقة في النوع. فإنَّ الحار إذا عرض له حركة بالقسـر إلى أسـفل ؛ وشاكل الحجر في ذلك ، كان نوعا الحركتين لا يختلفان في ذاتيهما ، إنّما يختلفان بالقسر والطبع ، والقسر والطبع لا يجعل الشيء مختلفاً . فإنَّ الحرارة التي تحدث في جسمٍ

Also, the black that comes to be by force and the one that is by nature are [both] a black that produces one and the same effect; they differ only in that this one is accidental, while that one is natural. The same holds for natural and non-natural[3] shapes and the like. Moreover, if the contrariety of motions were due only to force and nature, two forced motions would not be contraries, nor would two natural motions be contraries. Clearly, then, one motion does not become the contrary of another solely from the fact that the two agents of the motion are contraries. In the same way, you also know that one motion does not become the contrary of another on account of the fact that the two movers are contraries. It likewise is not on account of the time, because the nature of time has no contrary; and, even if it were to have some contrary, the contrariety would involve something accidental to the motion, not the motion's nature, since time is an accident of motion. Also, motions are not contraries [just] because that with respect to which there is one motion is the contrary of that with respect to which there is another motion. [That follows] since that with respect to which there is motion [might] be the same, while the motions are contraries. So [for example] the course from white to black, as well as from large to small, is the very same course as that from black to white, and from small to large, and, in general, between the intermediary things themselves—as, for example, the distance in descending is [the same as] the distance in ascending. In short, these intermediary things have no contraries, because they are things in the middle [of two contraries]. So how could it be on account of their contrariety that motions become contraries?

---

3. Literally, "forced."

بالقسر، والتي تثور بالطبع، حرارة متفقة الفعل. والسواد الذي يحدث بالقسر والذي يحدث بالطبع، سواد يؤثر تأثيراً واحداً، إنّما يختلف بأنَّ هذا عرضي وذاك طبيعي، وكذلك الأشكال الطبيعية والقسرية وغير ذلك. ولو كان تضاد الحركات أيضاً إنّما هو للقسر وللطبع، لما كانت حركتان قسريتان متضادتين، ولا طبيعيتان متضادتين. فبيّنٌ أنّه ليس تصير الحركة مضادّة للحركة لنفس أنَّ الفاعلين للحركة متضادان. وبمثل ذلك تعلم أيضاً؛ أنَّ الحركة ليست تصير مضادّة للحركة لأجل أنَّ المحركين متضادّان، ولا أيضاً لأجل الزمان؛ لأنَّ الزمان لا تتضادّ طباعه، ولو كانت تتضادّ لكان يكون التضادّ في أمر يعرض للحركة لا لطبيعة الحركة، فإنَّ الزمان عارضٌ للحركة. ولا أيضاً تكون الحركات متضادّة لأجل أنَّ الذي فيه الحركة مضادّ للذي فيه حركة أُخرى، فإنَّ الذي فيه الحركة يكون متفقاً، والحركات تتضادّ؛ فإنَّ الطريق من البياض إلى السواد، ومن الزيادة إلى النقصان، هو بعينه الطريق من السواد إلى البياض، ومن النقصان إلى الزيادة، وبالجملة هي بين المتوسطات بأعيانها، كما أنَّ المسافة في النزول هي المسافة في الصعود. وبالجملة فإنَّ هذه المتوسطات لا أضداد لها لأنّها متوسطات، فكيف تكون هي التي لتضادّها تصير الحركات متضادّة؟

(3) So nothing remains now but the termini *ad quem* and *a quo;*[4] for, when they are contraries, like black and white, the motions are contraries, but not in whichever way chance may have it. [That] is because the motion *from black* is not a contrary of the motion to black simply because it is from black, but because it necessarily follows from the fact that there is a motion *toward white* together with that, just as its being a motion *toward black* necessarily follows from its being a motion *from white.* There is no transition from black except toward white, nor is there transition toward black except from white. (As for from becoming transparent and toward becoming transparent, that is not a motion, but, rather, something that occurs all at once.) Were it the case that the motion from black was not directed toward white, then these two motions would not be contraries, just as something can be moved away from the right while not [being moved] to the left, but upward.

(4) So contrary motions are those whose limits are opposites. This is understood in two ways (which [in fact] are traced back to three ways). One of them is that their limits are opposite as a result of a real contrariety in the things themselves—as, for example, black and white, and the largest and smallest volume with respect to a given thing's nature. The second is that, although their limits do not oppose each other in themselves and essentially, they do oppose each other in two ways, one of which is relative to the motion, and the second relative to factors outside the motion. An example would be, for instance, the two limits of gazing along the continuous distance between the Heavens and the Earth, whether as two points or two places. The natures of the two points in the two places are neither contraries nor opposites in the way that black and white are, but

4. Following **Z** and **T** and rejecting **Y**'s suggested addition *tataḍāddu* and secluding *al-ḥarakah,* which appears in only two of the MSS consulted by **Y**. If retained, the sense would be "So nothing remains *contrary to the motion* but. . . ."

(٣) فلم يبق الآن إلاّ الأمور التي إليها وعنها ، فإنّها إذا كانت متضادّة ، كالسواد والبياض ، كانت الحركات متضادّة ، ولا كيف اتفق . فإنَّ الحركة من السواد ليس بضدّ للحركة إلى السواد ، لأجل أنّه حركة من السواد فقط ، بل لأجل ما يلزمه من أنْ يكون مع ذلك حركة إلى البياض ، كما يلزم كونها حركة إلى السواد من كونها حركة من البياض . فإنَّ الانتقال من السواد لا يكون إلاّ إلى البياض ، والانتقال إلى السواد لا يكون إلاّ من البياض . فأمّا من الإشفاف وإلى الإشفاف فذلك ليس بحركةٍ بل أمرٌ يقع دفعة ، ولو كانت الحركة من السواد قد تتوجه لا إلى البياض ، لم تكن هاتان الحركتان متضادتين ، كما أنّه يجوز أنْ يتحرك الشيء من اليمين لا إلى اليسار ، بل إلى فوق .

(٤) فالحركات المتضادّة هي التي تتقابل أطرافها ، وهذا يتصوّر على وجهين يرجعان إلى وجوهٍ ثلاثة : أحدها أنْ تكون أطرافها تتقابل بالتضادّ الحقيقي في ذواتها ؛ مثل السواد والبياض ، ومثل أكبر حجمٍ في طبيعة الشيء ، وأصغر حجمٍ في طبيعة ذلك الشيء والثاني أنْ تكون أطرافها لا تتقابل في ذواتها وفي ماهياتها ، بل تتقابل من جهتين إحداهما بالقياس إلى الحركة ، والثانية بالقياس إلى أمورٍ خارجةٍ عن الحركة ؛ مثل أنَّ طرفي النظر في المسافة المتصلة بين السماء والأرض مثلاً نقطتان أو مكانان ، وطباع النقطتين في المكانين لا تتضادّ ولا تتقابل تقابل السواد والبياض ، بل تتقابل لأمرٍ آخر خارج ، وذلك الأمر ، إمّا غير

[they] are opposites owing to some outside factor, where that factor either is or is not dependent upon the relation to motion. What is outside the relation to the motion occurs [for example] in that one of the two limits is at the maximal degree of proximity to the celestial sphere, whereas the second is at the maximal degree of remoteness from it. So [the gaze falls] necessarily upon one of [the celestial sphere's] limits if [that limit] is up and another if it is down. As for that which is dependent upon the relation to the motion, an example would be that one of the two limits is accidentally the beginning of one single motion and the other is accidentally the end. So the relation of each one of them to the motion is something different and opposite of the other's relation. Even if the relation of each one of them to the motion is a relation of relative opposition—since the beginning is a beginning of something that has a beginning and the end is an end of something that has an end, and vice versa—the opposition between beginning and end will not be this opposition. [That] is because the beginning is not the opposite of that end in that it is said relative to it. [That] is because it does not follow that when motion has a certain beginning, then it must be understood to have a certain end. Perhaps (if not necessarily) it will be known by some proof and an additional middle term. The same holds in the case of ending. Now, with two correlatives, to know either one is necessarily to know the other. The distance's beginning, however, is not something essentially understood relative to its end, nor is its end essentially understood relative to its beginning. So, between them, there is not the opposition of being correlative. [Still], there is inevitably some opposition between them (I mean, when they are through what is rectilinear), since it is impossible that the starting point and endpoint be joined together at one and the same

متعلق بالنسبة إلى الحركة، وإمّا متعلق بها. أمّا الخارج من النسبة إلى الحركة فبأنْ يكون أحد الطرفين في غاية القرب من الفلك، والطرف الثاني في غاية البُعْد منه، فيكون طرفٌ منه يلزمه، إنْ كان علوّاً، والآخر ⟨ي⟩لزمه إنْ كان سُفْلاً. وأمّا المتعلّق بالنسبة إلى الحركة، فمثل أنْ يكون أحد الطرفين عرض له أنّه مبدأ الحركة الواحدة، والآخر عرض له أنّه منتهى لتلك الحركة، فقياس كل واحدٍ منهما إلى الحركة مخالفٌ ومقابلٌ لقياس الآخر. فإنّه وإنْ كان قياس كل واحدٍ منهما إلى الحركة؛ قياس المقابل بالإضافة، إذ المبدأ مبدأ لذي المبدأ، والمنتهى منتهى لذي المنتهى، وكذلك بالعكس في الأمرين، فليس مقابلة ما بين المبتدأ والمنتهى هذه المقابلة، فإن المبدأ لا يقابل المنتهى بأنّه مقول بالقياس إليه، فإنّه ليس يلزم أنّه إذا كان للحركة مبدأ ما، وجَبَ أنْ يفهم من هذا بعينه أنَّ لها منتهى، عسى - إنْ كان ولا بُدّ - فسيعلم بدليلٍ ووسطٍ من خارج، والأمر في المنتهى كذلك. والمضافان، أيّهما علم، لزم العلم بالآخر، فليس ابتداء المسافة متصور الماهية بالقياس إلى منتهاها، ولا منتهاها متصور الماهية بالقياس إلى مبتداها، فليس بينهما تقابل المضاف، وبينهما لا محالة تقابل، أعني إذا كانا في المستقيمة. إذ يستحيل أن يكون المبدأ والمنتهى مجتمعين في شيءٍ واحدٍ

moment of time in one and the same thing to which they are related as a starting point and endpoint. Also, one of them is not some privation of the other, such that the endpoint would be the privative notion of the starting point, except by contrariety. There simply is no other kind of opposition but contrariety. It is not improbable, however, that, with respect to what is not rectilinear, there is some one and the same thing that is [both] a starting point and endpoint for the nonrectilinear motion. So, in that case, there would be neither contrariety nor opposition with respect to the starting and ending points.

(5) The first class, without doubt, makes motions to be contraries, whereas there seems to be doubt about the latter two classes. That is because the things possessing those limits do not oppose each other essentially, but only do so by happening to have some accident. So, when there is nothing that is really contrary, why should we make the motions contraries? We say that this premise is false. When *x* is dependent upon *y*, and [when] the contrariety that *y* happens to have is not[5] in its substance but owing to some accident that it happens to have, it does not necessarily follow that the contrariety in *x* is accidental. That is because this thing that is accidental to *y* might be some factor internal to the substance of *x*. So being delimited by [a given] limit is not essential to wax, but it is essential to the shape that is in the wax—that is, it is dependent upon the wax and subsists through it. Similarly, hot and cold bodies are contraries by virtue of their accidents and activities, and the [motions of] heating and cooling that proceed from them are not accidental, but real contraries. [That] is because, even if hot and cold are accidents relative to the body, they are essential or exist necessarily, such that [the motions of] heating and cooling are realized and have this form. Motion, then,

5. **Y** seems inadvertently to have omitted *laysa,* which appears in **Z** and **T**.

همـا بالقياس إليه مبتدأ ومنتهى، اجتماعاً في زمانٍ واحد، وليس أحدهما معنى عدمياً للآخر، حتى يكون المنتهى عدم المبتدأ إلاّ بالتضاد، ولا وجه من وجوه التقابل إلاّ التقابل بالتضادّ. وأمّا في غير المسـتقيم فلا يبعد أنْ يكون شـيءٌ واحدٌ مبتدأ ومنتهى للحركة، التي ليست على الاستقامة، فلا يكون في المبدأ والمنتهى هناك تضادّ أو تقابل.

(٥) وليس يقع الشـك في أنَّ القسـم الأول يجعل الحركات متضادّة، وأمّا القسمان الآخران فيشبه أنْ يقع هذا الشك فيهما، وذلك لأنَّ ذوات تلك الأطراف لا تتقابل لذاتها، بل تتقابل بعارضٍ عرض لها، فإذالم تكن متضادّة حقيقية، لِمَ نجعل الحركات متضادّة حقيقية؟ فنقول: إنَّ هذه المقدمة باطلة؛ فإنّه ليس إذا كان الشيء متعلقاً بشيءٍ ويكون ذلك الشيء ليس يعرض له التضادّ في جوهره، بل لعرضٍ يعرض له، يجب أنْ يكون التضادّ، في المتعلق بذلك الشيء، تضادّاً بالعرض؛ وذلك لأنّه يجوز أنْ يكون هذا الذي هو عارضٌ للمتعلق به أمراً داخلاً في جوهر المتعلّق. فإنَّ التحدّد بالطرف أمرٌ غير ذاتي للشمع، وذاتي للشكل الذي هو في الشمع، وهو ممّا يتعلق بالشمع ويتقوّم به. وكذلك الجسم الحار والجسم البارد، يتضـادّان بعرضيهما وفعلاهما؛ وهو الاسـخان والتبريد الصـادران عنهما، لا يتضادّان بالعرض بل بالحقيقة، لأجل أن الحار والبارد - وإنْ كان عارضاً بالقياس إلى الجسـم - فإنّه ذاتي، أو واجب الوجود حتى يكون الاسخان والتبريد متحقّقاً وعلى هذه الصورة.

does not depend upon the distance's limit inasmuch as it is merely a limit, however it might be. So, when [the motion] at the limit happens to have some accident, neither is [that accident] internal to the subsistence of the motion nor is its inclusion necessary. Very much to the contrary, the motion depends upon the limit only inasmuch as it is starting point and endpoint. [That] is because every motion, in its very substantiality, includes the posterior and the prior, because the very substance of motion is to be a departing from and tending toward. So the very substantiality of motion includes a starting point and endpoint, whether in actuality or in the potentiality proximate to actuality, which we have indicated.[6] So it is only insofar as distance's limits are a starting point and endpoint that motion is dependent upon them, and insofar as they are a starting point and endpoint that they are opposite each other, and insofar as they are opposites of each other that they are constitutive of the motion, even though that is not constitutive of them.

(6) It is patently obvious that the motion, which has a designated starting point and endpoint, where the two are actually different from one another [and] one of which cannot be brought to the other, but is as we described, is essentially from one contrary to another. Now, the two contraries are essential to it but are not essential to the subject, which is the limit. Someone might ask: How can the starting point be a contrary of the endpoint when motion's starting and ending points are sometimes in a single body, whereas contraries are not joined together in a single body? The answer is that contraries might be joined together in a single body when the body is not their primary, proximate subject. Only in the primary, proximate subject are contraries not joined together simultaneously. The subject functioning as the starting point

6. See 2.1.22.

فإنَّ الحركة ليست تتعلق بطرف المسافة، من حيث هو طرفٌ فقط كيف كان، حتى إذا عرض للطرفية عارضٌ كان غير داخلٍ في تقويم الحركة، أو لا يجب دخوله، كلا بل إنّما تتعلق الحركة بالطرف من حيث هو مبدأً ومنتهى. فإنَّ كل حركةٍ بجوهريتها تتضمن التأخّر والتقدّم، لأنَّ الحركة جوهرها مفارقة وقصد، فجوهرية الحركة تتضمن المبدأ والمنتهى، إمّا بالفعل، وإمّا بالقوة القريبة من الفعل التي أشرنا إليها. فالأطراف التي للمسافة، إنّما تتعلق بها الحركة من حيث هي مبدأ ومنتهى، وهي - من حيث هي مبدأ ومنتهى - متقابلة، وهي من حيث هي متقابلة؛ فهي مقوّمة للحركة، وإنْ كانت ليست متقوّمة بذلك.

(٦) وظاهرٌ بيّنٌ أنَّ الحركة التي يتعيّن لها مبدأ ومنتهى متغايران بالفعل لا يجوز أنْ يؤدي أحدهما إلى الآخر، بل يكون على النحو الذي وصفنا، فهي لذاتها من ضدّ إلى ضدّ، والضدّان ذاتيان لها، وليسا ذاتيين للموضوع الذي هو الطرف. ولقائلٍ أنْ يقول: كيف يكون المبدأ مضادّاً للمنتهى؛ ومبدأ الحركة ومنتهاها قد يكونان في جسمٍ واحدٍ، والأضداد لا تجتمع في جسم واحد؟ فيقال له: الأضداد قد تجتمع في جسمٍ واحدٍ، إذا كان الجسم ليس موضوعها الأول القريب، إنّما لا تجتمع الأضداد معاً في الموضوع الأول القريب. وموضوع المبدائية والمنتهائية ليس هو الجسم بل الطرف، ولا يجتمع في طرفٍ بالفعل أنْ يكون مبدأ حركةٍ مستقيمةٍ واحدةٍ بالاتصال ومنتهاها، وهذا كما قد يجتمع في

and endpoint is not the body, but the limit. In a limit, however, there is no actual joining together [such] that there is [both] a starting and endpoint of a single, continuous, rectilinear motion. This is like opposite things that are sometimes joined together in a single body (and even at the same time) without there being contrariety—as, for example, a body in which there exists a convex and concave line, and whatever is like that. The one who thinks that it is no more fitting that rectilinear motions are contraries of one another than that they be contraries of circular [motions]—since the course and distance in contrary rectilinear [motions] is one and the same—is simply being obtuse. He might just as well say that black and white are not contraries because their subject is one and the same. Were it one of the conditions of being contraries that two contrary things should not share anything in common, then two contraries would not be joined together in a single genus, nor would their subject really be one and the same; for *contrariety* is to be different to the maximal degree possible along a single course. Blackening is undoubtedly the contrary of whitening, and the course between them, which is one, is the intermediate states; however, the two opposing processes with respect to [that course] are as different as can be.

(7) Having laid these foundations, let us return to our intended goal of explaining that circular motion is not the contrary of rectilinear motion. We say: If there is contrariety between the two, then that contrariety either is or is not on account of being circular and rectilinear. On the one hand, if it is on account of being circular and rectilinear, then

جسم واحدٍ أشياء متقابلة، وإنْ كانا معه، بغير التضادّ؛ كجسمٍ يوجد فيه خطّ محدّب وخطّ مقعّر، وما أشبه ذلك. والذي ظنّ أنّ الحركات المستقيمة ليست أوْلى بأنْ تتضادّ، من أنْ تضادّها المستديرة - إذ الطريق والمسافة في المتضادّات المستقيمة واحدة - فقد سها سهواً عظيماً، وكان يلزمه أيضاً أنْ يقول: السواد والبياض ليسا بمتضادين؛ لأنّ موضوعهما واحد. ولو كان من شرط التضادّ أنْ لا يكون للضدّين أمرٌ مشتركٌ، لما اجتمع الضدّان في جنسٍ واحدٍ، ولما كان موضوعهما واحداً بالحقيقة، فإنّ التضادّ هو اختلافٌ في طريق واحد على غاية ما يمكن. ولا يُشك أنّ التسوّد ضدّ التبيّض، والطريق بينهما هو الوسائط وهو واحد، لكن السلوكين المتقابلين فيه هما على غاية الخلاف.

(٧) وإذ قد بيّنا هذه الأصول، فلنرجع إلى غرضنا من تبيين أنّ الحركة المستديرة لا تضادّ الحركة المستقيمة، فنقول: إنْ كان بينهما تضادّ، فإمّا أنْ يكون ذلك التضادّ لأجل الاستدارة والاستقامة أو لا يكون، فإنْ كان لأجل الاستدارة والاستقامة، كانت الاستدارة والاستقامة متضادّتين؛ لأنّ الشيء الذي به الاختلاف بين الأضداد المتفقة في الجنس

being circular and being rectilinear are two contraries, because the thing by which contraries agreeing in genus differ is a contrary. As was noted,[7] however, the proximate subject of the circular and rectilinear is not one and the same, nor, as we said,[8] can any part of the subjects undergo alteration from being circular to being rectilinear without its undergoing corruption. So, they are not contraries because of the contrariety of the motions, but neither is that with respect to which there is motion the cause of the motions' contrariety. So if their contrariety is not owing to that with respect to which [there is motion], it remains that it is owing to the limits. Now, if it is because of the limits that circular motion is contrary to the others, one and the same motion would have an infinite number of different motions contrary to it. [That] is because it is possible that the rectilinear line (which has been picked out and denoted as that along which there is this rectilinear motion) is a chord for a potential infinity of dissimilar arcs. There is, however, only one contrary of this one—namely, that which is at the maximal degree of remoteness from it. In this [way], we can also explain that the form of the rectilinear and the circular are not generically contrary to one another. [That] is because, if the rectilinear absolutely is a contrary of the circular absolutely, then, equally, *this* rectilinear thing is contrary to *this* particular circular thing, since this one thing can be an opposite of only one particular thing. [That is] because what is at the farthest degree of difference from this one in a given nature is itself one, and if it is not at the farthest degree, it is not a contrary. Since this individual is not numerically many, neither can its contrary be something common to many.

7. See 4.3.6
8. Ibid.

متضادّ. لكن الاستدارة والاستقامة - كما قيل - ليس موضوعهما القريب واحداً، ولا شيء من الموضوعات يجوز أنْ يستحيل من الاستدارة إلى الاستقامة إلاّ بفساده، على ما قلنا، فليس بضدّين بسببي تضادّ الحركات، بل ليس ما فيه الحركة هو السبب لتضادّ الحركات، فإنْ لم يكن تضادّهما لما فيه؛ بقي أنْ يكون للأطراف. ولو كانت مضادّة المستديرة لغيرها بسبب الأطراف، لكانت الحركة الواحدة بعينها تضادّها حركاتٌ لا نهاية لها مختلفة، لأنَّه يمكن أنْ يكون الخطّ المستقيم المعيّن المشار إليه الذي عليه هذه الحركة المستقيمة وترا لقسيٍّ غير متشابهة لا نهاية لها بالقوة، لكن ضد هذا الواحد واحدٌ فقط؛ وهو الذي في غاية البُعْد عنه. ويمكن أنْ تتبين بهذا أيضاً أنَّ صورة الاستقامة والاستدارة لا تتضادّ تضادّاً جنسياً، لأنّه إنْ كان مطلق الاستقامة مضاداً لمطلق الاستدارة؛ كان أيضاً هذا المستقيم يضادّ هذا المستدير بعينه. إذ لا يجوز أنْ يكون هذا الواحد مقابلاً إلاّ لواحدٍ بعينه، لأنَّ ما هو أبعد عن هذا الواحد في طبيعة الخلاف، فهو واحدٌ، فإنْ كان لا أبعد؛ فلا ضدّ. وهذا الشخص، لما لم يكن متكثّراً بالعدد، لم يجز أنْ يكون ضدّه معنى عامياً متكثّراً.

(8) Undermined in the same way is the claim of those who say that the many motions along the arcs can be the contraries to the one rectilinear motion. It is said that even if the contrary of one thing is one thing, these many things are like one thing inasmuch as they are circular. This claim is, in fact, mistaken. That is because the contrary of the single thing that is common to many things taken individually is [likewise] some single thing taken in common. It is not the case that the contrary of the single thing taken in common is one thing taken individually. So this single rectilinear line taken individually is not the contrary of all of those circular lines that are alike in being circular. The fact is that it seems more fitting that those circular lines are not like individuals of a single species, but, rather, [that] each one of them is an arc of some other circle whose curvature and protraction are different. It is not unlikely that there are numerically many circles that agree in species and do not differ in having a curved shape, but which in no way can be made to coincide with one another. In some way like this, what is rectilinear and what is circular differ, even while being alike inasmuch as they are extended lines. So it is not unlikely that two arcs that cannot be put in exact correspondence with one another will differ specifically, while agreeing in being two curve-shaped circular things. So how can all of those different arcs be some contrary for a single thing?

(9) Moreover, the demand of anyone who asserts that there is a generic contrariety between what is rectilinear and circular and a specific contrariety between two rectilinear things comes to naught by the [simple] fact that we do not deny that one and the same thing has a number of contraries from various perspectives, whether generically or specifically. That is because one thing may be the contrary of another with respect either to the nature itself or to certain accidents and states.

(٨) فسقطَ بهذا قول مَنْ قال إنَّ هذه الحركات القوسية الكثيرة، يجوز أنْ تكون مضادّة للمستقيمة الواحدة؛ قال وإنّه وإنْ كان ضدُّ الواحد واحداً، فهذه الكثيرة، من حيث هي مستديرة، كشيءٍ واحد. فإنَّ هذا القول خطأ، وذلك لأنَّ ضدّ الواحد بالعموم واحدٌ بالعموم، متكثّر بالشخص، ليس ضدّ الواحد بالعموم واحداً بالشخص. فليس ضدّ جميع تلك المستديرات المتفقة في معنى الاستدارة هذا المستقيم الواحد بالشخص، بل الأولى أنْ تكون تلك المستديرات ليست كأشخاص من نوع واحدٍ، بل كل واحدٍ منها قوسٌ من دائرةٍ أخرى، انعطافها وانجذابها إنعطافٌ وانجذابٌ آخر. ولا يبعد أنْ تكون الدوائر المتفقة في النوع هي التي تتكثّر بالعدد، ولا تختلف في الإحديداب، فيكون لا جواز مطابقة فيما بينهما بوجهٍ من الوجوه. وبمثل هذا ما اختلف المستقيم والمستدير، وإنْ اتفقا من حيث أنّهما خطّان ممتدان، فلا يبعد أنْ يختلف نوعا القوسين اللّذين لا ينطبق أحدهما على الآخر، وإنْ اتفقا في أنهما مستديران مُحْدَوْدبان، فكيف تكون تلك القسي المختلفة كلها مضادّة لشخصٍ واحد؟

(٩) ويسقط أيضاً سؤال مَنْ قال: ليكن بين المستقيم والمستدير مضادّة جنسية، وبين المستقيمين مضادّة نوعية، بأنْ يقال: إنّا لا نمنع أنْ يكون للشيء الواحد أضداد من جهاتٍ كانت جنسية أو كانت نوعية، وذلك لأنَّ الشيء يضادّ الشيء في طبيعة ذاته، وقد يضادّه في أعراض وأحوال. ونحن لا نمنع أنْ يعرض للحركات المستديرة أنْ يكون لها

Now, we ourselves do not deny that circular motions happen to have certain contraries that belong to rectilinear and circular things among their accidental features. We deny only that they have some contrary in themselves and essentially. This is like the fact that, in ethics, the mean is contrary to [both] excess and deficiency,[9] even though, in themselves, [excess and deficiency] are contraries as well. In fact, the contrariety of the excess and the deficiency is a real contrariety in the things themselves—namely, they are at the maximal degree of remoteness from one another. The contrariety of the mean and the two extremes, however, is not owing to the nature of the mean and the two extremes, but because the mean is a virtue, while the latter two are alike in being vices. Now, virtue is either some inseparable or accidental feature of that nature belonging to a mean, just like the latter two's being vices is something inseparable or accidental to them. So one limit is the contrary of the other as a result of their very substance, whereas the mean is a contrary accidentally. As for whether something can have one contrary due to its genus and another due to its species, you have already learned elsewhere[10] what there is to this when you discovered that the contrary is, in fact, the contrary of the thing itself and its specificity. So what is circular cannot be generically contrary to what is rectilinear, and one rectilinear thing [cannot be] specifically contrary to another. Also in this, one must not appeal to the generic contrariety of motion and rest to bolster one's case, and then the specific contrariety of two motions; for rest is a privative feature, not a contrary.[11] So it has become clear that rectilinear motion is not the contrary of circular motion.

9. Cf. Aristotle, *Nicomachean Ethics* 2.6.
10. The reference is probably to *Kitāb al-maqūlāt* 7.3.
11. See 2.4.

أضداد ، من المستديرات ومن المستقيمات في معانٍ تعرض لها ، وإنّما نمنع أن يكون لها ضدٌّ في ذاتها وماهيتها . وهذا ؛ كما أنَّ التوسّط في الأخلاق يضادّ التقصير والإفراط ، وقد يتضادّان هما أيضاً في أنفسهما ، ولكن تضادّ الإفراط والتقصير تضادّ حقيقي في الذات ، وهما المتباعدان غاية التباعد . وأمّا تضادّ التوسّط والطرفين ؛ فليس لطبيعة التوسّط والطرفين ، بل لأنَّ التوسّط فضيلة ، وذانك يجتمعان في الرذيلة ، والفضيلة معنى لازمٌ أو عارضٌ لتلك الطبيعة المتوسطة ، وأيضاً كون ذينك رذيلة معنى لازم لهما أو عارض ، وليس للفضيلة والرذيلة دخولٌ في ماهية هذه . فيكون التضادّ بين المتوسط والطرفين تضادّاً في عارض ؛ فالطرف يضاد الطرف بجوهره وذاته ، ويضاد الوسط بعارض . وأمّا هل يكون للشيء ضدٌّ من جهة جنسه ، وضدّ من جهة نوعه ، فقد علمت في موضع آخر ما في هذا ، وتحقّقت أنَّ الضدّ بالحقيقة هو ضدّ ذات الشيء ونوعيته . فلا يجوز أنْ تكون المستديرة تضادّ المستقيمة تضادّاً جنسياً ، وتضاد المستقيمةُ المستقيمةَ تضادّاً نوعياً ، ولا يجب أنْ يستعان في هذا بتضادّ الحركة والسكون تضادّاً جنسياً ، ثم بتضادّ الحركتين تضادّاً نوعياً ، فإنَّ السكون معنى عدمي لا مضاد ؛ فقد اتضح أنَّ الحركة المستقيمة لا تضادّ المستديرة .

(10) Likewise, you know that circular motions along arcs are not contrary to one another, because there can be infinitely many arcs agreeing in certain shared limits. Also, the motion from one limit of an arc to the other (and vice versa), when the arc is one and the same, has no contrary. You will understand that once you understand that the circular motion that involves [change] of position [rather than change of place], making a complete rotation, has no contrary whatsoever, because it has no actual limit.[12] When it is posited as having a limit at which a designated position comes to be actual by that act of positing, there comes to be, at [the designated position] (if there does), a starting point and endpoint. In doing so, the starting point and endpoint are not contraries on account of being the initial point of origin and ultimate point of termination, but because, whenever [that designated position] passes by you, there is a certain starting point and endpoint of a given motion. It does not happen no matter what, but because there is a starting point and endpoint of a given motion whose starting point is not itself the point at which its extension ends, such that the opposition between the starting point and end would, in fact, be on account of the relation to the motion. That happens only where the starting point and endpoint belong to a rectilinear motion whose extension is not such that one and the same point is appointed as both the starting and end point,[13] and so that [starting point and endpoint] are not together. When

12. See par. 8.

13. Literally, "the starting point is made an endpoint nor the endpoint a starting point."

(١٠) وكذلك لك أنْ تعلم أنَّ المستديرات التي على القسي لا تتضادّ؛ لأنّه يجوز أنْ تتفق في أطراف مشتركة قسيٌّ بلا نهاية. فأمّا الحركة من طرف قوس إلى طرفٍ آخر، والتي بالعكس – والقوس واحدة بعينها – فلا تكون مضادة لها أيضاً، تعلم ذلك إذا علمت أنَّ الحركة المستديرة الوضعية، التامّة الدوران، لا ضدّ لها بوجه، لأنّه لا طرف لها بالفعل. وإذا فرض لها طرف يكون فيه خروج وضْع معيْن إلى الفعل، بذلك الفرض اجتمع فيه، إنْ كان، مبدأ ومنتهى، إذ لم يكن المبدأ والمنتهى ضدّين لأجل المبدئية والمنتهائية، بل لأجل أنّهما – كما مرَّ ‹بك› – مبدأ ومنتهى حركة، ولا كيف اتفق، بل لأجل أنّهما مبدأ ومنتهى حركةٍ بصفةٍ لا يكون مبدؤها هو بعينه منتهاها في استمرارها، حتى يصح التعاند بين المبدأ والنهاية من جهة القياس إلى الحركة. وذلك إنّما يتفق حيث يكون المبدأ والمنتهى لحركةٍ مستقيمة، يكون الاستمرار فيها لا يجعل المبدأ منتهى ولا المنتهى مبدأ،

that is the case, you also know that the two motions along a single arc are not contraries. [That] is because the motion along those arcs, inasmuch as it is a curved motion, is not posited such that its starting point is essentially different from its endpoint; rather, that is accidental to some posited segment[14] or some chance pause. Otherwise, it would, in fact, continue toward the same starting point. In other words, it is single, continuous motion that does not reverse itself.

(11) Circular motions involving [change] of position—and especially those belonging to a body whose parts are homogeneous, whether taken as a body whose parts are homogeneous or in a body whose parts are homogeneous (where I mean homogeneous in nature and in the position of the parts)—are such that, if they are many and different, they are numerically many and different only because each one of the motions has been completed. [That] is because they begin from some position (once it is actually posited) and end at some position (once it is actually posited), where there is only numeric difference between the two. Also, in the middle it will have positions (once they are actually posited) that are only numerically different from the ones preceding them. Every one of the motions—indeed, its posited starting point, endpoint, and intermediary points—differ from another motion only numerically. So [circular motions involving change of positions] are only numerically different. Now, things that differ only numerically are not contraries, even if it is impossible that they be joined together. It is claimed that, just as the

14. Alternatively, the text's *qaṭ*^{c} might be understood as "traversal."

فذلك هو الذي لا يجتمع. وإذا كان كذلك، فقد عرفت أنَّ الحركتين اللّتين على القوس الواحدة لا تتضادّان، لأنَّ الحركة على تلك القوس لا يُفترض لها - من حيث هي حركة قوسية - أنْ يكون مبدؤها غير منتهاها مغايرة ذاتية، بل يعرض ذلك لقطْع يفرض ووقوفٍ يتفق، ولولا ذلك لصحَّ لها التوجه المستمر إلى المبدأ بعينه، وهي حركةٌ متصلة واحدة لا رجوع فيها.

(١١) والحركات المستديرة الوضعية، وخصوصاً ما يكون منها لجسم متشابه الأجزاء، موضوع على جسم متشابه الأجزاء، أو موضوع في جسم متشابهٍ الأجزاء، أعني المتشابه في الطبيعة وفيٍ وضع الأجزاء، فإنّها حركاتٌ - إنْ تكثّرت وتخالفت - فإنّما تتكثر وتتخالف بالعدد، لأنَّ كل حركةٍ منها تمت فإنّها تبتدىء من وضع، إذا فُرض بالفعل، وتنتهي إلى وضْع إذا فرض بالفعل، لا اختلاف بينهما إلاّ بالعدد، ويكون له في الوسط أوضاع، إذا فرضت بالفعل، لم تكن مخالفة لما قبلها إلاّ بالعدد. وكل حركة منها فإنَّ مبدأها المفروض ومنتهاها المفروض ووسطها المفروض، لا يخالف حركة أخرى إلاّ بالعدد، فهي لا تخالفها إلاّ بالعدد، ولا شيء ممّا لا يتخالف إلاّ بالعدد بأضداد، وإنْ كانت

circular differs from the rectilinear in that it has no actual limit, so likewise it differs from it in that its species of contrariety does not depend upon limits. [Such a claim] comes to naught given what we know—namely, that there is no way for there to be contrariety among motions except on account of the endpoints and limits. So, when the endpoints come to naught, so goes the way of contrariety, and so there is no contrary. From what we have said, then, you also know the situation concerning circular motion and have learned that rectilinear things have contraries. How could they not?! Indeed, descending and ascending are contraries in the aforementioned way, which belongs to motion inasmuch as it is a rectilinear motion, as well as in another way besides that—namely, that the two limits are sometimes contraries by way of being high and low as well. So the motion that has a contrary [call it *x*] is the one that is taken as most closely approximating a distance from some actual limit to some actual limit, and *x*'s contrary is that [motion] that starts at *x*'s endpoint and proceeds toward *x*'s starting point, and toward nothing else.

يستحيل أنْ تجتمع. وأمّا الذي قيل من أنّه كما أنّ المستديرة تخالف المستقيمة في أنها لا طرف لها بالفعل، فكذلك تخالفها في أنّ نوع تضادّها لا يتعلق بالأطراف، فيسقط - بما عرفناه - أنّه لا وجه لتضادّ الحركات إلاّ أنْ يكون بسبب النهايات والأطراف، فإذا سقطت النهايات، سقط وجه التضادّ، فلم يكن ضدّ. فقد علمتَ ممّا قلناه حال الحركة المستديرة. وأمّا المستقيمات فقد عرفت أيضاً أنّها تتضادّ - وكيف لا تتضادّ - وأنْ النازل والصاعد يتضادّ التضادّ المذكور الذي للحركة بما هي حركة مستقيمة، وتتضادّ تضادّاً خارجاً عن ذلك، وهو أنّ الطرفين قد يتضادّان من طريق أنّهما عُلو وسُفْل أيضاً. فالحركة ذات الضدّ هي التي تأخذ أقرب مسافة من طرفٍ بالفعل إلى طرفٍ بالفعل، وضدّها هو الذي يبتديء من منتهاها ذاهباً إلى مبتداها، لا إلى شيءٍ آخر.

Chapter Seven

Of the opposition of motion and rest

(1) The opposition between motion and rest is something I confirmed in what went before.[1] You have also learned that every genus of motion has a rest that is its opposite. We still need to explain the opposition of one rest to another *qua* rest and not inasmuch as a rest is natural and forced, and the other differences external to the substance of the two. So we say that rest is also something in which a certain opposition and contrariety occur on account of the things on which the state of rest depends. Now, when you closely consider the accurate account we presented in the chapter on motion's contrariety,[2] you'll easily understand that the thing producing rest, what is undergoing rest, and the time have nothing to do with that. You'll also know that rest does not depend upon either a starting point or endpoint that is in the [category] of place, although it does depend upon that with respect to which there is [rest].

(2) So it seems that the contrariety of that with respect to which [there is rest] makes rest a certain contrary. Now, that with respect to which [there is rest] admits of contrariety in two ways: contrariety that depends upon its being a space, direction, place, or other analogous terms (which, in general, is a contrariety that depends upon its essence); and

1. See 2.4.
2. See 4.6.

<الفصل السابع>

في تقابل الحركة والسكون

(١) أمّا تقابل ما بين الحركة والسكون، فأمرٌ قد تحققته فيما سلف، وعلمت أنَّ لكل جنس حركةٍ سكوناً يقابله. لكنه قد يجب علينا أنْ نعرف تقابل السكون للسكون، من حيث هو سكون، وسكون لا من حيث هو طبيعي وقسري، وغير ذلك من الفصول الخارجة عن جوهرهما، فنقول: إنَّ السكون أيضاً ممّا تقع فيه مقابلة ومضادّة ما، بسبب الأمور التي يتعلق السكون بها. وإذا تأملت ما اقتصصناه عليك في باب تضادّ الحركات، فعن قريب تعلم أنَّ المسكّن والمتسكّن لا دخل لهما في ذلك، ولا الزمان. وقد علمتَ أنَّ السكون لا يتعلق بمبدأ أو منتهى مكاني، ولكن يتعلّق بما فيه.

(٢) فيشبه أنْ يكون تضادّ ما فيه يجعل السكون متضادّاً، وما فيه بتضادّ على وجهين: تضاداً يتعلق بكونه حيّزاً وجهةً ومكاناً، أو أسماء أُخر ممّا يجري مجراه؛ هي وبالجملة تضادّاً يتعلّق بماهيته، وتضادّاً يتعلق بأمور أخرى مثل أنْ يكون مكاناً حاراً أو

contrariety that depends upon other things—as, for example, that it is a hot or cold place. The latter genus of contrariety is something foreign to rest, which does not bring about any change in the state of resting, such that if there were some body [*x*] in which [another] body [*y*] continues to be at rest, and [if] *x* happened to be heated or cooled or made white or black, [*y*'s] resting in *x* at one moment need not become the contrary of the state of rest in *x* at another moment. Instead, the resting in it continues to be one and the same, because this contrariety is not primarily in the very thing with respect to which there is rest, but in something else. When the contrariety is in the very thing with respect to which [there is rest] in that it is at rest upward at one time (and so where it rests is *up*) and it is at rest downward at another time (and so where it rests is *down*), then the former state of rest is aptly the contrary of the latter state of rest, where resting in the higher place is a contrary of resting in a lower place.

(3) It now remains for you to learn whether the state of rest that opposes moving away from above [that is, downward motion] is to rest above or to rest below. It has been said that resting above is contrary to moving away from above and not to moving above.[3] That is because resting above[4] is sometimes a perfection of the motion upward; but it would be absurd that the natural perfection oppose the thing, while the thing is led to an opposite and contrary. So this is what is said. As for myself, it is not clear to me why something doesn't lead to an opposite in the sense that some opposite immediately follows it. Were that the case, the existence of motion could not lead to the loss of [motion]; but who denies

3. Cf. Aristotle, *Physics* 5.6.230b15–16.

4. Secluding *ilá* (toward) with **Z** and **T**, which appears in only one MS consulted by **Y**.

مكاناً بارداً . وأمّا هذا الجنس من التضادّ ، فهو أمرٌ غريب عن السكون ، لا يغيّر من أمر السكون شيئاً ؛ حتى أنّه لو كان جسم يسكن فيه الجسم سكوناً متصلاً ، وكان يعرض أنْ يسخن أو يبرد ، أو يبيضّ أو يسود ، لم يجب أنْ يصير السكون فيه ، وقتاً مّا ، ضداً للسكون فيه وقتاً آخر ، بل يتصل السكون فيه واحداً بعينه ، لأنَّ هذا التضاد ليس في ذات ما فيه الساكن أولاً ، بل في شيءٍ آخر . وأمّا إذا كان التضادّ في ذات ما فيه ؛ بأنْ كان مرّة يسكن فوق ، فيكون الذي يسكن فيه فوق ، ومرّة يسكن أسفل ، فيكون الذي يسكن فيه أسفل ، فبالحري أنْ يكون هذا السكون مضادّاً لذلك السكون ، ويكون السكون في المكان الأعلى ضدّاً للسكون في المكان الأسفل .

(٣) وقد بقي أنْ تعلم هل السكون الذي يقابل الحركة من فوق هو السكون فوق ، أو السكون أسفل؟ وقد قيل إنَّ السكون فوق ضدّ للحركة من فوق ، لا للحركة إلى فوق ، وذلك لأنَّ السكون فوق قد يكون كمالاً للحركة إلى فوق ، ومحالٌ أن يكون الكمال الطبيعي مقابلاً للشيء ، وأنْ يكون الشيء يؤدي إلى مقابل وضدّ ، فهذا ما يقال . وأمّا أنا فلم يتضح لي أنَّ الشيء لا يؤدي إلى مقابلة ، بمعنى أنّه لا يعقبه مقابلة ، ولو كان كذلك ، لما جاز أنْ يؤدي وجود الحركة إلى فقدانها . ومَنْ ينكر أنَّ الحركة بالطبع إلى فوق إنما

that the natural motion upward is a natural motion upward only so that a natural state of rest occurs at it? So[5] this motion undoubtedly leads to its own loss. It is also not clear to me why resting above is a perfection of the motion in the sense that the *motion* is perfected by that. Quite the contrary, it is a perfection of *what is undergoing the motion,* whereas the motion is corrupted and passes away as a result of it. In other words, it is not the perfection of the motion, but the corruption of the motion. It is a perfection only of what is being moved, which that moved thing comes to have through the motion.

(4) It seems to me that every state of rest that happens to belong to the mobile is an opposite of every motion that is, in fact, in it (should [the motion] in it be replaced by a state of rest). [That] is because [resting] is a certain privation of any motion that was in [the mobile] toward or away from that location. In fact, resting is not the privation of motion inasmuch as it is toward a certain direction; otherwise, whatever is moving in some direction that differs from that one would be at rest. Instead, resting is the absolute privation of motion that is in that genus. The same holds for what is at rest with respect to the categories[6] of place,[7] quantity, and quality. For example, when one and the same place is preserved, the thing is at rest with respect to that place; and when one and the same quality is preserved, the thing is at rest with respect to that quality; and when one and the same magnitude is preserved, the thing is at rest with respect to that magnitude. It is impossible that there be something [of such a nature] that it remains in one and the same place but then cannot undergo any sort of local motion—and the same holds for alteration and the other [types of motion]. If it does not under undergo local motion but does undergo a motion with respect to position—as, for

5. Reading *fa* with **T**, for **Y** and **Z**'s *wa* (and).

6. Literally, "species."

7. Literally, "where" (i.e., the Aristotelian category of *pou*), here and in the following.

هي حركة بالطبع إلى فوق، ليحصل منه سكونٌ بالطبع، فلا شك أنَّ هذه الحركة مؤدية إلى فقدان نفسها. ولم يتضح لي أنَّ السكون فوق ‹هو› كمالٌ للحركة، بمعنى أنَّ الحركة تستكمل بذلك، بل إنّما هو كمالٌ للمتحرك. وأمّا الحركة فإنها تفسد وتبطل به، وذلك ليس كمال الحركة بل فساد الحركة، إنما هو كمالٌ للمتحرك، يحصل للمتحرك بالحركة.

(٤) وعندي أنَّ كل سكونٍ يعرض للمتحرك فهو مقابلٌ لكل حركة تصح فيه، لو كانت فيه بدل السكون، لأنّه عدمٌ لكل حركةٍ تكون فيه إلى ذلك الموضع، أو عن ذلك الموضع. فإنَّ السكون ليس هو عدم الحركة من حيث هو إلى جهةٍ ما؛ وإلاّ لكان المتحرك إلى خلاف تلك الجهة ساكناً، بل السكون عدم الحركة التي في ذلك الجنس مطلقاً. وكذلك الساكن في نوع أينٍ أو كيفٍ أو كمّ، إذا حفظ مثلاً أيناً واحداً؛ فهو ساكن في ذلك الأين، وإذاً حفظ كيفاً واحداً فهو ساكن في ذلك الكيف، وإذا حفظ مقداراً واحداً فهو ساكنٌ في ذلك المقدار. ويستحيل أنْ يكون الشيء يحفظ أيناً واحداً، ثم يكون عادماً لنُقْلةٍ دون نُقْلة، وكذلك في الاستحالة وغيرها. وإنْ كان يجوز أنْ يكون عادماً لنُقْلة وغير عادمٍ

example, something like a celestial sphere that is in another celestial sphere—then, while it is at rest with respect to place, it is absolutely undergoing motion with respect to position. The same holds for quality, for that which is at rest relative to change in quality is that which does not undergo change in quality. Also, that which is at rest relative to change in quantity is that which does not undergo change in quantity. Still, if someone is eager to make every motion as such have some state of rest that is its opposite and is the privation of that motion *qua* that motion, then he has to make resting from motion downward [the opposite and privation of] undergoing motion upward. If he is then eager to make the opposing state of rest that which the estimative faculty imagines to come upon the motion all of a sudden, so as to bring about its privation, then, despite the fact that he is at liberty to be eager, it is not necessary. [That is] because not every privation is the last, and, rather, sometimes it is earlier, from which it follows that resting in some area below would be that which suddenly comes upon the downward motion.[8] If he is then eager to make the opposing state of rest that which suddenly comes upon the motion such that it is like the prior preparedness and privation joined to potency, resting above is the opposite of the upward motion. As for the natural and forced opposition, it seems that resting above is not the opposite of upward motion (since there are two natures), but, rather, it is downward [motion]. The remaining differences by which motions differ are to be met with analogously.

---

8. In other words, if some moving thing had initially been resting in some place *x* that is above certain other places *y, z, ... n,* then, should that thing move downward and rest at any of those places *y, z, ... n,* that later state of rest would be contrary to the state of having rested at *x*.

لحركةٍ في الوضع، مثلاً مثل الفلك الذي يكون في فلكٍ آخر؛ فإنّه من حيث الأين ساكن، ومن حيث الوضع متحرك مطلقاً. وكذلك الحال في الكيف، فإنّ الساكن بقياس التغيّر في الكيف هو الذي لا يتغيّر في الكيف، والساكن بقياس التغيّر في الكمّ هو الذي لا يتغيّر في الكمّ. لكنه إنْ نشط أحدٌ أن يجعل لكل حركةٍ، من حيث هي بصفة، سكوناً يقابلها يكون عدم تلك الحركة – من حيث هي تلك الحركة – لزمه أنْ يجعل المتحرك إلى فوق ساكناً عن الحركة إلى أسفل. فإنْ نشط أنْ يجعل السكون المقابل هو الذي يتوهم طارئاً على الحركة فيعدمها، فمع أنّه يُرخص له في هذا النشاط من غير وجوب، إذ ليس كل عدم يتأخر، بل قد يتقدّم، يلزمه أنْ يكون السكون في ناحية تحت، هو الذي يطرأ على الحركة إلى أسفل. فإنْ نشط أنْ يجعل السكون المقابل هو الذي تطرأ عليه الحركة، حتى يكون كالاستعداد المتقدّم والعدم المقارن للقوة، كان السكون فوق، مقابل الحركة من فوق. وأمّا اعتبار التقابل بالطبيعة والقسرية؛ فيشبه أنْ يكون السكون فوق لا يقابل الحركة إلى فوق لأنّهما طبيعتان، بل التي إلى أسفل. وعلى هذا القياس تورد سائر الفصول التي بها تتخالف الحركات.

Chapter Eight

An explanation of whether one motion can really be continuous with another or whether that is impossible for them, such that there must be a state of rest between them

(1) Having learned how motion is one, and how motions are conjoined, as well as how they are opposite, it is fitting that we note which motions are continuous with which [other] ones, and which ones are not [continuous] and, instead, follow immediately upon one another and in succession. We say without question that, when [motions] of differing genera occur successively in a single subject, they are not as a single, continuous motion. Those that are of the same genera, however, such as one alteration after another or one local motion after another, deserve our further investigation. [That] is because there are serious doubts as to whether a stone's ascending motion is continuous with its descending motion and whether the motion along an arc is continuous with the motion along its chord. In general, are there continuous motions, each one of which is assumed to have some [terminus] *a quo* and *ad quem* of the motion, such that one of them would be an ultimate endpoint and the other a starting point, whether like a point (which is a limit of some distance), or a quality (which is a motion's terminus *ad quem,* or the like)?

<الفصل الثامن>

في بيان حال الحركات في جواز أن يتصل بعضها ببعض اتصالاً موجوداً، أو امتناع ذلك فيها، حتى يكون بينها سكون لا محالة

(١) قد عرفنا أنَّ الحركة كيف تكون واحدة، وكيف تتضام الحركات، وعرفنا أنها كيف تتقابل، فحري بنا أنْ نعلم أن أي الحركات تتصل بأي الحركات، وأيّها لا تتصل، بل تتشافع وتتالى، فنقول: أمّا المختلفة الأجناس، فلا شك أنّها إذا تعاقبت على موضوع واحد لم تكن على أنّها حركة واحدة بالاتصال؛ وأمّا المتفقة الأجناس، كاستحالة واستحالة، ونُقْلة ونُقْلة، فخليق بنا أنْ نحقّق الأمر في ذلك. فإنّه مما يعظم فيه الشك أنّه هل تتصل حركة الحجر الصاعدة بحركته النازلة، والحركة على قوس بالحركة على وترها وبالجملة، هل تتصل الحركتان اللتان يُفرض لكل واحدٍ منهما شيء عنه وإليه الحركة، فيكون لأحديهما غاية، وللآخر مبدأ، كنقطةٍ هي طرف مسافة، أو كيفية هي نهاية حركة إليها، أو غير ذلك فإنَّ قوماً جوّزوا هذا الاتصال، وقوماً لم يجوزوا، وأوجبوا أنْ يكون

Now, there is one group that allows this continuity, while there is another that does not, but requires that there be a rest between instances of these motions; and both those who allow it and those who deny it have their arguments. So let us list and explain [those arguments] and then present our own view.

(2) One of the arguments of those who allow it is to say: Consider a millstone that is either hurled upward or is falling downward, and, during its course, a very small pebble is going in the opposite direction so that it comes into contact with it. Does that pebble[1] first come to rest and then, thereafter, start its contrary motion, or are there two motions that are continuous with one another? If there is a rest, it follows that an ascending pebble must arrest the millstone's descending motion, which is absurd, whereas, if the two motions [namely, the pebble's ascending and then descending] are continuous, then the position of those who deny that has been refuted.

(3) They further said that it would be absurd for that state of rest to occur without there being some reason or other for it. If there is, then, some reason for it, it is because of either something's absence or [its] existence. If the reason for it is an absence of something—namely, the absence of some cause that brings about the motion—then there does not have to be in that body that is thrown upward, for instance, some principle of moving downward. So it ought not to move [downward] until its substance underwent a change, but that is not the case. If the reason [for resting] is some existing thing, it prevents the motion by being either some external force, or nature, or some internal psychological act of the will, none of which is the case.

---

1. Reading *tilka ḥaṣāh* with **Z** and **T** for **Y**'s simple *tilka* (that).

بين أمثال هذه الحركات سكون، وللمجوّزين حجج، وللمانعين حجج، فلنعدّها ولنكشف عنها، ثم لنورد ما عندنا.

(٢) فمن حجج المجوّزين قولهم. أرأيتم حجر رحى يُرمى إلى فوق، أو ينزل إلى أسفل، وتعارضه في مسلكه حصاة صغيرة حتى تماسّه، أتسكن تلك الحصاة أولاً، ثم تأخذ في ضد حركتها، أم تتصل الحركتان معاً؟ فإنْ سكن، وجب من ذلك أنْ تكون الرحى تحبسها حصاة صاعدة، عن الحركة النازلة لها، وهذا محال، وإنّ اتصلت الحركتان فقد بطل مذهب مَنْ يمنع ذلك.

(٣) وقالوا أيضاً؛ إنَّ ذلك السكون من المحال أنْ يحصل من غير أنْ يكون له سببٌ بوجهٍ من الوجوه، ثم إنْ كان له سببٌ، فإمّا أن يكون سبباً عدمياً أو يكون سبباً وجودياً، فإنْ كان سببه عدمياً - وهو عدم سبب التحريك - فيجب أنْ لا يكون في ذلك الجسم المرمي إلى فوق مثلاً مبدأ حركةٍ إلى أسفل، فينبغي أنْ لا يتحرك إلاّ أنْ يتغيّر جوهره، وليس الأمر كذلك. وإنْ كان السبب وجودياً فهو شيءٌ مانعٌ عن الحركة، إمّا قسري من خارج، وإمّا طبيعي أو إرادي نفساني من داخل، وجميع ذلك ليس.

(4) They also say that it is *not impossible* that something should come in contact with and depart from some particular thing in an instant and not remain in contact with it for a period of time so as to be at rest at it. Thus, a mainstay of the argumentation used by those who want to establish that there is rest is unsound, for they depend upon its being *impossible* that one and the same thing be in contact and then depart in a single instant. This [possibility], they said, presents itself [in, for example, the case of] a sphere mounted on a rotating wheel. So [for example] when one supposes a two-dimensional surface[2] above [the sphere] such that, on ascending, [the sphere] meets [the surface] and then departs from it, [the sphere] will be in contact with that surface at a point, but after that it does not continue to remain in contact with it for some period of time.

(5) As for those who deny that,[3] one of their arguments is that it is impossible for one and the same thing to be in contact actually with some definite and distinct end, except at two instants. Between any two instants, however, there is a period of time, and during that time there is no motion, in which case there is rest during it.

(6) Furthermore, they said that if it were possible for ascending to be continuous with descending as a single thing, then, from the two motions, there would come to be a single continuous motion, because it is the continuity that [explains] there being a single motion. It would be necessary, then, that two contrary motions are a single motion, which is absurd.

(7) Likewise, they maintained that, if it were possible for the two motions to be continuous, then the end of the upward motion, which then turns downward, must always terminate at that from which it began. In that case, the starting point of the rectilinear motion that retreats from some space is the very [space] intended by that retreat.

---

2. Literally, "a simple surface," but Avicenna uses *saṭḥ basīṭ* regularly to identify the innermost containing surface that is a thing's place, which is not three-dimensional so as to be a body.

3. Most of the arguments for this group are derived from Aristotle, *Physics* 8.8.

(٤) وقالوا أيضاً ، إنَّه لا يمنع أنْ يكون شـــيء يماسّ شـــيئاً معينـــاً في آنٍ ويفارقه ، ولا يبقى مماسّـــاً له زماناً حتى يكون ســـاكناً فيه. فلا يصح ما هو عمدة احتجاج مثبتي الســـكون ، فإنّهم يتعلقون بأنَّه لا يجوز أنْ يقع في آنٍ واحدٍ مماسّة ثم مفارقة. قالوا ؛ وهذا مثل كرة مركبة على دولابٍ دائر ، فإنها إذا فرض فوقها ســـطح بســـيط ؛ بحيث يلقاه عند الصعود ، ثم يفارقه ، فإنّها تماس حينئذ ذلك الســـطح بنقطة ، ولا تبقى مماسّة له بعد ذلك زماناً لا يزول.

(٥) وأمّـــا المانعون عن ذلك ، فمن حججهم أنَّ الشـــيء الواحد لا يجوز أنْ يكون مماسّـــاً بالفعل لغايةٍ معيّنة ومبايناً ، إلاّ في آنين ، وبين كل آنين زمانٌ ، وذلك الزمان لا حركة فيه ، ففيه سكون.

(٦) وقالوا أيضاً ؛ إنَّه لو كان اتصال الصاعد بالهابط شيئاً واحداً ، لكانت الحركتان تحـــدث منهمـــا حركة واحدة بالاتصال ؛ لأنَّ وحدة الحركة هـــي الاتصال ، فكان يجب أنْ تكون الحركتان المتضادتّان حركة واحدة ، وهذا محال.

(٧) وقالوا أيضاً ، لو جاز اتصال الحركتين لكان يجب أنْ تكون غاية الصاعد العائد هابطاً هي أنْ ينتهي في حركته مســـتمراً إلى ما عنه ابتدأ ، فيكون مبدأ الحركة المستقيمة الهاربة عن حيّز هو بعينه المقصود بذلك الهرب.

(8) Finally, they said that when something is in a process of becoming white, and then [at some instant *x*] it [actually] becomes white but [immediately starts] a process of becoming black, then, inasmuch as it is becoming black, there is some blackness in it. As such, however, there is a potential to be white in it. In that case, then, despite the fact that [at the instant *x*] it has [actually] become white, it would also be potentially white, which is absurd.

(9) These and similar things are the basis upon which the two groups argue, but the argumentation of neither one of them is outstanding (even if the second school of thought is true). The fact is that they have entrusted us with no demonstration such that it either completely satisfies us or brings us to a level of understanding that removes all doubts. [Consequently], those who advocate a state of rest took to task the arguments of the former [group].

(10) Concerning the report about the pebble,[4] either the air being pushed in front of the millstone turns the pebble around before the two come into contact (in which case, that state of rest occurs in the air before the contact) or it is not turned around until it meets the millstone (in which case, even if [the notion is] loathsome, it is not impossible that the millstone does stop, owing to the impossibility that the [pebble's upward and downward] motions should be continuous). That is similar to what happens owing to the impossibility of the void.[5] Indeed, it is not improbable that the existence of something necessary will render inactive that which can be so rendered or oppose that which can be so opposed, where there will be an amount of time during which there is the rendering inactive and opposing corresponding with the relation of the acting and being acted upon.

4. See. par. 2.
5. See for instance 2.8.25.

(٨) وقالوا أيضاً ، إنَّه إذا كان الشيء يبيض فابيَضّ وهو يتسوّد ، فمن حيث يتسوّد ففيه سواد ، ومن حيث هو كذلك ففيه قوة على البياض ، فيكون مع أنَّه أبيض ، فيه قوة على البياض ، وهذا محال .

(٩) فهذه الأشياء وما يشبهها عمدة ما يحتج به الفريقان ، وليس - لا واحد منهما - حسن الاحتجاج ، وإنْ كان المذهب الثاني هو الحق . لكنهم لم يتركوا لنا برهاناً أقاموه عليه ، بحيث نقتنع به أو لم يفهموناه تفهيماً يتعرضون به لأنْ يقع على وجهٍ يزيل الشكوك ، فلهؤلاء القائلين بالسكون أنْ ينقضوا ما احتج به أولئك .

(١٠) أمّا حديث الحصاة ، فإنَّها لا تخلو إمّا أنْ يكون الهواء المندفع أمام الرحى يصرف الحصاة قبل أنْ تقع بينهما مماسّة ، فحينئذ يكون ذلك السكون واقعاً في الهواء قبل المماسّة ، وإمّا أنْ لا يكون بحيث يصرفه حتى يلقى حجر الرحى ، فحينئذ لا يستحيل ، وإنْ كان شنعاً ، أنْ تتوقف الرحى لاستحالة اتصال الحركتين ، كما يقع مثل ذلك لاستحالة الخلاء . فإنَّ الأمر الواجب وجوده لا يبعد أنْ يبطل ما من شأنه أنْ يبطل ، أو يمنع ما من شأنه أنْ يمتنع ، ويكون القدر من الزمان ، الذي فيه الإبطال والمنْع ، بحسب مناسبة الفعل والانفعال .

(11) Against the other argument,[6] they can say that the reason for [the resting] is a certain absence of something—that is, the absence of the inclination opposing the motive power, for this motive power brings about motion only by producing an inclination. It is known, however, that when [the motive power] is in its natural place, it has no inclination there toward a given direction at all, and [yet] that power does exist. Thus, with respect to some direction other than [a thing's natural direction] toward which it is thrown by a forced inclination, the accidental forced inclination can, at times, oppose the inclination that [the motive] power naturally produces, from which it necessarily follows that it would not be undergoing motion. This is like when water is vigorously heated by a foreign [source], after which the natural coolness of the water is prevented from arising from the water's nature; for we know that a foreign inclination can overpower and bring about the absence of the natural inclination and prevent the natural motion. At the end of the [forced] motion, then, there might be a remnant of the foreign inclination that is just enough to prevent the natural power from producing the natural inclination. Despite that, it will be too weak to be able to bring about motion in that direction [opposite of the natural motion]. In fact, it will be so weak from causing motion that it no longer produces motion, and, [yet] it is not too weak to prevent the nature from producing the inclination [that will move it toward its natural direction]. The foreign inclination will not be able to produce motion while overcoming the natural motion, but neither will the natural power be able to produce the natural inclination until that remnant of foreign inclination passes away, either through itself or through some other cause. Something like this might also be observed between two opposing things when they rival one another for

6. See par. 3.

(١١) وأمّا الحجة الأخرى، فيجوز أنْ يقولوا عليها: إنَّ السبب فيه سببٌ عدمي، وهو عدم حدوث الميل عن القوة المحركة. فإنَّ هذه القوة المحركة إنّما تحرّك بإحداث ميلٍ، وقد علم أنَّها إذا كانت في مكانها الطبيعي لم يكن لها هناك ميلٌ إلى جهةٍ البتة، وتلك القوة موجودة. فلذلك يجوز في الجهة الأخرى، التي ترامت إليها بميلٍ قاسر، أنْ تكون تارة ممنوعة عن الميل الذي يحدثه بالطبع بمعارضة الميل القسري، ويلزم ذلك أنْ لا تتحرك؛ وذلك كسخونة الماء الغريبة إذا كانت قوية بعد، فإنّها مانعة عن أنْ ينبعث عن طبيعة الماء برده الطبيعي. فإنّا نعلم أنَّ الميل الغريب يستولي على الميل الطبيعي ويعدمه ويمنع الحركة الطبيعية، فيجوز أن يكون عند انتهاء الحركة بقية من الميل الغريب بقدر ما يمنع القوة الطبيعية عن إحداث الميل الطبيعي، ويكون أضعف من أنْ يقوى مع تلك الممانعة، على التحريك في تلك الجهة، بل يضعف عن التحريك فلا يحرّك، ولا يضعف عن ممانعة الطبيعة من إحداث الميل. فلا الميل الغريب يقوى على التحريك غالباً للقوة الطبيعية، ولا القوة الطبيعية تقوى على إحداث الميل الطبيعي، إلى أنْ تبطل تلك البقية من الميل الغريب بنفسها، أو يبطلها سببٌ آخر. ومثل هذا قد يشاهد بين المتقاومين أيضاً إذا تنازعا في

other reasons. In that case, this sometimes prevents the motions, while at other times it is prevented because of the need to rest for a period of time, after which the natural inclination arises when there is the production of motion. So not every inclination, as it just comes to be *qua* inclination,[7] comes to be together with a motion. Instead, it might either be too weak for that or be tainted with the opposite so as exactly to equal each other out until the taint is removed. This is like the inclination that occurs in a heavy load that nine porters apply themselves to, but [only] when a tenth joins them, is it [able to] be carried. In this case, the nine necessitate one inclination in it while depriving it of another, except that that inclination does not complete what is needed to carry [the load], and, in fact, an additional [porter] is needed. It also may be said that some existing thing is the reason for it—namely, some accidental factor. That is, the mover provides a certain foreign power by which the body is moved and indirectly provides a certain power that will bring [the body] to rest. Again, it will be some factor like the contrary of the inclination and a certain contrary form. It will be some foreign factor by which the body remains in some place in which it is and, likewise, leaves its place as a result of the inclination, in some cases forcibly and in others naturally, just as some inclination is forced and some natural.

(12) It has been said against the wheel argument[8] that no point in the true sense belongs to spheres occurring in nature and that [spheres] are [only] contacted at a surface. This does not particularly impress me. The more appropriate response is, instead, that wherever there is a true sphere either only a sphere surrounds it or it is not surrounded at all as in

7. Reading *maylan* with **Z** and **T** for **Y**'s *mathalan* (for example).

8. See. par. 4.

معانٍ أخرى، فيكون الامتناع عن الحركة تارة لهذا، وتارة يكون الامتناع بسبب وجوب السكون زماناً، بعده ينبعث الميل الطبيعي إذا وجد التحريك. فليس كل ميل، كلّما حصل ميلاً، حصلت معه حركة، بل ربما كان أضعف من ذلك، أو مشوباً بالمقابل شوب المتوسطات إلى أنْ يصفو. وهذا مثل الميل الذي يحصل في حِمْلٍ يتناوله محركون تسعة، فإذا انضمّ إليهم العاشر إستقلّ، فإنّ التسعة قد أوجبوا فيه ميلاً ما وأعدموا ميلاً، إلا أنَّ الحاجة لا تتم بذلك الميل في الإستقلال بل تحتاج إلى زيادة. ويجوز أنْ يقال إنَّ السبب فيه معنى وجودي، وهو أمرٌ عرضي أيضاً، وهو أنْ يكون المحرّك يفيد قوة غريبة يتحرك بها الجسم، وبتوسطها يفيد قوة مسكّنة؛ وهو أمرٌ كالمضادّ للميل، وصورة مضادّة، إنّه أمرٌ غريب به يحفظ الجسم مكان ما هو فيه، كما بالميل يترك مكانه، فيكون منه قسري وطبيعي، كما يكون من الميل قسري وطبيعي.

(١٢) وأمّا الحجة الدولابية؛ فقد قيل عليها أنَّ الكرة الطبيعية لا نقطة حقيقية لها، وأنها تماسّ بسطح. وهذا لا يعجبني، بل الجواب الأصوب أنّه حيث تكون كرة حقيقية فلا تكون إلاّ محاطة بكرة، أو لا محيط لها كما في السماوات، ولا يمكن معها هذا العمل –

the case of the Heavens. This practical application cannot work with [a real sphere], and, in the case where it can work, there is not a real sphere. Even if it were [a real sphere], it may well be impossible that, instantaneously, there is contact and the cessation [of the contact], and, because of that impossibility, there would necessarily be a brief pause. Additionally, between the sphere and the flat surface there either is or is not a void. Now, there cannot be a void between the sphere and the flat surface, and so there must be a plenum between them. Since there is a plenum between them, there will be the surface of that plenum that meets the flat surface, which is a two-dimensional surface,[9] and another surface that meets the convex [surface] of the sphere. It is impossible, however, that, on the outside surface of [the plenum], there be some point belonging to another body that is foreign [to the plenum]. [That] is because the point on a two-dimensional [plane] has no determinate position distinct from that two-dimensional [plane]. Consequently, contact does not occur between the surface and the flat plane at a point; but it was assumed that there was contact. This is absurd. Even setting all this aside, this [argument of theirs] makes the laws of nature dependent upon certain mathematical abstractions of the estimative faculty, which is not right. In fact, beyond going outside the discipline [of physics], that [argument] doesn't even entail what [they] wanted it to prove, but only requires that the continuity of the two designated motions be in the estimative faculty. We, however, don't deny that that continuity is in the estimative faculty. We deny that only of the natural things that deviate from the abstractions of the estimative faculty.

9. Literally, "a simple surface," but see note 2.

من حيث يمكن هذا العمل - فلا تكون كرة حقيقية. ولو كانت، فربّما استحال أنْ تماسّ دفعة وتزول، ووجب أنْ تقف وقفة ما، لاستحالة ذلك. ومع ذلك فلا يخلو، إمّا أنْ يكون هناك بين الكرة والصفيحة خلاء، أو لايكون. ويستحيل أن يكون بين الكرة والصفيحة خلاء، فيجب أن يكون بينهما ملاء، فإن كان بينهما ملاء كان سطح ذلك الملاء الملاقي يلاقي الصفيحة وهو بسيط مسطّح، وسطح آخر يلاقي تقبيب الكرة، ولا يجوز أنْ يكون في وجهه نقطة غريبة من جسم آخر، فإنَّ النقطة لا يتعيّن لها في البسيط وضْعٌ متميّزٌ عن أنْ يكون من ذلك البسيط، وإذا كان كذلك، لم تقع مماسّة بين الكرة وبين الصفيحة بالنقطة - وفرضت مماسّة - وذلك محال. على أنَّ هذا تعليقٌ لأحكام طبيعية بأوهام رياضية، وهو غير صواب، فإنَّ ذلك - مع أنّه خروجٌ عن الصناعة - فليس يلزم منه المراد على ما بيّناه، إلاّ أنْ يوجب منه اتصال الحركتين المذكورتين في الوهم، ونحن لا نمنع من اتصال الحركتين المذكورتين في الوهم، إنّما نمنع ذلك في الأمور الطبيعية الخارجة عن الأوهام.

(13) In their turn, the former group [namely, those who denied that there must be a moment of rest between two contrary motions] can go on the offense and take the arguments of the latter group to task. As for the first one[10] it is sophistical. That is because one means by *the instant at which there is separation* the limit of the time at which there is separation. So it is the limit of the time of separation, where [the separation] is the motion. So that [limit] is, itself, the instant at which there [still] was contact. It is not impossible, then, that the limit of the period of time of the motion be something at which there is no motion and, in fact, be different from the motion, and that the limit of the period of time of the separation be, itself, the instant of separation at which there is no separation. If by [*the instant at which there is separation*] one means some instant at which it is correct to say that the thing is separated, then it is true that there is a period of time between the two [instants]. Still, that period of time is the time during which it was moved from being in contact to that interval [of separation], which is not the time that it is at rest. [That] is especially [the case], given that, according to their school of thought, motion, separation, and what is analogous have no first [part of] moving or separating.[11] The same holds [even] if they drop the term *separating* and replace it with *not contacting,* for there can be contact at the limit of the time during the whole of which there is no contact. (Earlier, we provided an explanation upon which the truth of this position relies, and so let it be consulted.)[12] Be that as it may, all of that falls away when that with respect to which something is moved (I mean the distance) happens to have certain actual divisions already in it, in that [for instance] some of it is black and some white, or its parts are, as it were, striated, so that there are actual limiting points. Still, it is not

10. See. par. 5.
11. Cf. 3.6.3–6.
12. See 2.12.4.

(١٣) ثم لأولئك أنْ يعودوا وينقضوا حجج هؤلاء . أمّا الأولى فلأنّها سوفسطائية؛ وذلك لأنَّه أمّا أنْ يعني بالآن الذي يكون فيه مبايناً ، طرف الزمان الذي يكون فيه مبايناً ، فيكون طرف زمان المباينة – التي هي الحركة ، – فيكون ذلك بعينه الآن الذي كان فيه مماسّاً ، فلا يمنع أنْ يكون طرف زمان الحركة شيئاً ما ليس فيه حركة ، بل فيه أمرٌ مخالفٌ للحركة . وأنْ يكون طرف زمان المباينة هو نفس آن المماسّة ، وليس فيه مباينة . وإنْ عني به آنٌ يصدق فيه القول إنَّ الشيء مباينٌ ، فحقَّ أنَّ بينهما زماناً ، لكنه الزمان الذي تحرّك فيه من المماسّة إلى ذلك البُعْد ، وليس ذلك الزمان زمان السكون ، وخصوصاً – ومن مذهبهم – أنَّ الحركة والمباينة وما يجري ذلك المجرى ، ليس له أول ما يكون حركة ومباينة . وكذلك إنْ تركوا لفظة المباينة وأوردوا بدلها لا مماسّة ، فإنَّه يجوز أنْ تكون في طرف الزمان الذي في كله لا مماسّة ؛ مماسّةٌ . وقد سلف منا بيان يتعلّق به تحقّق هذا المكان ، فليستعن به . وعلى أنَّ جميع ذلك ينتقض إذا كان المتحرك فيه – أعني المسافة – قد عرض فيه فصولٌ بالفعل ؛ بأنْ صار بعضه أسود وبعضه أبيض ، أو كأن أجزاءه منضودة على التماسّ ، فكان هناك حدود بالفعل . لكنه ليس يبعد أنْ يقال إنَّه إذا عرض ذلك ، وجب أنْ يقع عند

completely outlandish to say that, if that happens, brief pauses must occur at the actual divisions, and that the motion is slower than it would have been had there been none. I suppose that some of them said that what is traversed is like that. As for that with respect to which there are accidental termini—as, for example, between white and black—they do not have limiting points relative to what is undergoing the motion, but relative to those qualities, and relative to [the fact] that it is something continuous, as if there is neither white nor black in it. I don't find this too astonishing, for, if the obstacle that they mentioned is not something relative, but owing to the actual existence of some thing at which it arrives and from which it departs (and this status undoubtedly exists in it), then, between black and white, there is an actual limiting point. Granted, when that does not exist, then neither is there any actual limiting point at all, except for the distance's limit, whether absolutely (in which case it is its extreme) or inasmuch as there is some distance (in which case it is its extreme as well as something other than its extreme—I mean wherever it pauses, even if it has not reached the limit of the distance *qua* interval).

(14) As for the second argument,[13] the former group can say that the motion's being a single motion is not according to just any chance manner of continuity, just as the line's being a single line is not according to just any chance manner of continuity. Instead, the continuity is one that makes magnitudes and the like to be a single thing—that is, the continuity in which there exists no actual common division. As for the continuity in the sense of sharing a common limit, that [sort] does not make lines, motions, and the like a single thing in such a way that there is no multiplicity in them actually (although perhaps potentially). Otherwise, a

13. See par. 6.

الفصـــول بالفعل وقفاتٌ ، وتكون الحركة أبطـــأ منها لو لم تكن . وأظنُّ أنَّ بعضهم قال : أمّا المقطوع فكذلك ، وأمّا ما تكون النهايات فيه بالعرض - كما بين الســـواد والبياض - فإنَّ الشـــيء لا يكون بالقياس إلى المتحرك ذا حدود ، بل بالقيـــاس إلى تلك الكيفيات ، وهو بالقياس إلى ذلك ، متصلٌ كأنَّه لا بياض فيه ولا سواد . وهذا ليس يعجبني ، فإنَّه إنْ لم يكن المانع الذي أوردوه أمراً بالقياس إلى شيء ، بل كان لوجود أمر بالفعل موصل إليه ومنفصل عنه ، وها هنا ذلك الحكم موجودٌ لا شك فيه ، فها هنا حدٌّ بالفعل بين السواد والبياض. ومسلّم أنَّه إذا لم يكن ذلك ، لم يكن ذلك ، لم يكن حدٌّ بالفعل البتة إلاّ طرف المسافة ، إمّا على الإطلاق فهو آخره ، وإمّا من حيث هو مســـافة؛ فهو آخره وغير آخره أيضاً ، أعني حيث يقف عليه المتحرك ، وإنْ لم ينته إلى طرف المسافة من حيث هو بعد .

(١٤) وأمّـــا الحجة الثانيـــة ، فلأولئك أنْ يقولوا إنَّ الحركة الواحدة؛ ليســـت تكون واحـــدة على أي نمط من الإتصال اتفق ، كما أنَّ الخط الواحد ليس يكون واحداً على أي نمطٍ مـــن الاتصال اتفق؛ بل الاتصال الموحد للمقادير وما يشـــبهها ، وهو الاتصال المعدوم فيه الفصْل المشـــترك بالفعل . وأمّا الاتصال الذي يكون بمعنى الاشـــتراك في طرف ، فذلك لا يجعل الخطوط والحركات وغير ذلك شـــيئاً واحداً - الوحدة التي لا كثرة فيها بالفعل ،

single line would really surround the triangle. Earlier, we ourselves gave a complete identification of the ways that *continuity* is said, and you learned that some continuity is unified and some is discrete.[14] So, then, these two motions [namely, ascending and then descending] are not a single motion by [way of] the unified continuity, but are two motions between which there is a discrete continuity. This continuity is a continuity of one thing with another through an actually existing limit that is commonly shared between them; and whenever there is not an actual two-ness, there will not actually be this continuity. This continuity is, instead, like two lines that meet at some angle that is itself an actual point. Therefore, this continuity is not the unified continuity, but the discrete continuity, where the status of this continuity is like black's being continuous with white. Through this, you also know the error in the next argument[15]—namely, that the final end would be, as it were, the same point from which its starts only if there were a unified continuity, not a discrete one. Things that are discrete and follow one another successively might be able to have a number of final ends successively.

(15) The last[16] argument[17] is fatuous. That is because, at the moment [when something actually] becomes white, it is not said to be in a process of becoming black; rather, that follows thereafter in a period of time whose limit is that instant at which it is white. Additionally, their argumentation is unsustainable if someone should maintain that this actual white is likewise potentially some other [shade of] white, because another white different from this white inheres in it potentially. Between the two, then, there would be a period of time that differentiates them. So relative to this existing white, there is no potentiality to be it, whereas, relative to some future white, there is.

14. See 3.2.8–10.
15. See par. 7.
16. Reading *akhīrah* with **Z** and **T** for **Y**'s *ukhrá* (other).
17. See par. 9.

بل عســى بالقوة، وإلاّ فالمثلث يحيط به خطٌّ واحدٌ بالحقيقة. وقد فرغنا نحن سالفاً عن تحقيق وجوه ما يقال عليه الاتصال، وعرفت أنَّ الاتصال منه موحّد ومنه مفرّق، فلا تكون إذن هاتــان الحركتان حركة واحدة بالاتصال الموحد، بل حركتــان إثنتان بينهما الاتصال المفرّق. فإنَّ هذا الإتصال هو اتصال شــيء بشيءٍ بطرفٍ موحودٍ بالفعل مشتركٍ بينهما، وما لم تكن إثنينية بالفعل، لم يكن هذا الاتصال بالفعل، بل هذا الاتصال يكون مثل خطّين ملتقيــين على زاوية ذات نقطة بالفعل؛ فهذا الاتصــال إذن ليس هو الاتصال الموحّد، بل الاتصال المفرّق، وحكم هذا الاتصال كاتصال السواد بالبياض، وبهذا تعلم أيضاً الغلط في الحجّــة التي تتلوها. وأنَّه إنّما، كأنْ تكون الغاية هي بعينها المبدأ، لو كان اتصالٌ موحّد لا مفرّق، والأشياء المتفرّقة المتتالية قد يجوز أنْ تكون منها غايات بعد غايات.

(١٥) وأمّا الحجّة الأخيرة فهي ســخيفة، وذلك أنَّــه عندما صار أبيض لا يقال إنَّه يتســود، بل ذلك بعده في زمان، طرفه هو ذلك الآن الذي هو فيه أبيض. ومع ذلك فلا يستمر احتجاجهم إذا قال قائل: إنَّ هذا الأبيض بالفعل هو بالقوة أبيض آخر أيضاً، لأنَّه في قوتــه أنْ يحل فيــه بياض آخر غير هذا البياض، وقد تخلّلهما زمانٌ يفصل بينهما، فيكون بالقيــاس إلى هذا البياض الموجود لا قوة له عليه، وبالقياس إلى بياضٍ ينتظر له قوة عليه.

(16) Having indicated and explained the arguments of these groups, we should reveal the argument for why we hold to the school of thought that we do. So we say that every motion in the proper sense proceeds from an inclination that is confirmed by either the repulsion of the thing standing in the way of the moving thing or the power needed [by the thing standing in the way] to hinder [the moving thing]. This inclination is, in itself, one of the things by which [the moving thing] reaches the motions' limiting points—namely, by freeing itself of anything that clings to it, pushing away whatever is in the path of the motion, and advancing toward something. Now, it is absurd that what arrives at a given limiting point should do so without some existing cause that makes it arrive. It is also absurd that this cause should not depart from the initial place of rest. This cause is proportional to what causes the departure and what pushes away, and it is that proportion that is called *inclination.* Indeed, this thing, inasmuch as it is what makes it arrive (even if the subject is one and the same), is not called an *inclination,* whereas this thing, as something that might exist at a single instant (while the motion is only that which might need a continuous time in order to exist), is called an *inclination.* Now, as long as the inclination is not constrained, repressed, or corrupted, the motion that necessarily results from it exists, whereas when the inclination is corrupted, its corruption is not itself the existence of another inclination, but, rather, that is something else that is perhaps joined with it. Also, when two motions come to be, the result is from two inclinations. Now, when the other inclination toward a different direction exists, [the inclination] is not itself this thing that brings about the arrival such that there would simultaneously be one and the same cause for [both] arriving at and

(١٦) وإذ قد أوضحنا حجج هؤلاء؛ فبالحري أنْ نعرف نحن الحجّة التي لأجلها تمسكنا بأحد المذهبين؛ فنقول: إنَّ كل حركة بالحقيقة فهي تصدر عن ميلٍ يحقّقه اندفاع الشيء القائم أمام المتحرك أو احتياجه إلى قوةٍ تمانعه. وهذا الميل، في نفسه معنى من الأمور به يوصل إلى حدود الحركات، وذلك بإبعاده عن شيءٍ يلزمه مدافعة لما في وجه الحركة، وتقريب من شيء. ومحال أنْ يكون الواصل إلى حدّ ما واصلاً بلا علّة موجودة موصلة، ومحال أنْ تكون هذه العِلّة غير التي زالت عن المستقر الأول، وهذه العلّة يكون لها قياسٌ إلى ما يزيل ويدافع، وبذلك القياس يسمّى ميلاً، فإنَّ هذا الشيء - من حيث هو موصل - لا يسمّى ميلاً، وإنْ كان الموضوع واحداً. وهذا الشيء الذي يسمّى ميلاً قد يكون موجوداً في آنٍ واحدٍ؛ وإنّما الحركة هي التي عسى أنْ تحتاج في وجودها إلى اتصال زمان، والميل ما لم يقسر ولم يقمْع، أو لم يفسد، فإنَّ الحركة التي تجب عنه تكون موجودة. وإذا فسد الميل، لم يكن فساده هو نفس وجود ميلٍ آخر، بل ذلك معنى آخر ربّما قارنه. فإذا حدثت حركتان فعن ميلين، وإذا وجد ميل آخر إلى جهةٍ أخرى، فليس يكون هو هذا الموصل نفسه؛ فيكون هو بعينه علّة للتحصيل وللمفارقة معاً. بل يحدث -

departing from [some point, like at the top of a projectile motion's trajectory]. Instead, some other inclination must come to be that has some first [instant] of coming to be. Now, [that inclination] exists at that first [instant], since its existence does not depend upon some time that no longer is. [In this respect, inclination] is not like motion and rest, which do not have a first [instant] of coming to be,[18] since they never exist except during or after a period of time. [That] is because they require an instant before which the body did not exist at it and after which it will not exist at it. So they require[19] a prior and posterior time. [Inclination], rather, is like the nonmotion that is at each instant. So, just like that instant that might function as motion's limit, [inclination] might itself be the limiting point of the nonmotion, such that there is no motion, [and yet] it exists at some instant as a limit of the continuous motion that will exist after it. In that case, between the motion and the nonmotion, there won't need to be one instant after another, but a single instant is enough. No absurdity arises because, at that instant, there is not simultaneously motion and rest, but only one of them.

(17) The instant at which there is the first existence of the second inclination, however, is not the instant at which there is the last existence of the first inclination, since, as we have explained, it is the last existence of the first inclination at which it exists when bringing about the arrival. So, if, on the one hand, it is found to be something that brings about the arrival over a period of time, then there would, in fact, have been a state of rest. If, on the other hand, it is found to be something that brings about the arrival only in an instant, that instant will be the last only if it is the last of what exists at it, since the last of it is something bringing about the arrival, and it does not bring about the arrival when it does not occur. The two instants were not one [and the same] precisely because it is not in something's nature to be that which

18. See 3.6.3–6.

19. Reading *yaqtaḍī* with **Z** and **T** for **Y**'s *qanqaḍī* (are completed).

لا محالة - ميلٌ آخر له أول حدوث ، وهو في ذلك الأول موجود ، إذ ليس وجوده متعلقاً بزمانٍ ليس ؛ كالحركة والسكون اللّذين ليس لهما أول حدوث ، إذ لا يوجدان على وجهٍ ما إلاّ في زمان ، وإلاّ بعد زمان ، إذ هي مقتضية لآنٍ لم يكن الجسم قبله فيه ولا يكون بعده فيه ، فيقتضي تقدماً وتأخراً زمانياً ، بل هو كاللاحركة التي تكون في كل آن . فكذلك الآن الذي قد يحدّ طرف الحركة ، يجوز أنْ يكون هو بعينه حدّ اللاحركة حتى يكون لا حركة ، موجوداً في آنٍ هو طرف حركة مستمرة الوجود بعده ، فلا يحتاج بين الحركة وبين اللاحركة إلى آنٍ وآن ، بل يكفي آنٌ واحدٌ ، ولا يعرض محال . لأنَّ ذلك الآن لا تكون فيه الحركة والسكون معاً ، بل واحد منهما .

(١٧) وأمّا الآن الذي فيه أول وجود الميل الثاني ، فليس هو الآن الذي فيه آخر وجود الميل الأول ، إذ هو آخر وجود الميل الأول الذي بيّنا أنّه يكون فيه موجوداً عندما يكون موصلاّ . فإنْ كان يوجد موصلاّ زماناً ، فقد صحّ السكون ، وإنْ كان لا يوجد موصلاّ إلاّ آناً ، فليس ذلك الآن آخر ، إلاّ أنْ يكون ما هو له آخر موجوداً ، فيه ، إذ ما هو له آخر هو موصل ، والموصل لا يكون موصلاً وهو غير حاصل . وإنّما لم يكن الآنان شيئاً واحداً ، لأنَّ الشيء لا يكون في طبيعته ما يوجب الحصول وما يوجب اللاحصول معاً ،

simultaneously necessitates occurring and not occurring such that its nature requires that, at [that single instant], there be some necessity in actuality that is not in actuality. Hence, the last instant of the first inclination is not the first instant of the second inclination.

(18) Pay no attention to whoever says that the two inclinations are together! How could there be something in which there is actually a resistance to a certain direction, or clinging to it while withdrawing from it? So one should simply not think that there is some downward inclination in a stone that is thrown upward; rather, there is a principle whose character is to produce that inclination when the obstacle [to it] is withdrawn or perhaps overcome. In the same way, there is a power and principle in water to produce coolness in the substance of the water when the obstacle is withdrawn or perhaps overcome, just as you have learned. So it has become evident that the two instants are distinct; and, between, every two instants, there is a time. It is most likely that what brings about the arrival remains, doing so for a period of time. Still, we take it as a something that brings about the arrival instantly in order that it be closer to that which requires the absence of rest. So the sophistry has been resolved, and you yourself are entrusted with basing the arguments of the First Teacher[20] upon this foundation.

20. The reference may be to Aristotle, *Physics* 8.8.262a21ff.

فتكــون طباعه تقتضي أنْ يكون فيه اقتضاء بالفعل وأنْ لا يكون اقتضاء بالفعل ، فإذن آنُ آخرِ الميل الأول ، غير آن أول الميل الثاني .

(١٨) ولا تصغ إلى مَنْ يقول إنَّ الميلين يجتمعان ، فكيف يمكن أنْ يكون شـــيء فيه بالفعـــل مدافعة جهةٍ أو لزومهـــا ؛ وفيه بالفعل التنحي عنها ؟ فلا يظنَّ أنَّ الحجر المرمي إلى فوق فيه ميلٌ إلى أســـفل البتة ، بل مبدأ من شـــأنه أنْ يُحدث ذلك الميل ، إذا زال العائق – وقـــد يغلب . كما أنَّ في الماء قوة ومبدأ يُحدث البرد في جوهر الماء إذا زال العائق ، وقد يغلـــب ، كما علمت . فقد بانَ أنَّ الآنين متباينان ، وبين كل آنين زمانٌ ، والأشـــبه أنْ يكون الموصـــل يبقى موصـــلاً زماناً ، لكننا أخذناه موصلاً آناً ؛ ليكـــون أقرب من الموجب لعدم السكون . فقد انحلت الشُبه ، وتول أنت بنفسك بناء حجج المعلم الأول على هذا الأصل .

Chapter Nine

On the motion that is naturally prior and a catalogue of the specific differences of motions

(1) Having reached this point of our discussion, we should complete the account of motions by determining which motions most deserve to be prior. So we say, first, that motion with respect to place or position is the prior motion. That is because augmentation inevitably involves a certain motion with respect to place that accompanies quantitative motion. That is, the growing thing inevitably acquires something new, [and that new thing] is moved toward and into [the growing thing], whereas motion with respect to place and position are free of that. Also, rarefaction and condensation inevitably involve alteration. Alteration, however, exists only after the existence of some motion with respect to place and position. [This is] because a single alteration does not always exist, since it is between contraries and inevitably has some cause that previously had not been actual but then became a cause. In that case, that cause must either be something that comes to that which is to be effected, or not. If it had not yet come and then arrives and so brings about the alteration, then a motion with respect to place and position occurred. If it has come but is not acting and so, consequently,[1] needs some alteration with

1. Reading *idhan* with **Z** and **T** for **Y**'s *dhā* (possessor).

<الفصل التاسع>

في الحركة المتقدمة بالطبع
وفي إيراد فصول الحركات على سبيل الجمع

(١) وإذ قد بلغ الكلام بنا هذا المبلغ، فبالحري أنْ نختم القول في الحركات؛ بأنْ نعرف أي الحركات أولى بالتقدّم؛ فنقول: أمّا أولاً، فإنَّ الحركة المكانية أو الوضعية أقدم الحركات، وذلك لأنَّ النمو لا يخلو عن حركةٍ مكانية مع الحركة الكميّة، ولا يخلو من واردٍ على النامي، يتحرك إليه وفيه، والمكانية والوضعية تخلو عنه. والتخلْخل والتكاثف لا يخلو عن استحالة، والاستحالة لا توجد إلاّ بعد وجود حركةٍ مكانيةٍ أو وضعية، إذ كانت الاستحالة الواحدة لا توجد دائمة، إذ هي بين الأضداد ويكون لها - لا محالة - علّة لم تكن من قبل علّة بالفعل، ثم صارت علّة. فلا يخلو، إمّا أنْ تكون تلك العلّة واصلة إلى المعلول أو لا تكون، فإنْ لم تكن واصلة فوصلت حتى أحالت، فقد حصلت حركة نقليةٌ أو وضعيةٌ، وإنْ كانت واصلة، ولكن ليست بفعل، فهو إذن يحتاج إلى استحالةٍ في

respect to its act of willing or the like in order to act, the question concerning *that* alteration still remains. If it does not require [anything else in order] to come or to undergo an alteration—that is, it and its object both exist, and yet there is no action—then it is simply not something that causes alteration. So the question about alteration still remains. Our question, however, concerns corporeal alterations by corporeal causes—that is, the one that acts after not having acted only by coming into proximity after having been remote. This question also concerns finite, rectilinear locomotions as well, for they do not continue infinitely.[2] Thus, certain motions must be prior to them in order that they exist. Positional and circular local [motion] (if it exists), however, does not have this form, but it is enough that it have one single permanent mover. It also turns out that the various kinds of relations produced between the moving thing and other bodies are causes from which other motions and alterations arise.

(2) From this, it is obvious that circular motions are prior, for motions with respect to place and position are prior. Also, the latter kind of motion is prior in nobility, because only the perfection of the substance as something actualized precedes its existence. [This substance] in no way, however, excludes [this kind of motion] from its substantiality but is something that, in itself, does not cease, while only a certain relation that it has to something external ceases. Also, circular motion is unique in that it is something complete that is not susceptible to increase. There does not need to be in it intensification and weakening in the way that, among natural and forced [motions], there are those that are the fastest; and, likewise, what is called a *mean* [*speed*]; and those that are undoubtedly slowest. The body having natural circular motion is prior to [all other] bodies, and, through it, the directions of the natural motions belonging to other bodies are defined.

---

2. Avicenna's point here is not to deny that there has been rectilinear motion from the infinite past and that there will continue to be so into the infinite future. Rather, his claim is that any particular rectilinear motion cannot continue infinitely. That is because there is only a finite amount of distance that any particular rectilinear motion can cover before it must reverse itself, that finite distance being the interval of the sublunar sphere. But after a reversal, there is another distinct rectilinear motion.

إرادته أو غير ذلك؛ حتى تفعل، والكلام في تلك الاستحالة ثابت. وإنْ كان لا يحتاج إلى وصولٍ ولا إلى استحالةٍ - وهو موجودٌ والموضوع موجودٌ وليس يفعل - فليس بمحيلٍ أصلاً، فالكلام في الاستحالة ثابت. على أنَّ كلامنا في الاستحالات الجسمانية عن عللٍ جسمانية، وهي إنّما تفعل بعدما لَمْ تفعل بالقرب بعد البُعْد، والكلام في الحركات النقلية المتناهية المستقيمة هذا الكلام، فإنّها لا تكون متصلة بغير نهاية، فتحتاج أنْ تتقدمها حركات حتى توجد. وأمّا الوضعية والنقلية المستديرة - إنْ كانت موجودة - فليس الأمر فيها على هذه الصورة، بل يكفي لها محرّكٌ واحدٌ ثابت. ويصلح أنْ تكون أصناف ما يحدث من المناسبات المختلفة، بين ذلك المتحرك وبين الأجسام الأخرى، أسباباً لانبعاث حركات واستحالات أخرى.

(٢) فبيّنٌ من هذا، أنَّ أقدم الحركات ما كان على الاستدارة، فإنّها أقدم الحركات المكانية والوضعية، وهذا الصنف من الحركات أقدم من سائر الحركات الأخرى وبالشرف أيضاً، لأنّه لا يوجد إلاّ بعد استكمال الجوهر جوهراً بالفعل، ولا يخرجه عن جوهريته بوجهٍ من الوجوه، ولا يزيل أمراً له في ذاته، بل يزيل نسبة له إلى أمرٍ خارج. وتُخص المستديرة بأنّها تامة لا تقبل الزيادة، ولا يجب فيها الاشتداد والضعف كما يجب في الطبيعية أن تشتد أخيراً في السرعة، والقسرية أنّ تشتد، كما يقال وسطاً، ولا شك أنها تضعف أخيراً. والجرم الذي له الحركة المستديرة بالطبع هو أقدم الأجرام، وبه تتحدّد جهات الحركات الطبيعية للأجرام الأخرى.

(3) Since we have exhausted the investigation of these notions, it is fitting that we summarize the specific differences that belong to motions. Firstly, we say: Whatever is related to a certain characteristic might have that characteristic predicated of it [in one of three ways. It might be predicated] essentially, in that that characteristic exists in it entirely—in the way, for example, that white is said of snow. Alternatively, [that characteristic] might not really exist in it entirely but only partially, in the way that seeing is said of man and black [is said] of the eye. Finally, [that characteristic] might be predicated accidentally without qualification, in that it is not in it but in something to which it is joined, just as the builder is said to write and whiteness is said to move when the white thing moves. So that is predicated of the mobile and the mover[3] essentially, either absolutely or partially—as when someone is said to write, but it is only his hand that writes, or [when] someone is moving, but it is only his hand that is moving. As for being predicated accidentally without qualification, in the way that something standing still on a ship is said to move, some are such that they simply cannot be characterized by that (like whiteness when it is said to move), while others are such that they can (like the nail that is firmly in the ship). Similarly, the mover might be accidental with or without qualifications, in the way mentioned in some previous chapters.[4]

3. Reading *muḥarrik* with **Z** and **T** for **Y**'s preferred *mutaḥarrik* (mobile).

4. For a general discussion, see 1.12.2. For the case of the physician curing himself as a specific example of an accidental mover, see 1.5.6 and 1.9.2.

(٣) وإذ قد استوفينا تحقيق هذه المعاني ، فبالحري أنْ نجمع الفصول التي للحركات ، ونقول : أولاً ؛ كل ما ينسب إليه صفة ، فإمّا أنْ يقال تلك الصفة له بذاته ، بأنْ تكون الصفة موجودة فيه كله ؛ مثل ما يقال إنَّ الثلج أبيض ، وإمّا أنْ لا تكون بالحقيقة موجودة في كله ، ولكنها بالحقيقة في جزئه ؛ مثل ما يقال إنَّ الإنسان يرى وأنَّ العين سوداء ، وإمّا أنْ تقال بالعرض على الإطلاق بأنْ لا تكون فيه ، بل في شيءٍ يقارنه ؛ كما يقال إنَّ البنّاء يكتب ، وكما يقال للبياض إنّه ينتقل عندما ينتقل الأبيض . فالمتحرّك والمحرّك إمّا أنْ يقال له ذلك لذاته مطلقاً أو للجزء ، كما يقال فلان يكتب ، وإنّما تكتب يده ، أو فلان يتحرك ؛ وإنّما تتحرك يده ، وأمّا أنْ يقال بالعرض مطلقاً ، كما يقال للساكن في السفينة إنّه يتحرك فمنه ما ليس من شأنه البتة أنْ يوصف بذلك ؛ كالبياض إذا قيل إنّه يتحرك ومنه ما من شأنه ذلك ؛ كالمسمار المسمّر في السفينة ، وكذلك المحرك قد يكون بالعرض مطلقاً وغير مطلق ، على ما قيل في أبواب سلفت .

(4) When the motion is in the very thing itself, then sometimes it arises from the nature of the thing, like the stone's descent, and is not from some external thing or by an act of volition or some intention. At other times, it arises from [the thing] by an act of volition; and, at still other times, [it arises] because of some external force, like the stone's ascending. Natural and volitional [motions] always share in common the fact that the term *spontaneous motion* applies to them, and that is because it does not come from something external (although that is perhaps properly said of that which is through an act of volition). Natural and forced motion might be in something other than [the category] of place and position. [That] is because there is natural alteration, like the health of someone who recovers on the natural critical day, and the cooling of hot water when, through its nature it undergoes alteration so as to become cool; and there is forced alteration, like water's undergoing alteration so as to become hot. There is also natural generation as, for example, the fetus from sperm and the plant from seeds, as well as forced generation, as, for instance, making fire by striking flint. Again, there is natural corruption, as in the case of someone's dying from old age, and forced corruption, like dying by murder and poison. There is a natural increase in the magnitude of the body, such as the growth of the young boy, while another is forced, like the growth brought on by fatty diets. There is also natural deterioration, like [that occuring] during old age, and forced deterioration, as [brought about] through illness.

(٤) والحركة إذا كانت في ذات الشـيء ، فقد تنبعث عن طبيعته لا من خارج ولا بإرادة ولا قصد ، كنزول الحجر ، وقد تنبعث عنه بالإرادة ، وقد تكون بسبب قسري من خارج ، كصعود الحجر . والطبيعي والإرادي يشتركان دائماً في أنْ يطلق عليهما لفظة الحركة الكائنة من تلقاء المتحرك ، وذلك لأنها ليست من خارج ، وربّما قيل ذلك خاصة للذي يكون بإرادة . والحركة الطبيعية والقسرية قد تكون في غير المكانية والوضعية ، فإنّ ها هنا استحالة طبيعية ، كصحة مَنْ يصح بالبُحْران الطبيعي ، وتبرّد الماء الحار إذا استحال بطبعه إلى البرد ، واستحالة قسرية كاستحالة الماء إلى الحر . وها هنا كَوْنٌ طبيعي ، مثل تكوّن الجنين والنبات من المني والبزور ، وكَوْنٌ قسريٌ مثل إحداث النار بالقدح ، وفسادٌ طبيعيٌ مثل الموت الهرمي ، وفسادٌ قسريٌ كالموت عن القتل والموت عن السُّم . وها هنا زيادة في مقدار الجسم طبيعية كنمو الصبي ، وأخرى قسرية كالنمو الذي يُستجلب بالأدوية المسمّنة ، وها هنا ذبولٌ طبيعي كما في الهرم ، وذبولٌ قسريٌ كما بالأمراض .

(5) You should know that by *natural motion* we do not mean the motion that simply proceeds from the nature when the nature itself has that which belongs to it. The nature is a stable, fixed entity, and what proceeds from it essentially is, likewise, a stable, fixed thing that subsists and exists together with the existence of the nature. Motion, however, (that is, the motion involving a traversal) always comes to an end and always involves renewal without being stable. The motion that we are investigating undoubtedly does as well, for it requires leaving something behind. Now, when the nature essentially requires that something be left behind, it undoubtedly requires that what is left behind be something outside of the nature. In that case, whenever nothing outside of the nature happens to occur, then some intention of [the nature] to leave will not naturally occur. Therefore, natural motion does not proceed from the nature unless some unnatural state happened to occur. There won't be an unnatural state, however, unless it parallels a natural one; since *this one* is not *that one,* that one is natural. So the unnatural [state] is left behind in such a way that something is directed toward the natural [state]; and, when any natural motion is not impeded, it terminates at some natural end. When that end is fully realized, it is impossible that the mobile should undergo natural motion, because motion is a sort of leaving behind and retreating, whereas the natural end is not something naturally left behind and retreated from. So, therefore, every natural motion is for the sake of seeking a state of rest, whether in a *where,* quality, quantity, or position.[5]

5. Literally, "in a *where,* a how, a how much, or a position," all of which correspond with the noted Aristotelian categories.

(٥) ويجب أنْ تعلم أنَّ قولنا حركة طبيعية ليس نعني به أنَّ الحركة تصدر البتّة عن الطبيعة، والطبيعة بحالها التي لها، فإنَّ الطبيعة ذاتٌ ثابتةٌ قارّةٌ، وما يصدر عنها لذاتها فهو أيضاً ثابتٌ قارٌّ قائمٌ موجودٌ مع وجود الطبيعة. والحركة، التي هي الحركة القطعية، تعدم دائماً وتتجدّد دائّماً بلا استقرار، والحركة التي حقّقناها - لا محالة - فإنّها تقتضي ترك شيء، والطبيعة إذا اقتضت لذاتها ترك شيءٍ فتقضي، لا محالة، ترك شيء خارجٍ عن الطبيعة. وإذا كان كذلك، فما لم يعرض أمرٌ خارجٌ عن الطبيعة، لم يعرض قصد تركٍ لها بالطبع. فإذن الحركة الطبيعية لا تصدر عن الطبيعة؛ إلاّ وقد عرضت حال غير طبيعية، ولا تكون حال غير طبيعية، إلاّ وبإزائها حال طبيعية إذ كانت هذه غير تلك؛ فتلك طبيعية، فتكون غير الطبيعية تترك تركاً متوجهاً إلى الطبيعية. وكل حركة طبيعية إذا لم تُعق، فهي تنتهي إلى غايةٍ طبيعية، ويستحيل إذا حصلت تلك الغاية أنْ يتحرك المتحرك بالحركة الطبيعية، لأنَّ الحركة تركٌ ما وهربٌ، والغاية الطبيعية ليست متروكة ولا مهروباً منها بالطبع. فكل حركةٍ طبيعيةٍ إذن فهي لأجل طلب سكون؛ إمّا في أيْنٍ أو في كيفٍ أو في كمّ أو في وضْعٍ.

(6) So every motion that does not come to rest is not a natural one. Hence, continuous circular motion is a non-natural [motion]. How could it be [natural]? That motion has nothing to do with the positions and places that are assumed to be something retreated from, unless it itself is what is naturally intended by that motion; but it is absurd that it would naturally retreat from something toward which it is naturally going. So circular motions either results from external causes or some non-natural and, in fact, volitional power. Now, it might be possible that the [end] result of some volitional power does not vary when the motives, obstacles, ends, and goals do not vary. In that case, there would not come to be new volitions, but they would be one and the same through which the object of the volition is reached during the motion. Also, the fact that the circular motion belongs to a simple body does not prevent that body from possessing a soul, although there are some who are troubled by [this], saying that the Peripatetics require that the soul belong only to a composite body. [The Peripatetics], however, go on to say that a simple circular motion proceeds from a soul and that it belongs to a simple body. That is because the Peripatetics did not deny that no simple body has a principle of animation, but denied only that that [simple animate] body would be one of the simple elemental ones underlying composition. So, as long as these simple [elemental] bodies neither undergo a certain composition, nor obtain a certain proper proportion, nor lessen the overpowering instances of contrariety, then neither will they be susceptible to life. If, however, a given simple body has no contrary in its nature, then it is most susceptible to life.

(٦) فكل حركة لا تسكن فليست بطبيعية. فالحركة المستديرة المتصلة إذن لا تكون طبيعية، وكيف تكون؟ وليس شـيء من الأوضاع والأيون التي تُفرض مهروباً عنه بالطبع بتلك الحركـة، إلاّ وهو بعينه مقصودٌ إليه بالطبع بتلك لاحركـة ومحالٌ أنْ تهرب الطبيعة عن أمرٍ تؤمّه بالطبع. فالحركات المستديرة تكون إمّا ‹عن› أسبابٍ من خارج، وإمّا عن قـوةٍ غيـر الطبع، بل قوةٍ إرادية. وقد يجوز أنْ لا يختلف ما يكون عن القوة الإرادية، إذا لـم تختلف الدواعي والموانع والغايات والأغراض، فلم تتجدّد الإرادات، وكانت الواحدة منها مبلوغاً بها المراد في الحركة. ولا يمنع كون الحركة المستديرة لجسم بسيطٍ أنْ يكون ذلك الجسم ذا نفس على ما يشكَل به بعضهم قائلاً إنَّ المشائين يوجبون أنْ لا تكون النفس إلاّ للجسم المركّب، ثم يقولون بحركةٍ مستديرة بسيطة هي صادرة عن نفس، وأنّها لجرم بسيط. وذلك لأنَّ المشائين لم يمنعوا أنْ يكون في البسائط كلها متنفسٌ، بل إنّما منعوا أنْ يكون ذلك الجسم من البسائط الأُسْطُقسّية الموضوعة للتركيب، فإنَّ هذه البسائط ما لم تركّب ولم تعتدل ولم تُسقط غلبات التضادّ، لم تقبل الحياة. فإنْ كان جسم بسيط لا ضدّ له في طبعه، فهو أقبل للحياة.

(7) Here we should define the number of ways [in which] *natural* is used, according to what is useful for the subject with which we are concerned, and then afterwards complete the discussion about natural motion. So we say that *natural* is sometimes said of the thing to which the natural feature belongs alone, while at other times it is not said relative to it alone, but relative to the natures of the whole thing taken collectively. An example of this latter sort is that the Earth is not really round when the water is removed from it. [This, in fact] is not something natural relative to earth itself, for the nature of every simple thing requires that there be no difference in it and, instead, that it must be homogeneous. So the natural simple shape must be a sphere. Still, when the feature that the nature of earth requires from its preparedness and action together is joined with the nature of the whole, the shape really belonging to it is natural—that is, a feature necessarily resulting from its nature and the nature of the whole,[6] and analogously for particular features of the whole (as will become clear in its proper place).[7] Similarly, the nutritive faculty's administration of the nutriment is unnatural to the nutriment itself [when] considered independently; but when it is related to the common nature of the whole [living thing], it is natural.

(8) [The notion of] *natural* that is proper to the thing is something proceeding from a natural power in it alone. By *natural power* we mean every power belonging to the thing itself that produces motion but is not by volition, whether it is a nature absolutely or one like the plant's soul. One of the two sorts falling under this heading is like the stone's being moved downward, which does not result from an act of volition, nor is it something whose direction varies. The second is like the growth of a

6. Avicenna's point is that, given the nature of the element earth alone, it should form a perfect sphere; however, when one considers the planet Earth with its water removed, the Earth obviously does not form a perfect sphere, but has canyons, mountains, and the like. Still, once one takes into account the actions and dispositions of earth when coupled with those of water, which jointly make up the planet Earth, the fact that the element earth does not make a perfect sphere is completely natural.

7. The reference is probably to *Kitāb al-maʿādin wa-l-āthār al-ʿulwīyah,* Avicenna's work mostly closely associated with the tradition surrounding Aristotle's *Meteorology.*

(٧) ويجب أن يعرف ههنا أن الطبيعي على كمّ وجه يقال، بحسب ما ينتفع به في الموضع الذي نحن فيه، ثم نتمم الكلام في الحركة الطبيعية فنقول: إنَّ الطبيعي قد يقال بالقياس إلى الشيء الذي له الأمر الطبيعي وحده، وقد يقال لا بالقياس إليه وحده، بل بالقياس إلى طباع الكل بالشركة. مثال هذا القسم هو أنَّ كون الأرض غير حقيقية التدوير، وانكشافها عن الماء ليس طبيعياً بالقياس إلى طبيعة الأرض نفسها. فإنَّ طبيعة كل بسيط لا تقتضي اختلافاً فيه، بل تقتضي التشابه، فيجب أنْ يكون الشكل الطبيعي البسيط كُرياً. ولكن الأمر الذي تقتضيه طبيعة الأرض من استعدادها وفعلها معاً، إذا قرنت به طبيعة الكل، كان وجود هذا الشكل له طبيعياً؛ أي أمراً يجب عن طباعه وطباع الكل، وما عليه مجرى الأمر الجزئي في الكل، على ما سنوضح هذا في موضعه. وكذلك تصرف الغذاء، بحسب تدبير القوة الغاذية، هو لنفس الغذاء غير طبيعي، ولكن إذا قيس إلى الطبيعة المشتركة للكل، كان طبيعياً.

(٨) وأمّا الطبيعي، الخاص بالشيء، فهو أنْ يكون صادراً عن قوةٍ طبيعيةٍ فيه وحده، ونعني بالقوة الطبيعية ها هنا كل قوة من ذات الشيء تحرّك لا بإرادة، كانت طبيعية صرفاً، أو كانت كنفس النبات. فيكون أحد قسمي هذا الباب على نحو تحرّك الحجر إلى أسفل؛ وهو الذي يكون لا عن إرادة، ولا أيضاً مختلف الجهة، والثاني على

growing thing, since that is not by an act of volition, but the directions do vary.[8] Also, there might be a motion to which the term *natural* is applied only equivocally (namely, that which is by volition but [which] does not vary in direction—as, for example, the motion of the first [celestial sphere]). Natural motion in this context is that which results from some power in the body itself that is directed toward the end that belongs to the nature of that body and in the way that the nature of that body requires when there is no obstacle. For example, the human hand will have five fingers that are similar at a time that they are generated and in a way that does not diverge from the required limits. Indeed, there might[9] be a motion that results from the nature but is not toward some natural end, as is instanced in the generation of the additional finger and unequal teeth. Also, a motion might not result from the nature but be toward a natural end, like one who throws a stone downward in a straight line, such that what results from the throw is that motion that would have resulted from the nature that is in the stone alone. Also, it might be uniform from beginning to end, but it is impeded. So, for instance, its motion is slower than necessary, or it has some quality that is not conducive to continuing toward the end, and this might be said to belong to it naturally (however, the proper sense is what we said initially). Also, sometimes the motion is natural, not relative to the nature proper to the thing, but relative to some external factors, for it is natural for sulfur to ignite when it meets fire,[10] and it is natural for iron to be attracted when it is near a magnet.

---

8. For instance, when I was young, I grew upward; but now that I am old, I am growing outward.

9. Reading *qad* with **Z** and **T**, which is (inadvertently) omitted in **Y**.

10. Reading *nār* with **Z** and **T** for **Y**'s *ghāz* (gas), which simply cannot be correct, since it is a French loan word that entered the Arabic language relatively late.

نحو تحرّك النامي إلى نموه، فإنَّ ذلك ليس بإرادة، ولكنه مختلف الجهة، وقد تكون حركة بإرادةٍ غير مختلفة الجهة، ولا تسمّى طبيعية إلاّ باشتراك الإسم؛ كالحركة الأُولى. فالحركة الطبيعية بحسب هذا الموضع هي التي تكون عن قوةٍ في الجسم نفسه تتوجه إلى الغاية التي لطبيعة ذلك الجسم، وعلى الوجه الذي تقتضيه طبيعة ذلك الجسم، إذا لم يكن عائق، مثل تكون يد الإنسان ذات خمس أصابع في مدةٍ في مثلها تتكون، وعلى نحوٍ من التوجه غير زائغ عن الحدود الواجبة، فإنّه قد تكون حركة عن الطبيعة ولكن لا إلى غايةٍ طبيعية، مثل تكوّن الأصبع الزائدة والسّن الشاغية. وقد تكون حركة لا عن الطبيعة، ولكن إلى الغاية الطبيعية، كمَنْ يرمي حجراً إلى أسفل على خطٍ مستقيم، رمياً لا تصدر منه مثل تلك الحركة التي فيه عن الطبيعة التي في الحجر وحدها. وقد يتفق أنْ يكون من المبدأ إلى الغاية، ولكن معوّقاً مثل أنْ تكون حركته أبطأ من الواجب، أو ذات كيفية غير موافقة للاستمرار إلى الغاية. فهذه قد يقال لها طبيعية، ولكن الحقيقي هو ما قلناه أولاً. وقد تكون الحركة طبيعية لا بالقياس إلى الطبيعة الخاصّة بالشيء، بل بالقياس إلى أمورٍ من خارج؛ فإنَّ الاحتراق طبيعي للكبريت عند ملاقاة النار، والانجذاب طبيعي للحديد عند مقاربة المغناطيس.

Chapter Ten

The way in which space and other things are natural to the body

(1) We shall prove that every body requires a certain space proper to it,[1] and it does so on account of its form by which it becomes a substance or the form of that which is predominant in it. It also frequently requires a certain quantity, quality, position, or the like. So, if the space that [the body] requires is something dependent upon it that does not depart from it, [the body] would have no natural motion to the space. The same would hold if its quality or quantity has this description. So, if its space is some space that it could depart by being forcibly removed from it, then its return is natural (if it is not forcibly prevented). Alternatively, if [the body] is such as not to be removed from its space, but [if] when it initially came to be, it was not in its [proper] space, it naturally moves toward it (if it is not forcibly prevented). If its [natural] quality is something that can be forcibly taken from it, like the quality of water (I mean its coolness), then, when [that] force ceases, the thing naturally turns toward [its natural quality]; and so the heated water, for instance, undergoes alteration so as to be cool. If its [natural] quantity is something that can forcibly be taken from it—as, for example, the air that is forcibly rarefied, so that [its volume] becomes greater, or compressed, so that it becomes smaller (as we reported in the chapter on the void)[2]—then, when the force ceases, the substance moves to its [natural] volume.

1. See 4.11.1–6.
2. See 2.9.7, 20–21.

<الفصل العاشر>

في كيفية كون الحيّز طبيعياً للجسم
وكذلك كون أشياء أخرى طبيعية له

(١) فنقول، إنَّ كل جسم، فسنبيّن أنّه يقتضي حيّزاً يخصه، والمقتضى لذلك صورته التي بها يتجوهر، أو صورة الغالب فيه، وقد يقتضي كمّاً أو كيفاً أو وضعاً، أو غير ذلك. فإنْ كان الحيّز الذي يقتضيه موقوفاً عليه لا يفارقه، لم تكن له حركة طبيعية ناقلة إلى الحيّز، وكذلك إنْ كان كيفُه بهذه الصفة أو كمُّه. فإنْ كان حيّزه حيّزاً يمكن أنْ يفارقه، بأنْ يزال عنه قسراً، فإنَّه يكون له عَوْدٌ بالطبع - إنْ لم يُمنع قَسْراً - أو كأنْ يكون لم يزل عن حيّزه، بل كان أول حدوثه في غير حيّزه، فإنَّه بالطبع ينتقل إليه إنْ لم يُمنع قَسْراً. فإنْ كانت كيفيته ممّا يجوز أنْ يسلب بالقَسْر ككيفية الماء - أعني برودته - فإنَّه إذا زال القاسر توجه إليها الشيء بالطبع، فاستحال الماء المسخّن مثلاً بارداً. وإنْ كانت كميته ممّا يجوز أنْ تُسلب بقَسْرٍ مثلاً كما يُخلْخل الهواء بالقَسْر حتى يصير أعظم، أو يضغط بالقَسْر حتى يصير أصغر - على ما أُخبرنا عنه في باب الخلاء - فإنَّه إذا زال القاسر انتقل الجوهر

Alternatively, its quantity will reach a level that it did not have when it first came to exist, but its initial existence was imperfect; and it is only by [a type of] replenishing that it becomes perfected and, indeed, through being nourished, naturally moves toward the perfection of its volume. Or again if the position of its parts is acted upon by force, like forcibly bending a straight piece of wood, then, if it is let go before breaking or splintering, it returns through its motion to the original position.

(2) Still, space may raise certain worries that other issues do not, for the body undergoes motion in a given direction that raises certain concerns for it. One of [these] is that it is moving toward that *direction;* another is that it is moved toward a certain *place;* and, again, another is that it undergoes motion to where its *collective kind* is. So the issue [of space] raises certain doubts and concerns, since it is not known to which one of these things it is moved. If water were to seek the direction and terminate its descent [only] at the lowest depths, it would neither stop short of the limit [where] earth rests, nor rise to the surface, nor settle within earth. The same would hold in the case of air, were the estimative faculty to imagine some part of it being forced toward the space of fire. It is found, however, to move from the space of fire toward its own space. You will learn that no two bodies belong by nature to one and the same space,[3] such that you could say that both earth and water seek one and the same direction and one and the same space, but [rather] it is just that earth is the more dominant and moves the farthest; and, likewise, air and fire both seek one and the same direction and one and the same place, but fire is the more dominant and moves the farthest. Now, if air were

3. See 4.11.7–10.

إلى حجمه، أو أتت كميته مما لا تحصل له في أول وجوده، بل كان أول وجوده وجوداً غير مستكمل، وإنّما يستكمل بالاستمداد، فإنه يتحرك إلى كماله في حجمه بالغذاء طبعاً. أو كان وضع أجزائه وضعاً مَقْسوراً؛ كما يحنى الخشب المستقيم بالقَسْر، فإنَّه إذا خلي سبيله من غير كسرٍ أو رضّ، رجع بحركته إلى الوضع الأول.

(٢) لكنه قد يَشْكل في أمر الحيّز ما لا يَشْكل في أمر غيره، فإنَّ الجسم المتحرك في جهةٍ ما تُعرّض له أمور؛ من ذلك أنّه متحرك إلى تلك الجهة، ومن ذلك أنّه يتحرك إلى مكانٍ ما، ومن ذلك إنه متحرك إلى حيث كليته، فيشتبه الأمر ويُشْكل، فلا يدري أنّه إلى أيّ واحدٍ من هذه الأشياء يتحرك. ولو كان الماء يطلب الجهة والنهاية في نزوله إلى أسفل، لما وقف دون حدّ وقوف الأرض، ولما طفا على الأرض، ولما رسب في الأرض، وكذلك حال الهواء لو توهم جزء منه مقسوراً إلى حيّز النار، فوجد ينتقل من حيّز النار إلى حيّز نفسه. وستعلم أنّه لا يكون لحيّزٍ واحدٍ جسمان بالطبع، حتى يكون لك أنْ تقول إنَّ الأرض والماء يطلبان جهة واحدة وحيّزاً واحداً، لكن الأرض أغلب وأسبق، وكذلك الهواء والنار يطلبان جهة واحدة وحيّزاً واحداً، لكن النار أغلب وأسبق. ولو

to seek what fire seeks but [not attain it because] it just does not have the strength to keep up with [fire], then, if we placed our hands over a portion of air, we would sensibly feel it rushing upward, just like when we hold [air] underwater in a bladder. If the mobile were to seek only the place—where *place* is the [inside] surface of the surrounding body and the *natural* [*place*] is the surface of the body that surrounds by nature—then water would stop at the air, wherever it was, because it would be at the surface of the natural body that surrounds it, and ascending fire would seek to be contained by a place that is the surface of the [lunar] sphere. It is absurd that it seek this, because it touches only a portion of the surface of the sphere on one side [and so is not surrounded]. Were it to seek [wherever] the collective kind [is], the stone released at the top of a well would just hang at [the well's] upper rim and not go on to the bottom, for it would be in contact with the collective kind here in the shortest distance. Also, the stone would ascend, if our estimative faculty were to imagine that its collective kind left its location. In that case, one of two things must follow. [The first is that the ascending] would be natural, one direction being distinguished from another—but this is absurd. Alternatively, [the stone] would be acted upon by the collective kind in some other way from some other direction [than down], such that its motion toward[4] the collective kind would not be from its nature, but would be by the collective kind's attracting it. We had supposed, however, that its motion was natural; and yet it is impossible that something act upon something of the same kind *qua* same kind such that a given action and influence are natural. If it is not accidental, then, a small amount of earth, like a clod, would be attracted more quickly than a larger amount.

4. Reading *ilá* with **Z** and **T** for **Y**'s *ʿan* (away from).

كان الهواء يطلب ما تطلب النار، ولكنه يعجز عن مساوقتها إليه، لكنّا إذا وضعنا أيدينا على شطرٍ من الهواء أحسسنا باندفاعه إلى فوق؛ كما إذا حبسناه في إناء تحت الماء. ولو كان يطلب المتحرك المكان فقط - والمكان هو سطح الجسم الذي يحويه، والطبيعي هو سطح الجسم الذي يحويه بالطبع - لكان الماء يقف في الهواء حيث كان، لأنَّه في سطح الجسم الطبيعي الذي يحويه، ولكانت النار المتصعدة تطلب أنْ يشتمل عليها مكانٌ هو سطح فلك، وهذا الطلب محال، لأنَّه إنّما يماس طائفة من سطح الفلك من جهة. ولو كان يطلب الكلّية لكان الحجر المرسل من رأس البئر يلتصق بشفيرها ولا يذهب غوراً، فإنَّ الاتصال بالكل هناك أقرب مسافة، ولكان الحجر يصعد، لو توهمنا أنَّ كلّيته زالت عن موضعها، فكان حينئذ لا يخلو إمّا أنْ يكون بالطبع يميّز جهة دون جهة وهذا محال، أو يكون قد انفعل عن الكلّية انفعالاً آخر من جهةٍ أخرى، فتكون حركته إلى الكلية ليست عن طباعه؛ ولكن بجذب الكلّية إياه، وقد فرضنا حركته طبيعية. وعلى إنَّه يستحيل أنْ يفعل الشيء في شبيهه فعلاً وأثراً بالطبع، من حيث هو شبيهه إلا بالعرض، ولكانت الأرض الصغيرة كالمدرة أسرع انجذابا من الكبيرة.

(3) What ought to be believed about this is that natural motion seeks the natural space and flees from the unnatural one—not absolutely, but together with a certain proper, ordered position of the parts of the universe and a certain proper position of the body acting on account of the directions (for the direction is not itself what is intended, except for the sake of this account's being in it). [It should, moreover, be believed] that [the *where* of] the collective kind belonging to every simple [element] is not what is intended in the natural motion belonging essentially to [the collective kind's] particular instances, but it is a subject where there is what is intended—and, in fact, what we mentioned [namely, the proper order and position] is what is intended. So the seeking is directed toward only this definite end and, in fact, nothing else, whereas the flight turns out to be away from what corresponds with [that end], whatever it chances to be. So, when the place is not a natural one, even if the ordered position is natural, it flees from it. An example is the air that is absorbed and trapped in baked bricks[5] that are then placed high in the air. In this case, the brick absorbs water from below owing to [three factors:] the air's intense flight from a foreign surrounding [body], the impossibility of there being a void in it, and the tiles' necessarily sticking together. So the water replaces [the air] in the pores, as it is drawn into them owing to the air's fleeing from them, even if the ordered position in remoteness and proximity is closer to what it should be, and is like water's fleeing from the air, even if the place is natural, but the ordered position is not realized.[6]

5. Reading *ājurrah* with **Z** and **T**, here and below, for **Y**'s *jarrah* (jar).

6. In the example, Avicenna has us imagine a situation where air becomes trapped within some earthen bricks, and then those bricks are placed high in the air. In this case, the air trapped within the bricks is, as it were, in its natural place—namely, in the air; however, there is not the proper order and position between the air in the bricks and the earth of the bricks, where that ordering and position is that earth be down, followed by water, then air. In order to achieve this order (or at least to get closer to it) the trapped air flees the surrounding earth of the bricks; however, if the air were to depart without anything filling the pores or spaces that are being left, then there would be a void, which Avicenna takes to be impossible. Consequently, moisture from the air surrounding the bricks is drawn into the pores to fill them. While the water's being surrounded by earth also does not represent the natural order and position, it is closer in proximity to that order.

(٣) فالــذي يجــب أنْ يعتقد في هذا هو أنَّ الحركــة الطبيعية تطلب الحيز الطبيعي وتهــرب من غير الطبيعــي لا مطلقاً ، ولكن مع ترتيبٍ من أجزاء الكل مخصوص ، ووضْع مخصوص من الجســم الفاعل للجهات . فإنَّ الجهة عينها غير مقصودة إلاّ لأجل كون هذا المعنى فيها ، وأنَّ الكلّية التي لكل بسيط ليست مقصودة في الحركة الطبيعية التي لأجزائها بذاتها ، ولكنها موضوعة حيث المقصود ، بل المقصود ما ذكرناه ، فالطلب يتوجه إلى هذه الغاية المتحقّقة فقط ، ولا يصحّ إلى غيرها . وأمّا الهرب فيصح عن مقابلاتها أيّها اتفق ؛ فإنّه إذا كان المكان غير طبيعي – وإنْ كان الترتيب طبيعياً – هرب منه ، مثل الهواء المنتشف المحصور في آجرّة مرفوعة في الهواء ، فإنَّ الجرّة تنشّــف الماء من أسفل لشدة هرب الهواء عن محيطٍ غريب ، واستحالة وقوع الخلاء فيه ، ووجوب تلازم الصفائح ، فيخلفه الماء في مســام الجرّة متصعداً فيها – لهرب الهواء عنها ، وإنْ كان الترتيب في البُعْد والقرب قريباً مــن الواجب ، وكَهرب الماء من الهواء إنْ كان المــكان طبيعياً ، وليس الترتيب حاصلاً .

(4) It is worth knowing whether the fleeing or the seeking is that which produces [the element's] motion. Now, were it only the fleeing and not a seeking, then a certain direction toward which [the element is moving] would not distinguish the fleeing from the seeking. The case of water is an example in that its nature produces a certain inclination in its substance, and that inclination produces an inclination and repulsion in that which it encounters, which if it were not to produce it in itself, it would not have produced the inclination in the other. So, for example, water's natural form produces coolness in another only by a certain coolness that emanates from [water's natural form] into the body in which there is [that form], which if that [coolness] were not first emanating with respect to [the form], it would not cool another, even if its form remains. When it is provided with a certain foreign heat, it produces the contrary of its action and so burns. Likewise, when [the water] is violently heated, there arises in it the accident that necessarily brings about a fire-like form, such that it produces the action of fire involving burning and ascending, and so it burns and ascends. That, however, does not require that there be, in this body, two powers that are necessarily contraries; one of which is that form and the other of which is this accident. That is because that form does not primarily require the motion and burning, but it is by means of a certain accident—namely, that which passed away when its contrary, from which this action primarily proceeds, comes to be. Equally, the form is the principle of the upward motion only by means of some accident that appears to be related to [the motion] as a momentary disposition—namely, the inclination. One must not think that that is not on account of the accident but is, instead, on account of, for instance, those fiery

(٤) وبالحري أنْ نعرف هل الهرب هو الذي يحرّكه أو الطلب، لكنه لو كان ليس إلاّ الهرب ولا طلب، لم تتعيّن جهة إليها الهرب دون الطلب. وحال الماء مثلاً في أنَّ طبيعته تُحدث ميلاً في جوهره، وذلك الميل يُحدث ميلاً واندفاعاً فيما يلاقيه، لولا أنَّه أحدثه في نفسه، لم يحدث الميل عنه في غيره، كحال الماء في أنّه إنّما تفعل صورته الطبيعية التبريد في غيره، بما يفيض عنها من بردٍ في جسمها التي هي فيه، لو لم يفض ذلك أولاً فيها لم يبرد غيره - وإن بقيت الصورة - وإذا استفاد حرارة غريبة، فعل ضدّ فعله، فأحرق. وكذلك إذا اشتدت سخونته، عرض فيه العرض الذي توجبه صورة النارية، ففعل فعل النار من الإحراق والصعود، فأحرق وصعّد، فلا يوجب ذلك أنْ يكون في هذا الجسم قوتان يتضادّ مقتضاهما، أحداهما تلك الصورة، والأخرى هذا العارض. وذلك لأنَّ تلك الصورة لا تقتضي الحركة والإحراق اقتضاء أولياً، بل بوساطة عارض، وهو الذي بَطَل وحصل ضدّه الذي هذا الفعل يصدر عنه صدوراً أولياً. فإنَّ الصورة أيضاً إنّما هي مبدأ للحركة إلى فوق، بوساطة عارض يشبه أنْ يكون بالقياس إليها ملكة وقتية وهو الميل. ولا يجب أنْ يظنّ أنَّ ذلك ليس لأجل العارض، بل لما يخالط الماء مثلاً من ناريات، تلك

components mixed with the water that get broken up, withdraw, and turn into vapor, while the water remains cold. If that were the case, then, when we set water and oil over a flame, the oil would necessarily turn into vapor first because it has a nature more susceptible to fire and, on account of being so thoroughly mixed with it, [the oil] would undergo alteration into it. Still, it is possible that some bodies, under the influence of force, do undergo motion contrary to the nature because of what dominates in the mixture, while others, of themselves, confirm this alteration as in water vapor, for, if it were owing to the fiery component, what we said would necessarily follow. You know that there is no cause or reason that prevents the fiery component from freeing itself of the water such that it would need the accompaniment of water, unless water is such that it undergoes motion so as to agree in being like [fire's] motion. We should now demonstrate that every body has some space proper to it.

الناريــات تنقص وتفــارق وتصعد ، ويبقى الماء بارداً ؛ ولــو كان كذلك ، لكان يجب إذا طبخنا الماء والدهن أنْ يتصعد الدهن أولاً ، لأنّه أقبلَ لطبيعة النار ، ولمخالطتها والاستحالة إليها . ١٨٩ – على أنّه من الجائز أنْ تكون بعض الأجســام المقســورة تتحرك إلى خلاف الطبيعة لمخالطٍ غالب ، وبعضها لنفس تأكد هذه الاســتحالة ؛ كما في البخار المائي ، فإنّه لو كان للنارية للزم ما قلناه . وأنت تعلم أنّه لا علّة ولا ســبب لا متناع النارية من التخلّص عن الماء ، حتى تحتاج إلى أنْ تستصحب الماء ، اللّهم إلاّ أنْ يكون الماء صار بحيث يتحرك نحو حركتها موافقة لها . لكنه بالحري أنْ نبرهن على أنَّ لكل جسمٍ حيّزاً يخصّه .

Chapter Eleven

On establishing that every body has a single natural space, and [on] the way space belongs to the body's collective kind and to its individual instances as well as to simple and composite [bodies]

(1) We say that every account and description of a body that simply must belong to that body is something belonging to it of [itself], naturally. An example of this is space, for there is no body that does not have space as one of its concomitants, whether as a place or as some position of an ordering. [Another] example is shape; for every body is something finite, and whatever is finite necessarily has a shape. Again, it inevitably has a certain quality or form other than the corporeality, because it is either easily susceptible to impressing and shaping, or [shaping] happens with difficulty, or [the body] is not so susceptible at all, all of which is something other than being corporeal. We might also be able to prove that certain other qualities are inseparable from the body as well.

(2) So we say that, among these things and what is analogous to them, there is certainly something necessary and natural belonging to the body. That is because what occurs by coercion and force is something accidental, which happens through some external cause. Now, the intellect might still be able to know the substance of the thing even when its

<الفصل الحادي عشر>

في إثبات أنَّ لكل جسم حيّزاً واحداً طبيعياً، وكيفية وجود الحيّز لكلّية جسم ولأجزائه، وللبسيط والمركب

(١) نقول إنَّ كل معنى وكل صفة للجسم لا بُدّ لذلك الجسم من أنْ يكون له؛ فانَّ له منه شيئاً طبيعياً، وهذا مثل الحيّز، فإنّه لا جسم إلاّ ويلحقه أنْ يكون له حيّزٌ، إمّا مكان وإمّا وضع ترتيب. ومثل الشكل، فإنَّ كل جسم متناهٍ، وكل متناهٍ فله شكل ضرورة، وإنْ كل جسم فله كيفية ما، أو صورة غير الجسمية لا محالة، لأنه لا يخلو إمّا أنْ يسهل قبوله للتأثير والتشكيل، أو يعسر أو لا يقبل، وكل هذا شيء غير الجسمية. وقد يمكن أنْ نبيّن ملازمة الجسم لكيفياتٍ أُخرى.

(٢) فنقول إنَّ هذه الأشياء وما يجري مجراها، لا بُدّ من أنْ يكون للجسم منها شيء طبيعي ضروري. وذلك لأنَّ الواقع بالقهر والقسرْ عارض بسببٍ يعرض من خارج، وجوهر الشيء قد يمكن أنْ يعقل ولا تعرض له الأسباب التي لوجوده منها بُدّ، إلاّ ما كان

causes that are nonessential for its existence do not occur and there is only whatever is necessary owing to [the thing's] nature, [since] it is not at all necessary that the intellect know the body only when some forced action in the body affects it. Consequently, the nature of the body might be able to be posited as existing—that is, according to what it is in itself—when there is no force acting upon it. When it is posited like that, then it and its nature remain; and when it remains like that, it inevitably has some shape and some [location] where it is. All of that belongs to it, either from its nature or from some external cause. We posited, however, that it was not an external cause. So it remains that it has it from its nature, and that which is from its nature belongs to it as long as its nature exists. Now, its nature is not by force, for, if it were such as to be susceptible to force, that could be forcibly removed from it, whereas, if its nature is not such as to be susceptible to force, then that will not be removed from it by force.

(3) Someone could say that it is possible that the appearance of each thing acting by force provides a certain shape and place, and that remains thereafter and is removed only by the appearance of something else acting by force. In that case, it would always involve a replacement of things acting by force in just the way that it involves a replacement of accidents. Now, from that, it would not necessarily follow that one of them is something essential and inseparable. Our response is that the body has non-necessary accidents in two ways: accidents that are concomitants of [the body] in itself, and accidents that follow upon it owing to its immediate vicinity, like its being up and down, being contiguous with, and facing. The accidents that follow upon it on account of its immediate vicinity do not necessarily belong to it, considered in itself, whereas,

منها لازماً لطباعه. وليس واجباً ضرورة أنْ يكون الجسم لا يعقل إلاّ ويلحقه فعل قاسر فيه. فإذا كان كذلك، فطبيعة الجسم قد يمكن أنْ تفرض موجوداً - وهو على ما هو عليه في نفسه - وليس يقسره قاسر. وإذا فرض كذلك، بقي وطباعه، وإذا بقي كذلك، لم يكن بدّ من أنْ يكون له أيْنٌ وشكلٌ، وكل ذلك لا يخلو إمّا أنْ يكون له من طباعه، أو من سببٍ من خارج، لكنا قد فرضنا أنّه لا سبب من خارج، فبقي أنْ يكون له من طباعه، والذي من طباعه يوجد له ما دامت طبيعته موجودة ولم تُقسر. فإنْ كانت طبيعته - بحيث تقبل القَسْر - أمكن أنْ يزول ذلك عنه بالقَسْر، وإنْ كانت طبيعته بحيث لا تقبل القَسْر، لم يزل ذلك عنه بالقَسْر.

(٣) فإنْ قال قائلٌ إنّه يجوز أنْ يكون كل قاسر يرد، فإنّه يعطي شكلاً ومكاناً، ثم يبقى ذلك فلا يزول إلاّ بقاسرٍ آخر يرد، فلا يخلو دائماً عن قاسر على التعاقب، كما لا يخلو عن الأعراض بالعاقب، وليس يلزم من ذلك أنْ يكون واحد منها ذاتياً لا يفارقه، فنقول: إنَّ الجسم تعرض له الأعراض التي ليست لازمة على وجهين: أعراضٌ تلحقه في ذاته، وأعراضٌ تلزمه من مجاوراته؛ مثل كونه فوق وتحت ومماسّاً ومحاذياً. والأعراض التي تلزمه لمجاوراته لا تصير ضرورية له باعتبار ذاته، والأعراض الأخرى فإنَّه لا يجب

while [the body] need not be devoid of the other accidents, their not being in it is, nonetheless, only possible. If its being devoid of them were impossible, such that [the body] subsists only by some of them existing in it, then they would be forms, not accidents. The fact is that accidents occur only after the thing exists as a substance such that the thing might exist, while each one of them does not. So it is possible to posit the substance of the body without any of them. As for [the accidents] owing to the immediate vicinity—namely, being contiguous with and what is analogous to that—they do not follow upon the body owing to its nature, but to its existing together with some other body. So, then, it is not at all necessary that the body be what essentially and actually bears some state that neither constitutes its essence nor entails what constitutes its essence. So the concern has been resolved.

(4) The case of things acting by force is just like the accidents, because the things acting by force do not constitute the essence of [the body]. What acts by forces, then, arises externally so as to provide a certain state without which that body would not have that state. Now, none of this is something necessarily belonging to the essence or what follows upon the essence. So the estimative faculty's imagining the body without anything acting upon it by force is not impossible relative to the nature of the body, whereas its imagining the body without possessing some *where* or space proper to it is impossible relative to the nature of the body. So the body, in its nature that belongs to it, necessarily entails that it have some space that it would have even if there were nothing acting by force (which might not occur). The same holds for the shape, the quality, and the like, as well as the position of the parts (if it has actual parts). So every body has a natural space; and if it has a place, then its space is a place.

أنْ يخلو منها ، بل يجوز أنْ يكون فيه عدمها فقط ، ولو كانت مما يستحيل خلوّها عنه ، بحيث لا يقوم إلاّ بوجود شيء منها فيه ، لكانت صوراً لا أعراضاً . بل الأعراض هي التي تعرض بعد تجوهر الشيء ، بحيث يجوز أنْ يوجد الشيء وكل واحدٍ منها معدوم ، فيمكن فرض جوهر الجسم دون شيء البتة منها . وأمّا المجاورات والمماسّات ، وما يجري مجرى ذلك ، فليس تلزم الجسم لطبيعته ، بل لوجوده مع جسم آخر . فليس إذن يجب - لا محالة - أنْ يكون الجسم لذاته حاملاً بالفعل لحالٍ ممّا لا يقوّم ماهيته ، ولا يلزم ما يقوّم ماهيته ، فقد انحل التشكّك .

(٤) وحال القواسر حال هذه الأعراض ، لأنَّ القواسر لا تقوّم ماهيته ، فإنَّ القاسر هو الذي يرد من خارج فيفيد حالاً ، لولاه لما كان لذلك الجسم تلك الحال ، وليس شيء من هذه واجباً أنْ يكون من الماهية أو لازماً للماهية . فتوهم الجسم - ولا قاسر يقسره ليس ممتنعاً بالقياس إلى طبيعة الجسم ، وتوهّم الجسم غير ذي أينٍ يخصه أو حيّزٍ ممتنع بالقياس إلى طبيعة الجسم . فالجسم يلزمه - في طبيعته التي له - أن يكون له حيز ، ذلك الذي لولا القاسر الذي يجوز أن لا يكون ، لكان له ، وكذلك الشكل والكيف وغير ذلك ، وكذلك وضع الأجزاء ، إنْ كان له أجزاء بالفعل . فكل جسمٍ فله حيّزٌ طبيعي ، فإنْ كان ذا مكان ، كان حيّزه مكاناً .

(5) One might say that earth is a simple body and [that] its nature requires that there be dryness in it. So [the dryness] either does or does not require that it have some shape. On the one hand, if it requires that it have some shape, then it must require a circular shape, owing to its simplicity. In that case, concerning[1] the dryness that assists what [earth's] nature requires, when some part of earth is deprived of the circular shape insofar as some other shape is imposed upon it, it would be necessary that[2] its nature return it to its circular [shape]. It is not found, however, to be like that. On the other hand, if the dryness prevents that and interferes with the nature of that part and what it requires, and [yet] the dryness proceeds from [the earth's] nature, then one and the same nature necessarily requires two mutually exclusive things; but this is impossible.

(6) Dryness, we say, emanates from it only in order to preserve the natural shape that its nature requires. So, from the fact that it preserves the shape of [the body], it necessarily follows that it preserves in each part the extension of what it is shaped into that the nature of [the dryness] primarily requires. So, when something acting by force disfigures a part of its shape, the rest of it cannot be sensibly aware of what happened, but needs to preserve what the nature [of the dryness] required. So, if the nature [of the dryness] turned around and required some yet different extension, the nature would be contradicting what it primarily requires [namely, to preserve any given shape]. In that case, then, it is doing what the nature primarily requires. So this scenario [namely, that the earth's dryness should preserve only a circular shape] would be the contrary of what [the nature] primarily requires and at variance with what the dryness requires, which is what the nature requires. It is not inconsistent that, in some accidental state, the nature requires that which

1. Reading *ammā* with **T** for **Z**'s and **Y**'s vowelled *immā* (either).
2. Reading a simple *an* with **Z** and **T** for **Y**'s *aw an* (or that).

(٥) ولقائـل أنْ يقول: إنَّ الأرض جرم بسـيط، وتقتضي طبيعته اليُبْس الذي فيه، فلا يخلو إمّا أنْ يقتضي له شـكلاً أو لا يقتضي، فإنْ اقتضى له شكلاً فيجب أنْ يقتضي شكلاً مستديراً لبسـاطته. فحينئذ أمّا أنْ يكون اليبس يساعد مقتضى طبيعته، فيجب أنْ تكون الأرض، إذا سلب جزء منها الشكل المستدير، بأنْ تشكّل شكلاً آخر، أنْ يعود بطبيعته فيسـتدير، وليس الموجود كذلـك. وإنْ كان اليبس يمنع ذلك، ويحول بين طبيعة ذلك الجزء ومقتضاه، واليبس صادرٌ عن طبيعته، فيجب أنْ تكون طبيعة واحدة تقتضي معنيين متفاوتين متقابلين، وليس هذا بجائز.

(٦) فنقول إنَّ اليبس إنّما يفيض عنه ليحفظ ما تقتضيه طبيعته من الشـكل الطبيعي حفظاً قوياً جداً، فإذا حفظ شكله لزم من ذلك أنْ يحفظ في كل جزء ما توجبه طبيعته إيجاباً أولياً، من انبسـاط الذاهب إلى شكله. فإذا ثُلم شيءٌ من شكله بقسْر القاسر لم يكن للباقي منه حسّ شعور بما حدث، بل كان عليه أنْ يستحفظ ما أوجبته الطبيعة. فإنْ عادت الطبيعة فأوجبت انبساطاً آخر، كانت الطبيعة هي المناقضة لموجبها الأول، فكان حينئذ مقتضى الطبيعة، بهـذه الحال ضدّ مقتضاها الأول، ومخالفاً لمقتضى اليبس الذي تقتضيـه الطبيعة. ولا يبعد أنْ تكون الطبيعة تقتضي – في حال عارض – أمراً مناقضاً

contradicts and opposes what it requires in its regular state. In that case, then, the two required things are not mutually exclusive contraries proceeding from one and the same power in one and the same state, such that there would be an absurdity. Instead, one of them proceeds from the power while it is in its natural state, while the other proceeds from it while it is in an unnatural state. That is, for example, like the state of rest that happens as a result of the nature when it is in a natural state, while motion happens to result from it when it is in an unnatural state. As for when a portion of some [elemental] component other than earth undergoes alteration into earth, its alteration is first into some nonspherical shape. That, then, is on account of external impediments and the difference of the parts in priority, posteriority, and immediate vicinity during the generation into earth.

(7) Having clarified this objective of ours, we ought to explain natural place, how it belongs to bodies, and how it belongs to what is simple and composite. We say: It is only fitting that we want to know whether some body or other could have two natural places or a single place, and whether two bodies can be at rest in it naturally. We would also like to know the state of simple bodies that have distinct parts: namely, does each one of [the parts] have some numerically different place that is absolutely proper to it, such that each one of them has some natural place different from that of another? How would the place of *this* become different from the place of *that* and proper to it to the exclusion of the other? How are those places related to the place that belongs to the whole? Again, we want to know the state of the composite body with respect to its natural where, for it certainly has some natural place. So what is that natural place? If it is a place of a single part, then the other parts won't be in their natural place.

ومقابلاً لما تقتضيه، في حال كونه سالماً. فليس إذن المقتضيان بمتضادين متمانعين صادرين عن قوةٍ واحدةٍ بحالٍ واحدة، حتى يكون محالاً، بل أحدهما يصدر عن القوة وهي على حالتها الطبيعية، والآخر يصدر عنها وهي بحالٍ غير طبيعية. وذلك مثل السكون يعرض عن الطبيعة إذا كانت على حالٍ طبيعية، ثم تعرض عنها الحركة إذا كانت بحالٍ غير طبيعية. وأمّا الجزء من عنصرٍ غير الأرض - إذا استحال إلى الأرض - فاستحال أولى استحالته إلى شكلٍ غير كُري، فذلك لموانع من خارج ولاختلاف الأجزاء في التكوّن أرضاً، اختلافاً في التقدّم والتأخّر والمجاورات.

(٧) وإذ قد أوضحنا غرضنا هذا، فبالحري أنْ نبيّن أنَّ المكان الطبيعي كيف يكون للجسم، وكيف يكون للبسيط منه وللمركّب، ونقول إنّه يخلق بنا أنْ نعرف هل يجوز أنْ يكون جسمٌ من الأجسام له مكانان طبيعيان، أو مكان واحد، وله جسمان يسكنانه بالطبع وأنْ نعرف حال الأجسام البسيطة التي لها أجزاء متمايزة، ولكل واحدٍ منها مكان آخر بالعدد يخصّه لا محالة، فيكون لكل واحدٍ منها مكان طبيعي غير الذي للآخر، وأنّه كيف يصير مكان هذا غير مكان ذلك، ويختص به دون الآخر وكيف نسبة تلك الأمكنة إلى المكان الذي للكل. وأنْ نعرف حال الجسم المركّب في أينه الطبيعي، فإنَّ له مكاناً طبيعياً لا محالة، فما ذلك المكان؟ فإنّه إنْ كان مكان جزءٍ واحدٍ، كانت الأجزاء الأخرى في غير مكانها.

(8) We say that it is impossible for a single body to have two natural places, except in the way that there are certain potential spaces in the entirety of the place of the whole. It occurs in whichever of [the spaces] that it does because of a special property that naturally belonged to it—like the clod of earth, for the space closest to that of the earth that is adjacent to [the clod] is natural to it. As for [having] two distinct places, that is impossible. [That] is because what a single individual requires is a single individual thing, whereas what the whole, composed of homogeneous parts, requires is what the entirety of all those parts requires. Now, it is not impossible for the bodies whose natures are alike to be continuous because of their nature. In fact, if it is impossible, it would be so only because of some accident that happens, whereas, in their nature they are such that it is possible that they should be continuous. Since their continuity is not impossible, how could their being contiguous be impossible? Should they be continuous or contiguous, nothing impossible has occurred. When they are continuous and contiguous, then when the entirety as something natural seeks the natural place, it is as a single thing, which is the entirety of these natures (in fact, this entirety consists of these natures). So it necessarily seeks the entirety of the space that is the space of this entirety. In fact, this space is due to this entirety as if it were an entirety that is a collection of each and every one of the spaces. So, then, the spaces of bodies that are alike in nature are the parts of a single space, and any particular body from that entirety has some space appointed to it from that entirety on account of the cause of that cause. As for the need of [a body] to exist in [some space] when it first comes to be and to be naturally suited to it, it is necessary, whereas [the body's] specific proximity—such as the fire that is moved upward to some part of the space of fire's collective kind itself—is only because of its being nearer to it.

(٨) فنقول: إنّه لا يجوز أنْ يكون لجسم واحدٍ مكانان طبيعيان، إلاّ على جهة أنَّ في جملة مكان الكل أحيازاً بالقوة، أيّها وقع فيه بسببٍ مخصّص كان طبيعياً له، كالمَدَرة فإنَّ أقرب حيّزٍ من حيّز الأرض يليها هو طبيعياً لها، وأمّا مكانان يتباينان فليس يمكن ذلك. فإنْ مقتضى الواحد بالشخص، من حيث هو واحد بالشخص، أمرٌ واحدٌ بالشخص، ومقتضى الكل المتشابه الأجزاء جملة مقتضى جميع الأجزاء. والأجسام المتشابهة الطبائع لا يستحيل عليها الاتصال لطبيعتها، بل - إنْ استحال - فإنّما يستحيل لعرضٍ يعرض وهي في طبيعتها؛ بحيث يجوز عليها، أنْ لو كانت متصلة، وإذ لا يستحيل اتصالها، فكيف يستحيل تماسّها؟ لو اتصلت أو تماسّت لم يعرض شيء مستحيل. وإذا اتصلت وتماسّت كانت الجملة - وهي تطلب المكان الطبيعي من حيث هي طبيعة واحدةٌ - هي جملة هذه الطبائع، بل هذه الجملة من الطبائع. فيجب أنْ تطلب جملة من الحيّز، هي حيّز هذه الجملة، بل هذا الحيّز لهذه الجملة، كأنّه جملة تجتمع من أحياز واحدٍ واحد. فإذن الأجسام المتشابهة الطبائع فإنَّ أحيازها كأنها أجزاء حيّز واحدٍ، ويكون لجسمٍ معيّنٍ من تلك الجملة حيّز يتعيّن له من تلك الجملة، لعلّة تلك العِلّة. أمّا وجوده فيه أولاً عندما حدث، وهو موافق له في الطبع، فوجب لزومه، وأمّا اختصاصه بالقرب، فإنَّ النار إنّما تتحرك إلى فوق وإلى جزءٍ من حيّز كلّية النار بعينه، لأنّه هو أقرب إليه.

(9) Someone might ask: If our estimative faculty were to imagine fire at the center of the celestial sphere, no part of it having an inclination toward some direction, what would happen to it with respect to its nature? Would it rest naturally—but that is absurd—or would it move toward some direction, no direction having been specified? We say that it would happen to be at rest, but by force. [That] is because a certain empty space would be formed at its center when [the fire] extends out equally from [the center] in [all] directions until each part of the extending mass encounters the natural place closest to it. In that case, however, the surrounding air and the like would not[3] permit [the fire] to interpenetrate it, since this penetration would not progress by piercing through [the air], because *piercing* is in one direction to the exclusion of another, whereas this would be an extending out in every direction. Thus, [the fire] would stay at rest by force. Furthermore, it is impossible that a void come to be at the center when it is pierced, whereas this force [namely, the fire's force to pierce through the air in order to move toward its natural place] would be a certain accidental force resulting from the nature.[4] This is really odd, since the nature would require something impossible to happen owing to certain accidents that occurred. So that would lead to something completely outside the natural order.[5] We do not know and deny the impossibility of this accident because we did not initially know and deny the impossibility of its effect on the subject. When the antecedent is possible, however, the consequent is as well; so, if the consequent is impossible, the antecedent is, too. So it has become apparent how the single body has a single place or single space by nature, and the way in which the space of the whole is related to the space of the parts, one to another, where these are the simple [elements].

3. Reading *lā* with **Z** and **T**, which has been (inadvertently) omitted in **Y**.
4. Reading simply *ʿan* with **Z** and **T** for **Y**'s preferred *min khārij ʿan* (external to).
5. Literally, "a foreign jurisdiction."

(٩) ولسائلٍ أنْ يسأل، إنّا لو توهمنا النار في مركز الفلك، لا ميل لجزء منها إلى جهة، فماذا كان يعرض لها في طبعها، أسكونٌ بالطبع - وذلك محال - أو حركة إلى جهة، ولا مخصّص لجهة؟ فنقول: كان يعرض لها السكون ولكن بالقَسْر، لأنّها كانت تقتضي أنْ يفرج عن فرجةٍ، في واسـطتها، تنبسـط عنها إلى الجهات بالسواء، إلى أنْ يلقى كل جزء من المنبسط ما هو أقرب إليه من المكان الطبيعي. لكن الهواء المحيط وغير ذلك، كان حينئذ لا يمكّنها من أنْ تداخله نافذة هذا النفوذ، إذ هذا النفوذ لا يتأتى بالخرق، لأنَّ الخرق يكون في جهةٍ دون جهة، وهذا انبساطٌ في كل جهة، فتكون ساكنة بالقَسْر وأيضاً فإنَّ الخلاء ممّا لا يجوز أنْ يحدث في الوسط عند انخراقه، وهذا القَسْر قَسْرٌ عارضٌ عن الطبع، وهو عجيب جداً، فإنَّ الطبع يقتضي أمراً صار غير ممكن لعارض عرض، فأدى ذلك إلى حكمٍ غريب. ونحن لا ندري استحالة هذا العارض ولا نمنعها، لأنّا لم ندر بعد استحالة المعروض له في الموضوع مقدماً ولا نمنعها، ولكن إذا جاز المقدّم جاز التالي، فإنْ امتنع التالي امتنع المقدّم. فقد ظهر أنّه كيف يكون للجسم الواحد مكان واحد بالطبع، أو حيّز واحد بالطبع، وأنّه كيف تكون نسـبة حيّز الكل إلى حيّز الأجزاء بعضها إلى بعض، وهذا للبسـائط.

(10) As for composite [bodies], their composition is from either two simple [elements] or more. If it is from two simple [elements], then either they are equal in power or one of them dominates. If they are equal in power but the position of one them does not chance to be opposite the side of the other, then they are not united, but are held together only by something acting by force to join them together. If their motions are in opposing directions and each one is equally distanced from its [natural] place, then they would balance each other out while each one forcibly acts upon the other. So they would be at a standstill, unless something breaks in and helps one of them. Alternatively, they may be at the shared limiting point between the two spaces, at which they might then naturally be at a standstill. If the power of one of them dominates and it brings the force to bear on the mix, the natural place is that of the dominant [element]. If [the compound] is from more than two simple [elements] but, among them, one dominates, then the [natural] space is that of the dominant [element]. If they are equal, then the two simple elements dominate whose direction is one relative to the subject[6] in which there is the composition. Also, the composite will be at the closest of the two spaces to the space where the composition occurs, but not going beyond it. [That follows] since the attraction from it toward the two sides is equal, and it does not overcome the check put on it by the simple [element] that seeks that space when the two attractions are at variance. Perhaps there is not a true mixture of the simple bodies by their [simply] being attached to one another except [in the following cases]: [(1)] There is a dominant one that joins and forcibly acts upon the other parts, preventing them from moving to their proper spaces. [(2)] Alternatively, the parts have become so small that they cannot act so as to pierce the bodies that are between them and their collective kinds. [(3)] Finally, there is some power forcibly acting upon the collection other than the powers of the simple [elements]. Let us now prove that every natural body has some principle of motion such that every body has some natural motion and that [that principle] is as a single species only.

---

6. **Z** and **T** read *mawḍiʿ* (location).

(١٠) وأمّا للمركّبات، فإنَّ تركيبها لا يخلو إمّا أنْ يكون عن بسيطين، أو عن أكثر من بسيطين، فإنْ كان عن بسيطين، فإمّا أنْ يكونا متساويين في القوة، أو أحدهما أغلب، فإنْ كانا متساويين في القوة، ولم يتفق أنْ كان وضع أحدهما بحذاء جهة الآخر؛ تفرّقا، ولم يحتبسا إلاّ بقَسْر جامع. وإنّ تواجهت حركتاهما، وبُعْد كل واحدٍ عن مكانه كبُعْد الآخر، تقاوما وقسر كل واحدٍ الآخر فوقفا، إلاّ أنْ يطرأ على أحدهما مُعينٌ، أو يكونا في الحدّ المشترك بين الحيّزين، فيجوز أنْ يقفا فيه بالطبع. وإنْ غلب فيه قوة أحدهما، والقَسْر على المزج حاصل، كان المكان الطبيعي مكان الغالب، وإنْ كان عن أكثر من بسيطين، وفيهما غالب كان الحيّز للغالب. وإنْ تساوت غلب البسيطان اللذان جهتهما واحد بالقياس إلى الموضوع الذي فيه التركيب، وحصل المركّب في أقرب الحيّزين من حيّز وقوع التركيب ولم يتجاوزه، إذ الجذب عنه إلى الجانبين سواء، والإمساك فيه عن البسيط الذي يطلب ذلك الحيّز لا يبطله، يخالف الجذبين. وعسى أنْ لا يصح امتزاجٌ من الأجسام البسيطة تتلازم به، إلاّ وهناك غالبٌ يجمع ويقسر الأجزاء الأخرى، مانعاً إياها عن الحركة إلى أحيازها الخاصة، أو تكون الأجزاء قد تصغّرت تصغّراً لا يمكنها أنْ تفعل في الأجسام التي بينها وبين كلياتها خرقاً، أو تكون قوة قاسرة على الاجتماع، غير قوى تلك البسائط. فلنبيّن الآن أنَّ لكل جسمٍ طبيعي مبدأ حركة، حتى يكون لكل جسمٍ حركة طبيعية، وأنّه على نوعٍ واحدٍ فقط.

Chapter Twelve

Establishing that every natural body has a principle of motion with respect to either place or position

(1) We say that every body either is or is not able to be forcibly moved from the location in which it is. If it is able to undergo locomotion away from the location in which it is, then, in its substance, either it has some inclination toward some space or it has no such inclination. Whatever the case, every body has some natural place or space at which its nature requires it to be, differing in that [respect] from the rest of the bodies, not by its corporeality, but only because there is some principle and power in it disposing it toward that place. So, when that power requires that place, and its corporeity[1] as such does not prevent it from undergoing locomotion and motion, then there is nothing in it contrary to its power or what its power requires demanding some other space. [That] is because, in a single body whose parts are not at variance, there cannot be two contrary powers. Otherwise, they would require two mutually exclusive actions, since the powers' being powers are in accordance with their action. When their actions are mutually exclusive, their

1. Normally, when speaking of corporeality, Avicenna uses the Arabic *jismīyah,* which is derived from *jism,* which is his standard term for the *physical* (or, perhaps, *composite*) body found in the sublunar realm; however, here he uses *jirmīyah,* derived from *jirm,* which is his preferred term for a *celestial* (or, perhaps, *simple*) body. His use of this different term is perhaps to note the fact that some of the celestial bodies, such as the outermost celestial sphere, are incapable of undergoing anything put positional (as opposed to local) motion. To mark the difference, I have used the admittedly artificial distinction between corporeality (*jismīyah*) and corporeity (*jirmīyah*).

<الفصل الثاني عشر>

في إثبات أن لكل جسم طبيعي مبدأ حركة وضعيةٍ أو مكانية

(١) نقول إنَّ كل جسم لا يخلو إمّا أنْ يكون قابلاً للنقل عن موضعه الذي هو فيه بالقَسْر، أو غير قابل، فإنْ كان قابلاً للنقل عن موضعه الذي هو فيه؛ فإمّا أنْ يكون له في جوهره مَيْلٌ إلى حيّزٍ، أو لا يكون له مَيْلٌ إليه البتة. لكن كل جسم فله مكانٌ طبيعي، أو حيّز طبيعي تقتضي طبيعته الكوْن فيه، وإنّما خالف سائر الأجسامَ في ذلك لا بجسميته، بل لأنَّ فيه مبدأ وقوة معدّة نحو ذلك المكان. فإذا كانت تلك القوة مقتضية لذلك المكان، وجرميته غير ممتنعة - بما هي جرمية - عن الانتقال والحركة، فلا مضادة فيه لقوته ولا لمقتضى قوته، تقتضي حيّزاً آخر. لأنّه لا يجوز أنْ يكون في جسم واحدٍ، غير مختلف الأجزاء، قوتان تتضادّان أو تقتضيان فعلين متمانعين، إذ القوى كونها قوى بحسب فعلها، وإذا تمانعت أفعالها تمانعت طبائعها، فاستحالت أن تكون معاً لجسم فإنَّ الجسم الذي فيه

natures are mutually exclusive, and so it is impossible that they simultaneously belong to some body. So the body in which there is a given power is that in which there is a principle of a certain action that proceeds inevitably, if there is no impediment. If the body is such that that action does not proceed from it (even if there is no external impediment), then that power is not in it. So, when there are two contrary powers in it, two contrary actions would, in fact, proceed, which is absurd. So, then, it is absurd that, in a simple, isolated body or in a dominant body of some compound, there should be two powers, one requiring a given place and the other impeding [the body] from it. Again, then, from what motion requires, the body is something able to move. So, when the body is forcibly removed from its natural place, it necessarily follows that it undergoes motion [back] toward its natural place once the external agent acting by force is removed. Something else that proves this is that any body in which there is no principle of inclination would undergo locomotion from a given *where* or position that it has instantaneously, which is absurd. The simple fact is that every body must be susceptible to the production of motion and a new inclination, and so there is a natural principle of inclination in it with respect to the very thing to which it is susceptible, whether a certain *where* or a position.

(2) Let us first single out for discussion the production of motion with respect to place as a way of making clear what is intended by what is most obvious, even if the explanation of motion with respect to place and position are doctrinally the same. We say that bodies found possessing inclination are like heavy and light: the heavy is what inclines downward, while the light inclines upward. Whenever there is an increase

قوة ما هو الذي فيه مبدأ فعلٍ ما ، يصدر لا محالة ، إنْ لم يكن عائق ، وإنْ لم يكون الجسم بحيث يصدر عنه ذلك الفعل – إنْ لم يمنع مانعٌ من خارج – فليس فيه تلك القوة . فإذا كانت فيه قوتان متضادّتان صحّ صدور فعلين متضادّين ، وهذا محال . فإذن من المحال أنْ يكون في جسمٍ بسيطٍ مفرد ، أو في غالب جسمٍ مركّب ، قوتان واحدة تقتضي مكاناً ، والأخرى تمنع عنه . ثم الجسم قابل للحركة – من مقتضى الحركة – فيلزم أنَّ الجسم إذا قسر على مفارقة مكانه الطبيعي أنْ يتحرك إلى مكانه الطبيعي ، عندما يفارق القاسر من خارج . ومّما يُبيِّن هذا أيضاً أنَّ كل جسمٍ ليس فيه مبدأ مَيْلٍ ما ، فإن نقله عما هو عليه من أيْنٍ أو وضْعٍ يقع لا في زمان ، وذلك محال ، بل يجب أن يكون كل جسم يقبل تحريكاً وإمالة طارئة ، ففيه مبدأ مَيْلٍ طبيعي في نفس ما يقبله ، كان أيْناً أو وضعاً .

(٢) ولنعيّن الكلام أولاً في التحريك المكاني على سبيل إيضاح المقصود فيما هو أظهر ، وإنْ كان المكاني والوضعي في مذهب البيان واحداً ، فنقول إنَّ الأجسام الموجودة ذوات المَيْل ، كالثقيلة والخفيفة . أمّا الثقيلة فما تميل إلى أسفل ، وأمّا الخفيفة فما تميل إلى فوق ، فإنّها كلّما ازدادت ميلاً كان قبولها للتحريك النقلي أبطأ ، فإنَّ نقل الحجر العظيم

in their inclination, their ability to produce locomotion is slower. So [for example] displacing a large, heavy stone is not like displacing a small, little one, and pushing a little air [such as a partially inflated bladder] under water is not like pushing a lot of air [such as a fully inflated bladder]. When small bodies are overwhelmed—for example, when a single mustard seed, a piece of straw, or a splinter of wood is thrown and does not pass through the air the way that the heavy body does—the reason for it is not that the heavier thing is more susceptible to being thrown and dragged. On the contrary, it is because some of these, owing to their smallness, do not receive from what propels them a power that can move both them and what is adjacent to them sufficiently to make them[2] capable of cutting the air. Moreover, the quick depletion [of that power] is due to the cause (which will be introduced in its proper place)[3] that depletes the acquired, accidental, and motive powers. For example, a single spark would be extinguished by the cause that depletes an acquired heat before a large fire would be so depleted. Also, some rarefied things are not able to pass through the air, but, instead, the air through which they pass interpenetrates them and is a cause for the depletion of their acquired power. You will learn that whatever opposes what passes through it is the thing that depletes the power of the motion.[4] This is like rarefied fire and water, for they are more susceptible to alteration. Now, if largeness and increased weight were the reason for the projectile's [capacity] to

2. Literally, "to have an effect on their strength," reading *shiddah* (strength) with **Z** and **T** for **Y**'s (inadvertent) *ḍiddah* (contrary?).

3. Although Avicenna had a somewhat detailed discussion of the causes that affect the speed of a mobile at 2.8.11–12, to which he appeals later in the same chapter to explain how an impressed power might be depleted, here he is clearly using the future tense, and so it would seem that he is referring to some future discussion. The reference may simply be to his passing remarks at 4.14.2.

4. Ibid.

الشديد الثقل أو جرّه، ليس كنقل الحجر الصغير القليل الثقل وجرّه، وزج الهواء القليل في الماء ليس كزجّ الهواء الكثير. وأمّا ما يعتري الأجسام الصغيرة، مثل الخردلة ومثل التبنة ونحاتة الخشب، من أنها لا تنفذ عند الرمي في الهواء نفوذ الثقيل، فليس السبب فيه أنَّ الأثقل أقبل للرمي والجرّ، بل لأنّ بعض هذه لصغرها لا تقبل من الدافع قوة محرّكة لها ولما يليها، تبلغ من شدّتها أنها تقدر بها على خرق الهواء، ومع ذلك فيكون سريع الاستحالة إلى البطلان من السبب الذي يُعرف في موضعه؛ وهو السبب الذي يبطل القوى المستفادة العرضية والقوى المحرّكة، كما أنَّ الشررة تطفأ قبل النار الكثيرة، من السبب الذي يبطل الحرارة المستفادة. وبعضها يكون متخلخلاً لا يقدر على خرق الهواء، بل يداخله الهواء الذي ينفذ فيه، ويكون سبباً لإبطال قوته المستفادة. فإنك ستعلم أنَّ مقاومة المنفوذ فيه هو المبطل لقوة الحركة، وهذا كالنار المتخلْخلة والماء المتخلْخل، فإنّه أقبل للاستحالة. ولو كان السبب في قبول الرمي الأبعد هو الكِبر وزيادة الثقل، لكان كلما ازداد المرميّ ثقلاً وكبراً،

be thrown farther, then whenever the projectile's size and weight were increased, it would be [capable of being] thrown even farther, which is contrary to fact. Instead, when only heavy and light, but no other causes, are considered, then the ability to move the smaller magnitude is greater, and there is a faster motion. So the ratio of the distances and times covered by things undergoing motion—both those moved by force and those having a natural inclination—is proportional to the relation between one inclination and another, except that the periods of time are inversely proportional to the distances. In the case of distances, the more intense the inclination, the greater the distance covered, whereas, in the case of time, the greater the inclination, the shorter the time. Now, if there is absolutely no inclination, and [if] the forcibly moved object is moved for a period of time, and [if] that period of time is proportional to a given time of a motion possessing a forced inclination—in which case it is proportional to a given ratio of one inclination (should it exist) to an inclination possessing the inclination of the forcibly moved object—then what has absolutely no inclination in it would be just as susceptible to the force as what does have a given inclination (should it exist). In that case, however, what has no impediment would be proportional to what does have some impediment (were it to exist), thus resulting in a contradiction exactly like the one we addressed in the case of the void, and for the very same reason.[5]

---

5. See 2.8.11–13. Roughly the same argument as appears here is also found in a slightly more developed form in *al-Ishārāt wa-l-tanbīhāt*, ed. Forget (Leiden, 1892), 109–110). "Indication: the body in which there is neither a potential nor actual inclination is not susceptible [110] to a forcible inclination by which it is moved, and, in general, it will not be forcibly moved. If this were not the case, then let x be forcibly moved in a given time [t_1] [and along] a given distance [d_1] and let y, for example, in which there is a given inclination and resistance [i_1], be moved. Clearly, then, y will be moved [d_1] in a longer time. Now, let z [have] an inclination [i_2] weaker than that inclination [i_1] which, as a result of the same mover, covers a [greater] distance [d_2] in the same time [t_1], whose ratio to the first distance [d_1] is the ratio of the time as the one possessing the first inclination [t_2] and the time of the one lacking the inclination [t_1] such that it is forcibly moved the same distance in the same time of the one lacking the inclination. Thus, there will be two forced motions [x and z], z having a resistance in it and x not having a resistance in it, that are of comparable states with respect to speed, which is absurd. Note: You must note here that there is not some indivisible time [i.e., 0 amount of time] such that, during, it a certain motion having no inclination might occur and would have no ratio to a given time of a motion possessing an inclination."

كان أقبل للرمي، والأمر بخلاف ذلك، بل إذا اعتُبر الثقل والخفّة ولم تعتبر أسبابٌ أخرى، كان الأقل مقداراً أقبل للتحريك القسري وأسرع حركة. فتكون نسبة مسافات المتحركات بالقَسْر ولها مَيْل طبيعي ونسبة أزمنتها على نسب المَيْل إلى المَيْل، لكن النسبة في المسافات بعكس النسبة في الأزمنة. أمّا في المسافات، فيكون الأشد ميلاً أطول مسافة، وأمّا في الزمان فيكون ذلك أقصر زماناً. وأمّا إذا لم يكن مَيْلٌ أصلاً، وتحرّك المقسور في زمان، ولذلك الزمان نسبة إلى زمان حركة ذي الميل بالقَسْر، ويكون على نسبة، لو وجد، إلى ميل ذي الميل المتحرّك بالقَسْر، فيكون قبول ما لا مَيْل فيه أصلاً للقَسْر، كقبول ذي مَيْلٍ مّا لو وجد، فيكون الذي لا مانع له، على نسبة ذي مانعٍ مّا لو وجد. ويعرض مثل ما قلنا في باب الخلاء من الخُلْف، وعلى ذلك الوجه بعينه.

(3) Something that proves that is that the influence upon what undergoes forced rectilinear or circular motion varies, being stronger and weaker. When that [influence] varies, the strong one is obviously obeyed, while the weak one hinders. Now, the hindering factors do not belong to the body *qua* body, but through something in it that seeks to maintain its state of place and position, where this is the principle that we are explaining. So, in every body undergoing forced locomotion, there is a principle of a particular inclination. It has already been proved in the case of locomotion with respect to place. In the case of forced locomotion with respect to position, if that body is able to undergo local motion from its place, then [the need for a principle of inclination] is obvious as well. If it is not able, then it certainly has some power by which it stays fixed in its place, which necessarily follows upon it, and is specific to it, and is not [due to] its corporeality.

(4) So we say that there is also a principle of motion in this body, which is evident when it is considered next to what was learned about the body that is able to move locally from its location. That is because [the body that moves with respect to position] has a certain numerically [distinct] position either in that which surrounds it or around that which it contains, whether in *that* or around *this*. So it has that, either from some cause in itself and from its natural form, or from some cause outside of the nature. It is absurd that it itself should require that. Indeed, none of the parts that are posited in it, and its various directions, and

(٣) ومّما يبيّن ذلك، أنَّ المقسور على الحركة المستقيمة أو المستديرة، يختلف عليه تأثير الأقوى والأضعف. وإذا اختلف ذلك، فظاهرٌ أنَّ القوي مطاوعٌ والضعيف مُعاوقٌ، وليست المعاوقات للجسم – بما هو جسم – بل بمعنىً فيه يطلب البقاء على حاله من المكان والوضع، وهذا هو المبدأ الذي نحن في بيانه. فكل جسم منتقل بالقَسْر ففيه مبدأ ميلٍ ما. أما الانتقال المكاني فقد بيّناه، وأمّا الانتقال القسري الوضعي ‹فإنَّ› ذلك الجسم، إنْ كان قابلاً للنقل عن مكانه فقد ظهر، وإنْ كان غير قابلٍ، فله لا محالة قوة بها يثبت في مكانه، وتلزمه وتختص به، وهي غير جسميته.

(٤) فنقول إنَّ هذا الجسم فيه مبدأ حركة أيضاً، ويستبين إذا اعتبر قريباً ممّا اعتبر به أمر الجسم القابل للنقل عن موضعه. وذلك لأن له وضعاً ما بالعدد فيما يحويه، أو حول ما يشتمل هو عليه، أو في ذلك أو حول هذا. فلا يخلو إمّا أنْ يكون له ذلك عن علّةٍ في ذاته وعن صورته الطبيعية، أو عن علّةٍ خارجةٍ عن الطبيعة. ومحال أنْ تقتضي ذلك ذاته، فإنَّ الأجزاء التي تفرض فيه، والجهات المختلفة التي تكون له، والأجزاء التي تفرض فيما

the parts that are posited in that with which it is in contact, are any more suited—I mean that some part of it is in a given direction—to being contiguous with any given particular part, since the whole does not vary. So the nature of the body does not require the same position, since there is nothing that some of the homoeomerous parts deserve by their nature that the other homoeomerous parts do not equally deserve; rather, all of that might belong to every one of [those parts]. This is not like what belongs to the parts of bodies that are able to be separated [from a given place], for we find that every posited part in [a body that can be so separated] can be specified by something proper to it. [That] is either because the existence of [the part] first occurs *there;* or because it is the nearest of the locations to the one at which it existed; or [because] it underwent locomotion to it, departing from its natural space. Having some existence at which it first is or undergoing forced local motion to it is not specific to every part inasmuch as it is in it, either by the nature considered alone or by force, but owing to the nature joined with some specifying feature.[6] As for that which is unable to depart from its place, neither does this status hold for it nor can this interpretation be brought in line with it, <because it can be said that this part of earth, as a result of its being an individual, required that it be in this place, since this part actually exists in [this place], whereas this status does not apply to the body in which there are no actual parts.>[7] When that is the

6. Avicenna's point is that when the parts of, for example, earth undergo local motion, such that one part of earth leaves behind some other parts, that departing part comes to have some specifying feature, which is a result of the local motion, by which it can be distinguished from some other part of earth. Such specifying factors, as Avicenna notes, are like the following: the earth, which was *here,* for example, in my compost heap, is now *there,* for example in my garden; or the water that is in a certain cloud was high in the sky but now lower and so closer to the region of water; or this clod of earth is different from that one thrown in the air since that one is not in its natural location. In contrast, when there is only positional motion, the parts retain their same relative positions vis-à-vis one another, and so there is no specifying factor that results from the positional motion by which one part can be distinguished from another.

7. *Because it can be said... no actual parts:* This clause, which occurs in **Y**'s text, is absent in **Z**'s text and appears as an addition to **T** in the margins.

يماسّه، ليس شيء منها أولى بشيء منها ؛ أعني أنّه ليس جزء يكون منه في جهة ؛ أولى بماسّة جزء بعينه، إذ الجميع غير مختلف. فطبيعة الجسم ليس تقتضي الوضع بعينه، إذْ المتشابهات لا يستحق بعضها بطبعه شيئاً من المتشابهات بعينه دون بعض، بل يكون جميع ذلك جائزاً لكل واحدٍ منها. وليس هذا كما يكون لأجزاء الأجسام القابلة للتفرّق، فإنَّ كل جزء يفرض نجده متخصصاً بما تخصّص به، لأنَّ أول وجوده وقع هناك، أو لأنّه أقرب المواضع من موضع وجد فيه أو نقل إليه خارجاً عن حيّزه الطبيعي، أمّا لوجودٍ يكون الأول فيه، أو لوقوعٍ الانتقال بقاسرٍ إليه فيكون اختصاص كل جزء بما هو فيه، لا بالطبع المجرّد ولا بالقَسْر، بل للطبع المقترن بمعنى مخصّص. وأمّا الذي لا يقبل مفارقة مكانه، فليس حكمه هذا الحكم، ولا يجري عليه هذا التأويل، لأنه يمكن أنْ يقال إنَّ هذا الجزء من الأرض اقتضى بشخصيته أنْ يكون في هذا المكان، إذ هذا الجزء موجود فيه بالفعل، وأمّا في الجسم الذي لا جزء فيه بالفعل فلا يستمر هذا الحكم فيه. فإذا كان كذلك، لا

case, none of the parts of that body can be specified by something proper to it, whether by the nature taken alone or the nature joined to some state involving force that some cause necessitates. Now, were there some impure mixture of a cause acting by force and what is required from its nature [such that there results] something that is required by certain causes proper to the parts of the elements by their spaces, then, if there were not that cause [acting by force] or [if] it ceased, it would not be in the nature of [that part] that it be specified. So, in its nature—in every state and however you wanted to dispose of the divisions—[the part] might or might not have that [instance of] facing and being contiguous with [some given direction that it is in]. Also, in its nature, it is able to move with respect to position.[8] We have already explained, however, that, in everything that is able to move from a certain thing, whether a *where* or a position, there is a principle of motion and a natural inclination. So, in this body, there must be a principle of inclination with respect to position.

(5) Know that what is intended by the explanation and examination we have presented is this: Every body in which there appears the production of some inclination whose principle is not in it naturally, but proceeds either from some external cause or a conjoined soul (which produces motion according to some intention and brings about a certain inclination that was not in the body), undergoes motion as a result of that [principle] only when there is a prior inclination in [the body]. The discussion

8. Avicenna's point, which is admittedly convoluted, might more easily be explained with a simple example. Imagine that you are holding a perfectly homogeneous sphere or ball in front of you. There is nothing about the nature of that sphere that requires that the "part" facing you have that and only that position. That very part could have been pointing toward your feet, toward the ceiling, or away from you. Consequently, there is nothing about the nature of the sphere that requires that its "parts," never change their relative position. So, by its nature, the sphere is able to rotate while never leaving the place or space in which it is rotating so as to have undergone locomotion.

يكــون جزء من أجزاء ذلك الجســم متخصّصاً بما تخصّص بــه بالطبع مفرداً ، ولا بالطبع مقارناً لحالةٍ قسرية أوجبها سببٌ . ولو كان هناك شوْبٌ من سببٍ قاسرٍ ومقتضٍ من طبعه أمراً اقتضى أســباب تخصّص أجزاء الأُسْطُقسّات باحيازها ، لكان في طبعه أنْ لا يكون متخصصاً به ، لو لم يكن ذلك الســبب أو زال . فيكون في طبعه على كل حال ، وكيف تصرّفت الأقسام ، جواز أنْ يكون على تلك المحاذاة والمماسّة وأنْ لا يكون ، ففي طبعه أنْ يقبــل نقلاً في الوضع . وقــد بيّنا أنَّ كل قابل نقل عن أمرٍ ما - أيْنٌ أو وضْعٌ - ففيه مبدأ حركةٍ ومَيْل طبيعي ، فيجب أيضاً أنْ يكون في هذا الجسم مبدأ مَيْل في الوضع .

(٥) واعلــم أنَّ المقصود فيما وضح بما شــرحناه من البيان والمكشــوف عنه ؛ هو أنَّ كل جســم تطرأ عليه إمالة لم يكن مبدؤها فيه بالطبع ، بل تصدر عن ســببٍ خارج ، أو نفسٍ مواصلة تحرّك بحســب القصد وتحدث ميلاً لم يكن في الجســم ، فليس يصح أنْ يتحرك الجسم عن ذلك ، إلاّ وفيه ميل متقدم . فإنّ الكلام في التحريك المبتدأ ، الواقع بقصد

concerning the principle for the production of motion that occurs through the intention of the soul is just like the one concerning its inclination that occurs though some external cause. So [for example] you see that the animal soul produces various motions in [the body] owing to the body of [the animal], while the power is the same depending on [how] more and less heavy the inclination in the body is (where you find a certain resistance owing to the heavier [inclination]). So you find that the argument stands. Also concerning this, you should consult the discussions in the things that will follow[9] and so find in them what is more to your liking, if you desire more details.

(6) Having clearly explained that there is a principle of motion in every natural body, and that there is a principle of circular, positional motion in the body that does not depart from its natural place, we say: It is impossible that, in a single body, there should be a principle of rectilinear motion and a principle of circular motion such that, when it is in its natural location, it is moved with respect to position, whereas when it is somewhere other than its natural place, it is moved toward [its natural place], whether or not there is in it a principle of inclination to move circularly.[10] If it does not [have an inclination for circular motion while it is away from its natural place], and then it comes to be in its natural place, and [yet] this inclination is not produced, then it would necessarily follow that there is no principle of circular motion in it, whether it is in or out of its place. If this inclination is produced in it, then this inclination is not something innate to it, following upon its

9. The phrase "the things that will follow" translates the Arabic *al-lawāḥiq,* which can also be translated *Supplements* or *Appendices* and so could refer to Avicenna's *al-Lawāḥiq*, an apparently lost commentary on the *Shifāʾ* written by Avicenna himself, which he conceived while writing the *Shifāʾ*. While I can find no convincing arguments one way or the other as to whether the reference in the present context is to that work, the issue of how the body effects the actions produced by the soul is taken up throughout Avicenna's *Kitāb al-nafs,* the psychological section of the *Shifāʾ* (e.g., 2.3, where he speaks of tactile sensation). For discussions of Avicenna's *al-Lawāḥiq,* see Dimitri Gutas, *Avicenna and the Aristotelian Tradition* (Leiden: E. J. Brill, 1988),140–44; and David C. Reisman, *The Making of the Avicennan Tradition* (Leiden: E. J. Brill, 2002), 247.

10. The position may be that of John Philoponus as found in his *Contra Aristotelem;* see, for instance, *Against Aristotle, on the Eternity of the World,* trans. Christian Wildberg (Ithaca, NY: Cornell University Press, 1987), frag. 12*.

النفس، كالكلام في ميله الواقع بسببٍ من خارج. فإنك ترى نفس الحيوان يختلف تحريكه لبدنه، والقوة واحدة بحسب ما في بدنه من الميل الثقيل، الزائد والناقص، وتجد للزائد مقاومة ما، فنجد الكلام قائماً، ثم في هذا مباحث يجب أنْ ترجع فيها إلى اللواحق، فتجد ما يقنعك فيها، إنْ كنت في الإسهاب أرغب.

(٦) فقد بانَ واتضح أنَّ كل جسم طبيعي ففيه مبدأ حركة، وأنَّ الجسم الذي لا يفارق مكانه الطبيعي ففيه مبدأ حركة وضعية مستديرة؛ فنقول إنّه لا يجوز أنْ يكون في جسمٍ واحدٍ مبدأ حركة مستقيمة ومبدأ حركة مستديرة، حتى يكون - إذا كان في موضعه الطبيعي - تحرّك في الوضْع، وإذا كان في غير مكانه الطبيعي تحرّك إليه على الاستقامة، لأنه عندما يتحرك إلى مكانه بعينه بالاستقامة لايخلو، إمّا أنْ يكون فيه مبدأ مَيْلٍ إلى حركةٍ مستديرة أو لا يكون، فإنْ لم يكن، فإذن حصل في مكانه الطبيعي ولم يحدث هذا الميل، وجب من ذلك أنْ لا يكون فيه مبدأ حركة مستديرة، لا في مكانه ولا

substance, but something it comes to have in its natural place, the cause of which is nothing but its being contiguous with its natural place according to a certain position, or its occurring in a natural space according to a certain position. That [instance of] being contiguous with as well as that of occurring, [at the natural place] does not require that there be some inclination from one state to a like state; but neither do [these instances] require some flight from itself to something like itself. So what necessitates[11] this inclination is not [the body's] arriving at the space. Regardless of whether you take the necessitation to be by means of some nature or not, when [the nature's] body is present in some natural space, this inclination, in that case, proceeds from [the nature]; for the point of investigation and the account concerning the whole of that is the same. Likewise, you cannot say that the soul that produces motion obtains there, in the production of the motion and inclination, some principle after not having it, because the intention and act of volition occur after not having occurred, since this has already been denied. It is clearly impossible that something like it occurs, unless there is some principle of inclination in the nature. So that inclination must be necessary even if it is from the soul, in which case its being necessary would be from a perpetual natural act of volition as long as that body exists. Accordingly, the state of what moves rectilinearly is necessarily not such that it sometimes undergoes motion and sometimes rests such that it is moved into some place different from its place and rests in its place, where both belong to it naturally. So, likewise, perhaps this body can

11. Reading *mūjib* with **Z** and **T** for **Y**'s *yūjibu,* whose sense would be "that inclination does necessitate [the body's] arriving at the space."

خارجاً عن مكانه. وإنْ حدث فيه هذا الميل، كان هذا الميل ليس غريزياً تابعاً لجوهره، بل أمراً يحدث له في مكانه الطبيعي، ولا تكون العلّة فيه إلاّ مماسّته لمكانه الطبيعي على وضْعٍ مّا، أو حصوله في حيّزٍ طبيعي على وضعٍ مّا، وتلك المماسّـــة وذلك الحصول لا يوجبُ ميـــلاً عن حالٍ إلى مثلها، بل لا يوجب هرباً عن ذاته إلى مثل ذاته. فليس إذن موجب ذلك الميل موافاة الحيّز، سواء أخذتَ الإيجاب إيجاباً بتوسط طبيعة، أو أخذته إيجاباً بلا توسّـــط طبيعة، إذا حصل جســـمها في حيّز طبيعي، صدر عنها حينئذ هذا الميل، فإنَّ البحث في ذلك كله واحدٌ والكلام واحد. ولا أيضاً لك أنْ تقول: إنَّ النفس المحرّكة تأخذ هناك في التحريك والإمالة أخذاً مبتدأ بعد ما لم يكن، لحدوث القصد والإرادة، بعد ما لم تكن، فقد مُنع هذا أيضاً. وبيّنٌ أنّه غير ممكن أنْ يقع مثله؛ إلاّ وهناك مبدأ مَيْلٍ في الطبع، فيجب أنْ يكون ذلك الميل لازماً، وإنْ كان عن نفسٍ فيكون لزومه عن إرادةٍ طبيعيةٍ دائمة، ما دام ذلك الجســـم موجوداً. ولا يلزم - على هذا - حال المســـتقيم من أنّه تارة يتحرك وتارة يســـكن، يتحرك في غير مكانه، ويسكن في مكانه، وكلاهما طبيعي له. فكذلك

move rectilinearly into some place other than its [own] while moving circularly in its place, where both would be natural with respect to the two states. This, however, is not necessary precisely because rectilinear motion is not absolutely natural, as we have explained it;[12] rather, the natural is the *where* that the thing's nature requires when there is no impediment. So, when it departs, this nature requires the return to [its natural place] and to some particular location of it, where the principle of the two is one and the same.

(7) As for circular motion, the principle that we established is [such] that it naturally requires [the circular motion], always and however it might be, if [the circular motion] is absolutely natural. If it is not absolutely natural, but is like the rectilinear [motion] that the nature requires when there is a certain accidental factor, which would be the loss of the natural position, then it must stop whenever it is found. [In this case,] some given particular position would be natural, but the situation is not like that. [That is] because it is not the case that, just as a certain *where* is more suited to the body than another, so, likewise, from the position [the body] has with respect to the *where* (which is like [any other *where*]), one position would be more suited to it than another. Clearly, then, this inclination does not come to be upon reaching the natural place; rather, if there is [this inclination], it is according to the other sort—namely, that it is always, together with [the body]. So, when there is some principle of rectilinear motion in the body, it must be possible for this body to depart

12. See 4.10.3.

ربّما جاز أنْ يكون هذا الجسم تستقيم حركته في غير مكانه، وتستدير حركته في مكانه، ويكونان كلاهما طبيعيين في اختلاف الحالين، وإنّما لا يلزم هذا؛ لأنَّ الحركة المستقيمة ليست طبيعية على الإطلاق على ما شرحناه، بل الطبيعي هو الأين الذي تقتضيه طبيعة الشيء، إذا لم يكن عائق، فإذا فارق اقتضت هذه الطبيعة الردّ إليه، وإلى موضعٍ معيّنٍ منه، ويكون المبدأ فيهما واحداً.

(٧) وأمّا الحركة المستديرة فإنّ المبدأ الذي أثبتنا أنّه يوجبها بالطبع، يوجبها كيف كان ودائماً، إنْ كانت طبيعية على الإطلاق، وإن كانت ليست بطبيعية مطلقة، بل هي كالمستقيمة التي تقتضيها الطبيعة عند عارض؛ كان ذلك عند فقدان الوضع الطبيعي، فيجب أنْ يقف عند وجدانه، وكان يجب أنْ يكون الطبيعي هو وضْعٌ ما بعينه، إلاّ أنّه ليس كذلك. فإنّه ليس كما أنَّ أيْناً أولى بالجسم من أين، فكذلك من الوضع الذي له في الأين المتشابه، وضْعٌ أولى به من وضْع. فبيّنٌ أنَّ هذا المَيْل لا يكون حادثاً عند الوصول إلى المكان الطبيعي، بل إنْ كان فيكون على القسم الآخر، وهو أنّه يكون معه دائماً، فإذا كان في الجسم مبدأ حركة مستقيمة، وجب أنْ تجوز مفارقة هذا الجسم لمكانه الطبيعي،

from its natural place in order to undergo rectilinear motion from the unnatural [place back] toward [its natural place]. Now, when a single, simple body is in some place other than its natural one, there must be two inclinations in it: an inclination to [move] rectilinearly, and an inclination from it to [move] circularly. In that case, there would simultaneously exist, in one and the same substance, opposing things. They, however, are not analogous to the opposing things that undergo mixing such that there is a mean among them, for means are like a certain mix of two extremes. Now, powers undergo mixture so as to produce a mean only when each one of them is such that it can to a greater or lesser [extent] be turned toward the other direction. The result, then, is not two powers, but there is one power that is weaker and less intense than the two extremes. Being rectilinear and circular, however, are not susceptible to increase and decrease such that the rectilinear quality of [the body] would gradually become circular, or, being circular, become rectilinear, and, during the time of that becoming and existing in the intermediary state be something neither rectilinear nor curved. The fact is that, if what is rectilinear can quit being rectilinear and [can] itself become something circular, it would quit being rectilinear and embrace the quality of circularity all at once, without it being said that it had quit being rectilinear and, voilà!, had gradually become circular, being a little too zealous!—or, likewise, it quit being circular to be rectilinear.

حتى يتحرك عن غير الطبيعي إليه بالاستقامة. وأن يكون في جسم واحدٍ بسيطٍ، إذا كان في غير مكانه الطبيعي، مَيْلان مَيْلٌ إلى الاستقامة، ومَيْلٌ عنه إلى الاستدارة فيكون في جوهرٍ واحدٍ أمور متقابلة موجودة معاً، وليست مما يجري مجرى متقابلاتٍ تمتزج حتى يكون بينها وسطٌ، فإنَّ الوسائط أُمور كأنها مَزْج من الطرفين. وإنَّما تمتزج القوى امتزاجاً يؤدي إلى الوسط، إذا كان من شأن كل واحد منها أنْ يقبل الأقل والأكثر قبولاً، يصرف إلى الجهة الأخرى، فيكون الحاصل ليس قوتين، بل قوة واحدة هي أضعف وأنقص من الطرفين. ولكن الاستقامة والاستدارة لا يقبلان الاشتداد والتنقص، بأن تأخذ الاستقامة قليلاً قليلاً إلى الاستدارة، أو الاستدارة إلى الاستقامة، وهو في زمان ذلك الأخذ والوجود في التوسّط لا في مستقيم ولا في مُنْحنٍ. بل المستقيم إنْ أمكنه أنْ يفارق الاستقامة، ويصير بعينه مستديراً، كانت مفارقته الاستقامة دفعة، ومواصلته الاستدارة دفعة، من غير أنْ يقال قد فارق الاستقامة وهو ذا قد استدار قليلاً وهو يمعن فيه، أو فارق الاستدارة إلى الاستقامة كذلك.

(8) The curve found in the arcs [of a circle] does not provide a pathway from being rectilinear to being circular, nor does being circular produce one of them. So, when neither being rectilinear nor being circular is able to become more or less so, then, likewise, neither are the two powers over them so susceptible. So no power is produced that is an intermediate between what is rectilinear and what is circular, and neither is this collection by way of being a mixture. So, obviously, in one and the same body, there is not simultaneously a principle of rectilinear motion and a principle of circular motion. From this and what went before, it comes together that, in the body that delimits directions, there is a principle of circular motion but no principle of rectilinear motion. [That] is because these two principles are not joined together, and [additionally] because it has been proven about that body that neither in its entirety nor in its parts can it turn out to depart from its natural location, whereas, among the bodies situated in it, there are principles of rectilinear motion away from it and toward it. So, inasmuch as there is some direction in nature, there are three kinds of motions: one is around the middle, another is away from the middle, and the third is toward the middle, <whereas, when you take into account the directions by supposition and convention, the motions exceed this number, but they are not natural.>[13] Having completed the explanation of natural motion, it is fitting that we investigate non-natural motion.

13. This sentence has apparently been transposed and made the last sentence of the chapter, but its placement seems more natural here.

(٨) وأمّا الانحناء الموجود في القطوع، فليس سبيلاً من الاستقامة إلى الاستدارة، ولا استدارة تؤدي إلى أحدهما . فإذا كانت الاستقامة والاستدارة لا تقبلان الأشد والأضعف؛ فكذلك لا تقبلهما القوتان عليهما ، فلا تحدث قوة متوسطة بين المستقيم وبين المستدير، ولا يكون أيضاً هذا الاجتماع على سبيل الامتزاج. فيظهر أنه لا يكون في جسم واحدٍ مبدأ حركة مستقيمة، ومبدأ حركة مستديرة معاً . ويجتمع من هذا وممّا قبله أنّ الجسم المحدّد للجهات فيه مبدأ حركة مستديرة وليس فيه مبدأ حركة مستقيمة؛ لأنَّ هذين المبدأين لا يجتمعان، ولأنَّ ذلك الجسم قد بانَ من أمره أنّه لا يصح على كلّيته، ولا على أجزائه، مفارقة موضعه الطبيعي. وأما الأجسام الموضوعة فيه؛ ففيها مبادىء حركات مستقيمة عنه وإليه، فتكون - حيث تكون جهة في الطبع - ثلاثة أصناف من الحركات؛ واحدة حول ‹ال› وسط، وأخرى عن الوسط، وثالثة إلى الوسط. وإذ قد بالغنا في تعريف حال الحركة الطبيعية، فحريّ بنا أنْ نحقّق حال الحركة غير الطبيعية، وأما إذا اعتبرت الجهات بالفرض والوضع، فتزيد الحركات على هذه العدة، ولكن لا تكون طبيعية.

Chapter Thirteen

Accidental motion

(1) Some unnatural motion is said to be essential and some accidental. The accidental [motion] of something *x* does not involve *x* itself primarily departing from a *where,* position, quantity, or quality. Instead, *x* is connected with some other thing *y* so as to accompany it, such that, when *y* has some state to which [*x*] is related[1] that is replaced, the [motion] of *x* is accidental. As you have learned,[2] [accidental motion] occurs with respect to the *where* and position in two ways. On the one hand, what is said to undergo accidental motion may, in itself, be in some place or possess some position and be susceptible to motion, but it does not depart its place or position. Instead, it is that in which it inheres that departs its place, while that one accompanies it. So, owing to the motion of that in which it is, it necessarily comes to be in some [definite] position toward which one can point that is different from the one in which it was, or it has some other position relative to the directions. On the other hand, it might not be of the sort that has a *where* or position and can undergo motion.

(2) Examples of what accidentally happens to move from a certain *where* or position (that is, the sort that can undergo motion) are [the following]. [An example of motion] with respect to the *where* would be like that which is at rest in a box, maintaining its place, but the box is being carried along; [or], again, someone standing still aboard a ship while the ship is being carried along. [An example] concerning position is when our estimative faculty imagines a sphere within a sphere—the

1. Reading *yansibu* with **Z** and **T** for **Y**'s *fa-nasaba* (and so/then is related to it).
2. See 4.9.3.

<الفصل الثالث عشر>

في الحركة التي بالعرض

(١) نقــول إن الحركة غير الطبيعية منها ما يقال بالذات، ومنها ما يقال بالعرض. أمّا الذي بالعرض فهو أن يكون الشيء لم يلحقه في نفسه مفارقة أينٍ أُولى، أو وضع أول، أو كيفٍ أو كم، بل هو مقارن لشيء آخر مقارنة لازمة، فإذا تبدل لذلك الشيء حالٌ ينسب إليــه كانت له بالعرض. أمّا فــي الأين والوضْع، فهو على وجهين على ما علمتَ؛ فإنّه إمّا أن يكون ما قيل إنّه متحركٌ بالعرض هو في نفسه في مكانٍ وذو وضْع وقابل للحركة، إلاّ أنّه لم يفارق مكانه أو وضعه، بل الشــيء الذي هــو محمول فيه قد فارق مكانه، وهذا ملازمٌ له. فيلزم أنْ يقع له – لأجل حركة ما هو فيه – حصولٌ في جهةٍ تقع إليها إشارة، غير الجهة التي كان يقع عليه الإشــارة، أو يقع له وضْعٌ آخر بالقياس إلى الجهات؛ وإمّا أنْ لا يكون من شأنه أنْ يكون له أيْنٌ ووضْعٌ، ولا من شأنه أنْ يتحرك.

(٢) مثــال الذي يعرض له ما يعرض للمنتقل من مفارقة أيْنٍ ووضْع، وهو من شــأنه أنْ يتحــرك؛ أمّا في الأين كالمنقول في الصندوق وهو ســاكنٌ فيه حافظٌ لمكانه، والقاعد في السفينة والسفينة تنقله. أمّا في الوضْع، فإنّا إذا توهمنا كرة في كرة، وقد ألصقت بها

one having been attached to the other by either rivets, glue, nature, or the like—and the outer sphere moves, such that the relation of its parts to the parts of that which surrounds it changes, such that it undergoes real motion with respect to position. The inner, attached sphere would, in fact, exactly follow [the outer sphere] in that every part of it is attached to some part whose relative position is changing. In that case, its relative position would be changing, albeit accidentally, since the relative position of the part of the inner sphere to what surrounds it is not changing in the way that the relative position of the parts of the surrounding sphere are changing in relation to the parts of its place. So, if the position is considered only relative to the parts of what surrounds that which is situated in it and what is surrounded by that which is situated around it—and, in general, to the parts of what is contiguous with that which has a position, whether being in contact with what surrounds (as in the case of the sphere within a sphere) or being in contact with what is surrounded (as in the case of the outermost celestial sphere relative to that which is inside it with which it is in contact)—then the inner sphere has not changed its position. If the position is not considered with respect to what is [immediately] contiguous, but, instead, with respect to the directions with which it is parallel and facing, then the inner one has essentially changed its position also. [That] is because the parts of [the inner sphere] did change with respect to the [directions] that they were facing when that which surrounds [those parts] changed. Indeed, it is fitting that the position it has by taking the whole into account does essentially change, whereas its position relative to what contains it does not change. Position is of two sorts: one position takes into account the whole, while the other takes into account a given thing.

بمسامير أو بغراء ، أو بالطبع ، أو بغير ذلك ، فحرّكت الكرة الخارجة حتى تغيّر نسبة أجزائها إلى أجزاء المحيط بها ، تغيّراً هو حقيقة الحركة في الوضْع ، فإنَّ الكرة الداخلة الملصقة بها قد تعرض لها متابعة في أنَّ كل جزء منها يلزم جزءًا ينتقل فينتقل ، ولكن بالعرض ، إذ لا تنتقل نسبة ما بين أجزاء الكرة الداخلة وأجزاء المحيط بها ؛ كما تنتقل نسبة أجزاء الكرة المحيطة مع أجزاء مكانها . فإنْ كان اعتبار الوضْع ، إنّما هو بحسب القياس إلى أجزاء المحيط الموضوع فيه والمحاط به الموضوع عليه ، وبالجملة إلى أجزاءٍ مّا تماسّ ذا الوضْع مماسّة محيط ، كما لكرةٍ في كرة ، أو مماسّة محاطٍ كما للفلك الأعلى بالقياس إلى ما يماسّه في داخله ، فلا تكون الكرة الداخلة قد تبدل وضعها - وإنْ كان الوضع ليس باعتبار المماسات ، بل باعتبار الموازيات والمحاذيات في الجهات - فتكون الداخلة قد تبدل أيضاً وضعها بالذات . فإنَّ الأجزاء منها قد استبدلت المحاذيات مع استبدال المحيط ذلك ، بل الأولى أنْ يكون قد تبدل الوضع الذي له بحسب الكل بالذات ، ولم يتبدل الوضع الذي له ، بالقياس إلى ما يحويه . والوضْع وضعان ، وضْعٌ بحسب الكل ، ووضْعٌ بحسب شيء .

(3) What we believe about the higher atmosphere's[3] moving together with the motion of the sphere of the Moon is of this sort, for that motion is not by force, as is supposed.[4] That is because, if there were this force, it would be the kind that moves the mobile owing to what it encounters and what pushes it. Now, when one sphere is around another and then is moved, but does not stick to any part of what is below it, but slides over a smooth surface,[5] not encountering any opposition in the direction of its motion such that something standing in its way must get pushed back, then nothing precludes the inner one from being at rest while the outer one undergoes motion around it, passing along its surface unobstructed. The cause of the former motion is that any part of the fire you posit can have some part of the celestial sphere assigned to it as the natural place toward which it moves naturally and at which it rests so as to be inseparable from it, [since] it is attached to it *naturally,* so that it must be inseparable from it (even though the attachment necessitated by glue or rivets is separable). So, when the place moves, there is, inseparably connected with it and naturally following it, that which has its natural place in it, holding on to what it encounters of it. So the motion of the higher atmosphere relative to the celestial sphere accidentally moves with respect to position. Now, if water, while it is in the air, were to attain the natural ordered position that we explained before[6] together with its attaining the natural position (I mean the natural surrounding surface) such that no shuffling about or inclination remains in it, nor

---

3. The Arabic *huwā*ʾ is Avicenna's preferred term for the element air; however, it also can mean *atmosphere.* Since, strictly speaking, it would be the element fire that is in contact with the sphere of the Moon such as to move together with it (a point that Avicenna will specifically make shortly), as well as the fact that he uses the specific locution *al-jaww al-ʿālī* (higher atmosphere) later, *higher atmosphere* seems preferable to *higher air* as a translation here.

4. Aristotle's discussion in his *De caelo* 2.7 (esp. 289a19–28) of the celestial sphere's motion producing friction in the upper atmosphere at least suggests this position; also see *Meteorology* 1.3 (esp. 341a1–2). It is far from clear whether Aristotle in fact held this position, and I have not had the opportunity to check the Arabic commentary traditions surrounding the *De caleo* and *Meteorology* to see if this position was ascribed to him.

5. Literally, "something simple on an extended ground."

6. See 4.10.3.

(٣) ومن هذا القبيل ما نعتقده من حركة الهواء العالي مع حركة فلك القمر ؛ فإنَّ تلك الحركة ليست كما يظنّ عن قَسْر ، وذلك لأنَّ هذا القَسْر إنْ كان ، كان من جنس تحريك المتحرّك لما يلاقيه ويدفعه . وإذا كانت كرة على كرة ، فإنها إذا تحركت ولم تتشبث بشيءٍ ممّا تحتها ، بل زحفت على بسيطٍ غير مقاوم في جهة حركتها ، حتى يلزم أنْ يندفع القائم في وجهها باندفاعها ، فلا مانع من أنْ تسكنَ الداخلة منهما ، وتتحرك الخارجة عليها ؛ ماضية على سطحها من غير انغلاق . فالسبب إذن في تلك الحركة أنَّ كل جزء تفرضه من النار قد تعيّن له جزء من الفلك كالمكان ، وهو بالطبع يتحرك إلى المكان الطبيعي له ويسكن عنده ، لازماً إياه ملتصقاً به التصاقاً طبيعياً يوجب من لزومه إياه ، وإنْ زال ما يوجبه الإلصاق بالغرى والمسامير . فإذا تحرّك المكان لزمه وتبعه ما هو بالطبع متمكنٌ فيه حافظٌ لما يلاقيه منه ، فتكون حركة الجو العالي ، بالقياس إلى الفلك ، حركة بالعرض في الوضع . ولو كان الماء – وهو في الهواء – مصيباً للترتيب الطبيعي الذي بيّناه قبل ، مع إصابته الموضع الطبيعي ، أعني السطح المحيط الطبيعي ، حتى لم يبق فيه ارجحانٌ

were there any variation in the parts of the earth over which it exists, then [the water] would follow the motion of the air in whichever direction it was moved. As it is, though, water does not, for the most, part attain the natural place according to the way that is natural; but, instead, most of it is still teeming in the lowest region. Also, there is variation from below in some of its parts. So, when [water's motion] follows the motion belonging to air, its parts in the higher region follow it, but with a rippling effect, whereas the noted cause accidentally affects the [water] in the lower region, from which it happens to become, as it were, differentiated [into parts]. The higher atmosphere, however, attains the natural place in the natural way. So its being inseparable and attached are appropriate to it. In contrast, because of mountains and strong winds, the air happens to have something additional that requires a differentiation of it into its parts. This explains the case of accidental motion. From this, then, the calumny that some[7] raise comes crashing back down. They said: If the motion belonging to fire is forced but is an everlasting motion, then an everlasting force has existed; but this is at odds with what you believe.[8] If this [circular] motion is natural, but the body [undergoing] it has some other natural motion, like rising, then a simple body has two natural motions; but you have denied that.[9] This is just an instance of a mobile that, [although] it is the sort that can be moved essentially, it is [currently] undergoing accidental motion.

(4) An example of something undergoing accidental motion that is not the sort that can be moved [essentially] would be something conjoined [to a body] not as one body to another but as something or other existing in the body, whether a form in its material or an accident in the body. So, because of the body, [these forms or accidents] happen to

7. The reference seems to be to part of Philoponus's argument in book 1 of his *Contra Aristotelem*.

8. See, for example, 2.9.18, where the denial of an everlasting force is, at least, presupposed.

9. See 4.12.6.

ومَيْلٌ، ولا اختلف أجزاء ما يقوم عليه من الأرض، لكان يتبع حركة الهواء في أي الجهات تحرّك. لكـن الماء ليس مصيباً – في أكثر الأمر – المكان الطبيعي على الوجه الذي هو طبيعـي، بـل في أكثر الأمر به انضغاط بُعْدٍ إلى السُـفْل، واختلاف في بعض أجزائه من تحـت، فإذا تبع الحركة الهوائية، تبع<ـتـ> ها أجزاؤه العالية في كثير من الأمر على سـبيل التموّج؛ وأمّا السـافلة فيعرض لها السبب المقول، فيعرض من ذلك كالتميّز. والجو العالي يصيـب المكان الطبيعي على الوجـه الطبيعي، فيحق عليه لزومه والالتصاق به، على أنَّ الهواء قد عرض له أيضاً – بسبب الجبال والرياح – أمرٌ أوجب تميّزاً مّا في أجزائه؛ فهذا بيان حال الحركة بالعرض. ويسـقط من هذا تشـنيعٌ أورده بعضهم فقال: إنْ كانت الحركة التي للنار قَسْـرية، وهي حركة دائمة، فقد وجد قَسْـرٌ دائم، وهذا خلاف لرأيكم، وإنْ كانت هذه الحركة طبيعية، ولجسـمها حركة أخرى بالطبع كالسمو، فيكون لجسمٍ بسيطٍ حركتان طبيعيتان، وقد منعتم من ذلك – فهذا مثال ما يكون المتحرك بالعرض من شـأنه أنْ يتحرك بالذات.

(٤) وأمّا مثال المتحرّك بالعرض الذي ليس من شـأنه أنْ يتحرك، فهو أن يكون هذا المقارن ليست مقارنته مقارنة جسم لجسم، بل مقارنة شيءٍ من الأشياء الموجودة في الجسم صورة في هيولاه، أو عرضاً في الجسم؛ فتصير له – بسبب الجسم – جهة تختص بها

have a [general] direction that is specified by pointing toward the [body] itself. [These forms or accidents] also have certain [accidental] parts like the parts of the body that are specified in that they are adjacent to what is conjoined with the body to which the body [that they are in] is adjacent.[10] So they happen to have, as it were, a *where* and position on account of the body's *where* and position. So, when the body comes to have some other place, the direction that occurred by pointing changes; or, when it comes to have some other position, a state of some given part changes, since it happened to that thing like the parts. So it is said that it has moved with respect to either the *where* or the position.

(5) Now, if the soul is a certain form that resides in the matter of the body, then, when the body happens to undergo accidental motion, the soul accidentally follows. The same also holds for the rest of the changes that happen to that part alone in which the soul resides. If there is some part of the soul that is not conjoined with [the body] in the sense of being impressed into the body in which it is, then it does not undergo motion, not even accidentally. <Because of this, we judge that the human soul is not impressed into matter, since it is not changed by the changes that happen to the body, whereas, if it were impressed into it, it would be changed by the body's changing, as is known.>[11] It has been asked:[12] Why is it said of the soul that it accidentally undergoes motion

10. In other words, since, for example, my body has certain parts, the form of humanity in me or the accident of whiteness in me has certain accidental parts. Also, since part of me, namely, my feet, are next to the ground, while another part of me, namely my head, is in the air, the form of humanity in me or the accident of whiteness in me will accidentally be next to the ground or in the air.

11. This entire sentence does not appear in either **Z** or **T**, and only appears in two MSS consulted by **Y**, one dating from the fourteenth and the other from the fifteenth centuries, and so both are rather late.

12. While Aristotle briefly discusses accidental motion in *Physics* 8.4, nothing like the present question is raised there. It is possible that one of Aristotle's later commentators raised the issue, but I have not be able to find anything like the present question in the extant commentaries available in Arabic. The issue of the soul's accidental motion is also taken up in Aristotle's *De anima* 1.3; but neither he nor Themistius (whose paraphrase of *De anima* would be the most likely source) takes up the question of the soul's undergoing accidental alteration, which is the issue raised here.

الإشـــارة الواقعة إلى ذاته. وتصير له أجزاء كأجزاء الجسم تختص بأنْ تلي ما يليه الجسم من الأجسام المقارنة له، فتصير له كالأين لأيْن الجسم، وكالوضع لوضع الجسم. فإذا حصل للجســـم مكانٌ آخر تبدلت الجهة المصابة بالإشـــارة، أو إذا حصــل له وضعٌ آخر تبدلت حــال جزء مّا، إذ صار لذلك الأمر كالأجزاء، فقيل إنه قد انتقل في الأيْن أو في الوضع.

(٥) وإن كانت النفس صورة قائمة في مادة البدن، فإذا عرض للبدن الحركة بالعرض لحقت النفس بالعرض، وكذلك سائر التغيّرات التي تعرض لذلك الجزء الذي تقوم فيه النفس وحــده، وإنْ كان من النفس ما ليس مقارنته بأنْ ينطبع في البدن الذي فيه، فإنّه لا يتحرك ولا بالعرض. فلهذا نحكم بأنَّ النفس الإنسانية غير منطبعةٍ في المادة، إذ لا تتغيّر بما يعرض للبدن من التغيّرات، ولو كانت منطبعة فيها لكانت تتغيّر بتغيّر البدن، كما علم. وقد سئل أنّه لِمَ كانت النفس يقال لها إنّها تتحرك بالعرض في الأيْن، ولا يقال لها تســوّد بالعرض في

with respect to where, but it is not said of it that it accidentally becomes tanned during the body's becoming tanned? Our own response is to say that the fact of the matter requires that when the former turns out to apply to the soul accidentally, then the latter one does as well (or, to be more precise,[13] when the tanning is with respect to the primary organ itself, in which the soul is), even if one of them is brought about in the normal course of things. As it is, though, the locomotion of that in which there is the soul (should the soul be impressed [into matter]) appears more frequent than the appearance of its other alterations. That is because people take it as the norm that when the body leaves some target area that is being pointed toward, what accompanies it leaves as well, but another instance of pointing toward [the body] will pick it out [again] (even should it be an imperceptible feature of the body). When a tan comes to be in the body and remains in it, [people] are less inclined to notice its coming to be in something else [other than the body], and being joined with it, when that [other] thing is imperceptible. It is as if, on account of the idle prattle that they hold—namely, that everything needs to occupy space—they have decided that whatever exists (whether perceptible or imperceptible) needs to occur in some space, but that a tan needs only what receives it, not acknowledging the existence of anything toward which one cannot point. So this is the reason that the two scenarios are different in the mind of the masses; but, since the reason is not necessary, neither is what it requires.

---

13. Reading *wa-dhālika idhā* with **Z** (and in close parallel with **T**'s *fa-dhālika idhā*) for **Y**'s *dhāka idh* (this one since).

اسـوداد البدن؟ ونحن نجيب فنقول: إن التحقيق يوجب أنه إذا صحَّ إطلاق ذلك على النفس بالعرض، صحّ إطلاق هذا وذلك إذا كان السواد في العضو الأول الذي فيه النفس بعينـه، وإن كان أحد الأمريـن أوقع في العادة. ولكن ظهور نقْلة ما فيه النفس - إنْ كان منطبعة به - أكثر من ظهور سائر استحالاته، وذلك لأنَّ الناس يحكمون بأن الجسم إذا زال عن إصابة إشارة مّا، زال ما معه، فصار إليه إشارة أخرى تخصه، ولو كان الشيء غير محسوس الجسم. وأما السواد فإنّه إذا حصل في الجسم واستقر فيه، لم يلتفتوا إلى حصوله في شيء آخر ومقارنته له؛ إذا كان ذلك الشيء غير محسوس. كأنهم يوجبون الحصول في حيّز لكل موجودٍ كان محسوساً أو غير محسوس، ولا يوجبون التسوّد إلاّ لقابله؛ لغلبة إيجاب التحيّز عندهم لكل شـيء ما، لا يؤمنون بموجود لا إشارة إليه، فهذا هو السبب الذي اختلف به الأمران عند الجمهور، ولأنّه سـبب غير واجب، فمقتضاه غير واجب.

(6) Since you have now learned the situation with respect to *where* and position, judge[14] the remaining categories in the same way. So something is, for instance, said to become black accidentally when it itself is not the subject of the blackness. Instead, [the subject] is some other body to which it is joined or mixed, whether it is some body in which it is an accident or some body that is in itself in the subject but is not being considered in itself—like when we say that the building has become black. [That] is because the primary[15] subject of blackening is not some substance accompanied by the structure. In fact, [the blackness] is accidental to the substance accompanied by the structure (if this substance is susceptible to blackening), but it might be said of the substance, even though it is not a primary subject of blackness. The fact is that [blackness's] primary subject is something in [the substance], but not like some part of it—namely, it is the surface. So it is believed that the primary subject of blackness is the surface, and [the blackness] belongs to the body on account of the surface. Since we have addressed motion that is accidental, let us discuss non-natural motion that is essential—namely, motion that is by force. We'll then follow that up with a note about spontaneous motion.

---

14. Reading *uʿḥkum* with **Z** and **T** for **Y**'s *fa-mā ḥakama* (what was judged).
15. Reading *al-awwal* with **Z** and **T**, which **Y** (inadvertently) omits.

(٦) وإذ قد علمت الحال في الأين والوضْع ، فاحكم بمثلها في سائر الأبواب ؛ فإنّه يقال إنَّ الشيء مثلاً يسوّد بالعرض ، إذا كان الموضوع للسواد ليس هو ، بل جسمٌ آخر يقارنه أو يخالطه ، أو جسمٌ هو عرض فيه ، أو جسمٌ هو بعينه في الموضوع ، وليس هو هو بعينه بالاعتبار . كقولنا إنَّ البناء قد اسوّد ؛ فإن السواد ليس موضوعه الأول جوهراً مع البنائية ، بل الجوهر مع البنائية عرض له ، إن كان هذا الجوهر ‹هو› القابل للسواد . وقد يقال للجوهر ، إذا كان ليس موضوعاً أولاً للأسود ؛ بل موضوعه الأول شيء فيه لا كجزء منه وهو السطح ، فإنَّ السواد يعتقد أنَّ موضوعه الأول هو السطح ، ولأجل السطح يوجد للجسم . وإذ قلنا في الحركة التي بالعرض ، فلننقل على الحركة غير الطبيعية التي بالذات ، وهي الحركة التي بالقَسْر ، ثم نقول في الحركة التي من تلقائها .

Chapter Fourteen

On forced motion and the mobile's spontaneous motion

(1) As for non-natural motion that nonetheless exists in the very thing described as [moving], there are those that are by force and those that are spontaneous. Let us begin by discussing those by force. We say that motion that is by force is that whose mover is external to what is being moved and is not something that its nature requires. This might either be something simply outside of the nature (as, for example, the motion produced in dragging a stone along the face of the Earth), or it might be contrary to that which is by nature such as, moving the stone upward or heating water. Motion also might be outside of nature with respect to quantity (as you have learned)[1] as, for example, the increase of size brought on by inflating or intentionally fattening something up, as well as the deterioration caused by sickness. As for the deterioration due to old age, in one way it is natural and in another not. It is natural relative to the nature of the universe for it is just part of the natural course of events of the universe and is necessary. It is not natural relative to the nature of that body, but is due to the weakening of that nature and its being ravaged. Also, it seems that the recovery that occurs on the critical day is a natural alteration, whereas that which is not in this way is by a non-natural alteration. Likewise, death owing to maturation is natural in one way, while [death] by disease and murder is not at all natural.

1. Cf. 4.9.4.

<الفصل الرابع عشر>

في الحركة القَسْرية وفي التي من تلقاء المتحرّك

(١) وأمّا الحركة غير الطبيعية، ولكنها مع ذلك موجودة في ذات الموصوف بها؛ فمنها بالقَسْر ومنها ما يكون من تلقائه. ولنتكلم أولاً في التي بالقَسْر، فنقول إنّ الحركة التي بالقَسْر، هي التي محركها خارج عن المتحرك بها وليس مقتضى طبعه. وهذا إمّا أنْ يكون خارجاً عن الطبع فقط، مثل تحريك الحجر جرّاً على وجه الأرض، وإمّا أنْ يكون مضادّاً للذي بالطبع، كتحريك الحجر إلى فوق وكتسخين الماء. وقد تكون حركات خارجة عن الطبع في الكم كما علمت؛ مثل زيادة العظم الكائن بالأورام، أو بالسمن المجتلب، والذبول الذي يكون بسبب الأمراض. وأمّا الذبول الذي للسّن فهو من جهة طبيعي، ومن جهةٍ ليس بطبيعي، فهو طبيعي بالقياس إلى طبيعة الكل، فإنه أمرٌ تجري عليه طبيعة الكل ويجب، وليس طبيعياً بالقياس إلى طبيعة ذلك البدن، بل هو لعجز تلك الطبيعة واستيلاء الغاصب عليها. ويشبه أنْ تكون الصحة التي بالبحران باستحالةٍ طبيعية، والتي تكون لا على تلك الجهة باستحالةٍ غير طبيعية، وكذلك الموت الأجلي طبيعي من وجه، والمرضي والقتلي غير طبيعي البتة.

(2) Forced motion with respect to place may be by either pushing or pulling, while carrying is more like accidental motion. Forced rotation is something composed of [both] pushing and pulling, while rolling sometimes results from two external causes and sometimes from a natural inclination together with a forced pushing or pulling. As for that which occurs when [the mover] becomes separated from the mobile—as, for example, the projectile or what is set rocking—the learned are divided into various schools of thought about it. Among them are those who think that [(1)] it is because the pushed air comes back around behind the projectile, at which [point] it comes together with a power that presses that which is in front of it forward.[2] There are others who say that [(2)] there is something pushing the air and the projectile simultaneously, but [that] the air is more susceptible to the pushing and so is pushed more quickly and thus pulls what is placed in it along with itself. Again, others think that [(3)] the cause of that is a power that the mobile acquires from the mover, remaining in it for a while until continuous collisions with that with which it comes into contact deplete [that power] and it becomes worn down.[3] So the more it becomes weakened by that, the greater the influence of the natural inclination and the collision is upon it, until the power becomes depleted; and then the projectile proceeds in the direction of its natural inclination.

---

2. Aristotle at *Physics* 8.10.266b27–267a20, and again at *De caelo* 3.2.301b23–30, suggests that projectile motion involves the mover's imparting an initial motion to the projectile and the surrounding medium, such as the air. The initially moved air, a^1, in turn moves the projectile a bit, as well as putting into motion some air adjacent to it, a^2, after which a^2 then moves the projectile, as well as setting into motion some more air, a^3, and so on until some final amount of air, a^n, can no longer set both the projectile and some further air into motion, at which point the motion ceases. I have not been able to find in Aristotle's physical writings anything as elaborate as the two suggestions, mentioned here and in the immediately following sentence, concerning how the air moves the projectile. They may both be attempts by later commentators to flesh out Aristotle's rather bare-bones explanation of projectile motion.

3. This appears to be John Philoponus's theory of *rhopē*, or impetus; see, for example, his *In Phys.* 641.13–642.20.

(٢) والحركات المكانية القسْرية، فقد تكون بالجذب وقد تكون بالدفع. وأمّا الحمْل فهو بالحركة العرضية أشبه، والتدوير القسري مركب من جذبٍ ودفعٍ، والدحرجة ربما كانت عن سببين خارجين، وربما كانت عن مَيْلٍ طبيعي، مع دفعٍ أو جذبٍ قسري. وأمّا الذي يكون، مع مفارقة المحرّك، مثل المرمي والمرجوج، فإنَّ لأهل العلم فيه اختلافاً على مذاهب. فمنهم مَنْ يرى أنَّ السبب فيه رجوع الهواء المدفوع فيه إلى خلف المرمي، والتئامه هناك إلتئاماً بقوةٍ تضغط ما أمامه. ومنهم مَنْ يقول إنَّ الدافع يدفع الهواء والمرمي جميعاً، لكن الهواء أقبل للدفع فيندفع أسرع فيجذب معه الموضوع فيه. ومنهم مَنْ يرى أنَّ السبب في ذلك قوة يستفيدها المتحرّك من المحرّك، تثبت فيه مدة إلى أنْ تبطلها مصاكّات تتصل عليه ممّا يماسّه وتنخرق به. فكلما ضعف بذلك، قوي عليه الميل الطبيعي والمصاكّة، فأبطلت القوة، فمضى المرمي نحو جهة ميله الطبيعي.

(3) The supporters of air's motion [that is, position (2),] say: It is no exaggeration at all that the motion of air can become powerful enough to carry along stones and large bodies. Indeed, a very loud sound sometimes brings about an avalanche, and there are mountains that, when yelled at, come crashing down to their very bases. Also, thunder levels large buildings and topples small mountains, and a deafening roar splits boulders. There are even those people who, by numerous and prolonged trumpet blasts, capture fortresses built on summits. [Against position (1), they ask]: Now, how can we say that the air comes back around to the rear, coming together in such as way that it presses what is in front of it forward? What is the cause of its forward motion when [the air] comes together so that it pushes what is in front of it? [Against position (3), they ask]: How could we maintain that the mover lends some power to the mobile? That is because the power must be either natural, psychological, or accidental. Yet it is none of these, since you [who support this position] have maintained that the power producing upward motion is in the substance of fire, in the sense of the form, whereas it is an accident when it is the stone [moving upward]. So how can a single nature be [both] an accident and a form? Also, were the mover to impart a power, the action of [the power] would be at its strongest at the beginning of its existence and then, thereafter, it should begin to fall off. What [we] find, however, is that its action is strongest during the middle of the motion. If the cause of this motion, however, is the air's carrying along the projectile, then a cause might be found for that—namely, that the air is attenuated by the motion, and so [the projectile] moves more quickly, [since] the motion belonging to the projectile more effectively pierces through the air through which it is passing. This cause, however, is not found in this case [of impressed power].

(٣) قال أصحاب القول بتحرك الهواء ، وليس يعظم أنْ تكون حركة الهواء تبلغ من القوة ما يحمل الحجارة والأجسام العظيمة ، فإنَّ الصوت العظيم ربّما دكَّ أنْفاً من الجبل ، وها هنا جبالٌ إذا صيح فيها انحطم أركانها . والرعد يهدّ الأبنية المشيدة ويقلب قلل الجبال ، ويفلق الصخور الصمّ . ومن الناس مَنْ يفتتح القلاع المبنية في القلل بتكثير البوقات والإلحاح عليها . وكيف يمكننا أنْ نقول إن الهواء الراجع إلي خلف التأم إلتئاماً ضغط ما قدّامه إلى قُدّام ، وما سبب حركته إلى قدّام عند الالتئام حتى يدفع ما قدّامه؟ وكيف يمكننا أنْ نقول إنَّ المحرّك أعار المتحرك قوة ، وذلك لأنّها لا تخلو من أنْ تكون إحدى القوى التي هي الطبيعية والنفسانية والعرضية؟ وليست طبيعية ولا نفسانية ولا عرضية ؛ لأنَّ القوة المحرّكة إلى فوق زعمتم أنّها في جوهر النار ، بمعنى الصورة ، وإذا كانت في الحجر عارضاً ، فكيف تكون طبيعة واحدة ، عرضاً وصورة؟ ولو كان المحرّك أفاد قوة لكان أقوى فعلها في ابتداء وجودها ، ثم كان يجب أنْ تأخذ في الانسلاخ ، والموجود هو أنَّ أقوى فعلها في الوسط من الحركة . وأما إنْ كان علّة هذه الحركة حمل الهواء للمرمي ؛ فقد يوجد لذلك علّة ؛ وهو أنَّ الهواء يتلطف بالحركة فيزداد سرعة وانخراقاً لما ينفذ فيه من الهواء الناقل للمرمي ، ولا توجد هذه العلّة هناك .

(4) There is a group who has maintained [a theory] of engenderment.[4] They claimed that it is of the nature of motion that [another] motion be engendered after it, and of the nature of tendency that a tendency be engendered after it. They, however, did not deny that the motion ceases and then is followed by a state of rest, and then, thereafter, a motion is engendered as a result of the tendency. This really is the most atrocious of the accounts. [That] is because what is engendered inevitably is something that comes to be after not being; but whatever comes to be after not being has some originator that is a cause of the coming to be. Now, if that cause is a cause inasmuch as it exists, then the first motion must exist together with the second, whereas, if it is [a cause] inasmuch as it does not exist, there would necessarily always be some cause of the motion. If, despite that, the cause is the tendency's continued existence, then why do you allow a subsequent state of rest when the principle of the motion actually exists as it should and there is no obstacle to the motion, either from the mobile or in the distance [to be covered]? If the tendency also does not exist, the discussion about it is the same as the one about motion.

(5) As it is, when we independently investigate the issue, we find that the most sound school of thought is of those who think that the mobile acquires an inclination from the mover [position (3)], where the inclination is what one sensibly perceives when one forcibly tries to bring to rest some natural or forced [motion]. In this case, one senses the pushing power, which is able to be more or less than another; for sometimes it is greater, while, at other times, less than what[5] undoubtedly

4. The doctrines of *tawallud* (engenderment) and *iᶜtimād* (tendency) were standard among many early Mutᶜazilī mutakallimūn; see, for instance, al-Ashᶜarī, *Maqālāt al-Islāmīn wa ikhtilāf al-muṣalīn,* 300–15. Al-Ghazālī also discusses the doctrine in his *Iqtiṣād;* for a translation and discussion, see Michael E. Marmura, "Ghazali's Chapter on Divine Power in the *Iqtiṣād,*" *Arabic Science and Philosophy: A Historical Journal* 4 (1994): 279–315; reprinted in Michael E. Marmura, *Probing in Islamic Philosophy: Studies in the Philosophies of Ibn Sina, al-Ghazali and Other Major Muslim Thinkers* (Binghamton, NY: Global Academic Publishing, 2005), 301–34.

5. Reading with **Z** and **T** *fa-marratan takūnu ashadd wa-marratan takūnu anqaṣ mimmā* (for sometimes it is greater, while, at other times, less than what), which is (inadvertently) omitted in **Y**.

(٤) وقد قال قوم بالتولّد، وقالوا لأنَّ من طبع الحركة أنْ تتولّد بعدها حركة، ومن طبع الاعتماد أنْ يتولّد بعده اعتماد، ولم يمنعوا أنْ تكون الحركة تعدم ثم يتبعها سكون، ثم يتولد عن الاعتماد بعد ذلك حركة، وهذا أشنع ما يقال، فإنَّ المتولّد لا محالة شيء حادث بعد ما لم يكن، ولكل حادث بعد ما لم يكن؛ محدثٌ هو علّة للحدوث. وتلك العِلّة، إنْ كانت عِلّـة بأنْ توجد؛ وجب أنْ توجـد الحركة الأولى مع الثانية، وإنْ كانت بأنْ تعدم؛ وجب أنْ تكـون دائماً علّة للحركة. وإن كان السـبب، مع ذلك، بقـاء الاعتماد، فلِمَ تجوّزون سكوناً يلحق، ومبدأ الحركة موجود على ما ينبغي بالفعل، وليس هناك مانع عن الحركة من المتحرّك ولا في المسـافة؟ وإنْ كان الاعتماد أيضاً يُعدم، فالكلام فيه كالكلام في الحركة.

(٥) لكنـا إذا حقّقنا الأمـر، وجدنا أصح المذاهب مذهب مَـنْ يرى أن المتحرك يستفيد مَيْلاً من المحرّك، والميل هو ما يُحسّ بالحسّ إذا حوول أنْ يسكّن الطبيعي بالقَسْر، أو القَسْري بالقَسْـر الآخر، فيحسّ هناك من القوة على المدافعة التي تقبل شـدّة ونقصاً فمرة تكون أشـدّ ومرة تكون أنقص مما لا يشـك في وجوده في الجسم، وإنْ كان الجسم

exists in the body, even if the body is at rest by force. Furthermore, the school of thought of those who think that the air is pushed forward and in turn pushes [the projectile] forward has missed the mark. How could it have hit the mark, when the discussion concerning the air is just the same as the one concerning the projectile? That is because this air that is pushed forward either continues to be moved at the same time that the mover comes to rest, or it does not. If it does not continue [to be moved], then how does it pass through [the air] carrying [the projectile] along? If it continues [to be moved], then the discussion concerning [how it continues to be moved] still remains. Next, if [the air] moves faster, and thus should penetrate a wall more forcefully than the arrow does—for in their view, the arrow penetrates only through the power of what causes it to penetrate: namely, from the motion of the air that is faster—and yet [if] the air is obstructed and deflected by those things that stand in its way, then why is the arrow not obstructed and deflected? If the reason is that what is adjacent to the tip of the arrow is obstructed, while what is adjacent to the arrow's notch still retains its power, then the arrow should be swifter than the air, whereas they maintained that the air is swifter. If the arrow is swifter, then the air that is adjacent to the arrow should not have enough pushing power to cause the arrow to penetrate the wall that stands in its way, if it were not being pushed from behind. [That] is because the arrow's penetrating the wall cannot be said to be like its penetrating the air, for, in their opinion, the air carries and pushes [the arrow] along by [itself] being pushed [forward]. If that is

ســاكناً مما قُســر . ومذهب مَنْ يرى أيضاً أنَّ الهواء يندفع فيدفع ، مذهبٌ غير ســديد . وكيف يكون ســديداً والكلام في الهواء كالكلام في المرمي؟ ذلك لأنَّ هذا الهواء المدفوع إمّا أنْ يبقى متحركاً مع ســكون المحرك ، أو لا يبقى ، فإنْ لم يبق ، فكيف ينفذ ناقلاً؟ وإنْ بقي ، فالكلام فيه ثابت . فإنْ كان أســرع حركة ، فيجب أنْ يكون نفوذه في الحائط أشــدّ من نفوذ السهم – فإنْ السهم إنما ينفذ عندهم بقوة مُنْفذةٍ ، هي من حركة الهواء الذي هو أســرع – والهواء يُحبس ويرد عن الأمور القائمة في وجهه ، فلِمَ لا يحبس الســهم ويرد؟ فإن كان الســبب فيه إنَّ الذي يلي نَصْل الســهم يُحبس ، والذي يلي فُوقه يكون بعد على قوته ، فقد وجب أنْ يكون السهم أسبق من الهواء ، وجعلوا الهواء أسبق ، فإنْ كان السهم أســبق فيجب أن لا يكون للهواء الذي يلي الســهم ، من قوة الاندفاع ما ينفذ السهم الممنوع بالحائط ، لولا دفعه من خلف . فإنَّ نفوذ السهم في الحائط لا يجوز أنْ يقال إنَّه كنفوذه في الهــواء ، فــإنَّ الهواء يحمله ويدفعه عندهم باندفاعه ، وإنْ كان ذلك من جذب الســهم ما

from the arrow's pulling what is behind it in such a way that it turns around and pushes what is pulling it, then what is being pulled would be pulled with a greater strength than that which necessarily belongs to what is doing the pulling. If this strength is a power or inclination, then the claim in favor of that [namely, of an acquired inclination] turns out to be the case, whereas, if it is a mere consequence [of its cause], then it ceases when its cause ceases. So, if [the strength] continues, then a power and inclination are the cause. Also, why is it that the things that happen to be in this air around the immediate vicinity of the arrow fall to the ground, and the air does not carry them along? Indeed, the air prevents heavy things that are carried along in it from falling to the ground[6] precisely because of the motion's strength, which [even] uproots the heavy thing; and when strong winds rush through tree limbs, they break them; and yet it does not carry an arrow along, should it be placed in it?! It would be fitting, then, that this air that moves the large stone should be empirically confirmed[7] by small bodies in the vicinity of what necessarily produces calm [winds]. These folks presume that when they say that the air is moved faster and so produces intervening motions in the parts of the air that are straight ahead and [in] the arrow that is situated in [those parts], they have said something of consequence; but it is not [so]. That is because this motion might be produced in one part after another of the air that is straight ahead. In that case, however, the ones that are being moved undergo motion after the mover quits acting; but you attacked [this] claim, even if their motion is together. Alternatively, [the intervening motions in the parts of the air] might be together, and the original mover is either moved together with them or

6. Reading *rusūb* with **Z** and **T** for **Y**'s *rasūl* (messenger), which probably is just a typographical error.

7. Reading *ikhtibār* with **Z** and **T** for **Y**'s *ikhtiyār* (choice).

خلْفـه جذباً يعود به دفعاً لجاذبه، فيكون المجذوب أشـدّ انجذاباً من الجاذب الملازم له. وهذه الشـدة - إنْ كانت قوة ومَيْلاً - فقد حصل القول بذلك، وإنْ كانت متابعة فقط؛ فتزول مع زوال سببها، فإنْ بقيت فيكون السبب القوة والمَيْل. وما بال الأشياء، التي يتفق حصولها في هذا الهواء اللصيق بالسـهم ترسـب ولا يحملها الهواء؟ فإنَّ الهواء إنّما يمانع الثقال المحمولة فيه عن الرسـوب بحركةٍ شـديدة يصير بها مقاوماً لخرق الثقل، والرياح إذا هبّت على أغصان الشـجر هشـمتها، مع أنها لا تحمل سهماً، لو وضع فيها. فهذا الهواء الـذي ينقل الحجـر الكبير، بالحري أنْ يكون إختباره بقرب الأجسـام الصغار ممّا يوجب كسـرها. وهؤلاء يظنّون أنهم إذا قالوا إنَّ الهواء يتحرك أسرع؛ فتحدث حركات متشافعة في أجزاء الهواء قُدُماً، والسهم موضوع فيها؛ إنّهم قالوا شيئاً، وليس كذلك. وذلك لأنّه لا يخلو إمّا أنْ تحدث هذه الحركة في أجزاء الهواء قُدُماً، شيئاً بعد شيء، فيكون المتحرك منها يتحـرك بعد هدوء المحرك؛ فقد انتقضت الدعوى، وإنْ كانت حركتها معاً، فإمّا أنْ تكون معاً والمحرك الأول يتحرك معها، أو هو واقف. فإنْ كان مع حركة المحرّك الأول فيجب

comes to a stop. If it is together with the motion of the original mover such that the arrow must stop after it, and if it is after its motion, then the doubt still remains—namely, that there is a motion and a cause by which the motion continues to exist that is different from the initial mover.

(6) As for the report that the forced mover increases in power in the middle, there is no harm in that when the power is a [mere] posit. Also, the motion of the air is of no use in [explaining the varying speed of the projectile], since the difficulty still stands. That is because the original objector can ask: Why is it that this air is faster only during the middle of the time of the motion? Indeed, if that is because [the air] acquires a greater rarefaction as a result of the motion, then it would be more appropriate that what is borne along in it is not affected by it. [That] is because [the air] will take up a larger volume and will become weaker in strength. Now, what is larger in volume and weaker in strength moves slower than what is not like that [when] the imparted motion is one and the same. If the rarefaction under consideration belongs only to the air that is passed through and not to what is passing through it, then why is this friction in the middle more capable of bringing about a decomposition and attenuation than the friction at the beginning? Certainly, if either what produces the friction or that which the friction acts upon were the same thing during the whole process, then that would make sense. [In that case,] either what produces the friction would be like a drill, since it becomes hotter according to how long it is used and so is more powerful according to the attenuation, or the influence on

أنْ يقف السهم بعده، وإنْ كان بعد حركته، فقد بقي الشك؛ وهو أنَّ هناك حركة وسبباً به تستمر الحركة، هو غير المحرّك الأول.

(٦) وأمّا حديث ازدياد المحرّك القسري قوة عند الواسطة، فليس يضير في ذلك فرض القوة ولا تنفع فيه حركة الهواء، وذلك لأنَّ الإشكال فيه قائم. وذلك لأنَّ للمتشكك الأول أنْ يقول إنَّ هذا الهواء ما باله إنّما يكون في أوسط زمان الحركة أسرع؟ فإنه إنْ كان ذلك لا ستفادته بالحركة تخلخلاً أكثر، فهو أولى بأنْ لا ينفعل عنه المنقول فيه لأنّه يصير أكبر حجماً وأضعف قواماً. والأكبر حجماً والأضعف قواماً، فإنّه يكون عن تحريكٍ واحدٍ بعينه، أبطأ حركة مما ليس كذلك، وإنْ كان التخلخل المعتبر إنما هو للهواء المنفوذ فيه لا للنافذ، فلِمَ كانت هذه المحاكّة في الوسط أقوى في التحليل والتلطيف من المحاكّة التي في الابتداء؟ نعم؛ لو دامت المحاكّة على شيءٍ واحدٍ يلقى إمّا الحاكّ وإمّا المحكوك، لكان لذلك معنى أمّا الحاكّ فكالمثقب، فإنّه على طول المزاولة يصير أسخن، فيكون على التلطيف أقوى، وأمّا المحكوك فلأنَّ دوام الحكّ عليه يكون ممّا يزيده تأثيراً بعد تأثير. وها

what is acted upon by the friction would increase more and more, owing to the continuous application of friction to it. In the present case, however, there is no single thing producing the friction or being acted upon by the friction. Instead, in their view and according to the logical conclusion of their theory, [the air] must be moved like a chain being pushed forward, where each part that is posited is some particular agent producing friction in some particular object. Perhaps the way of presenting this cause of the increase in the class associated with power is more obvious. So it could be that, when the friction on the projectile occurs over and over again, it is constantly becoming hotter and is still becoming hotter while the acquired power is becoming weaker. The attenuation acquired by the heating, however, either will correct for or counterbalance what is lost by the weakening, as long as something in the power remains. So, when the power is successively pounded upon and loses its strength, the friction also weakens and reaches a degree at which it does not correct for the pounding influence. Still, concerning that, we have not completely made up our mind about this cause, even if it might be one of the auxiliary causes of the increasing [speed] at the middle.

(7) Since it has become clear how there is forced motion, how many kinds [of it] there are, and that all motion results from a certain power in the mobile by which it is moved[8] whether forced, accidental, or natural—let us now discuss the motion that is said of what undergoes spontaneous motion. Among the theoreticians, there have been differences of opinions and bickering over this issue. [Yet] this topic does not merit the level of scrutiny and dispute that has occurred among them;[9] for [the whole of]

---

8. Reading *yandafi*^{c}*u* with **Z** and **T**, which would make "mobile" the subject of the verb, as opposed to **Y**'s *tandafi*^{c}*u,* which would make "motion" (or less likely, but still possible, "power") the subject of the verb.

9. Cf. Aristotle, *Physics* 2.4 for the early *endoxa,* and then 2.5–6 for Aristotle's preferred account of spontaneity.

هنـا لا الحاكّ ولا المحكوك واحد، بل عندهـم - وعلى قياس قولهم - يجب أنْ يتحرك كسلسـلةٍ مدفوعة قُدُماً، ويكون كل جزء يفرضه حاكّاً بعينه، لمحكوك بعينه. وعسى أنْ يكون وجه إعطاء هذه العلّة لهذا التزيّد في الباب المنسـوب إلى القوة أوضح. فعسـى أنَّ الحكّ إذا تكرّر على المرمي أكثر يسخن أكثر، فلا يزال يتسخن بالحكّ أكثر، والقوة المستفادة تضعف، إلاّ أنَّ التلطيف المستفاد بالتسخين يكون متداركاً أو موفياً على المعنى الذي يفوت بالضعـف، ما دام في القوة ثبات ما. فإذا ترادف الحكّ على القوة واسـترخت؛ ضعف أيضاً الحكّ، وبلغ مبلغاً لا يفي بتدارك تأثير الحكّ. على أنّا لا نعوّل في ذلك على هذه العلّة كل التعويل، وإنْ كان قد يجوز أنْ يكون ذلك من إحدى معنيات العلل المزيدة في الوسط.

(٧) فقد اتضح أنَّ الحركة القسـرية كيف هي، وعلى كم قسم هي، وانَّ كل حركة فعن قوة تكون في المتحرك، بها يندفع؛ إمّا قَسْـرية وإمّا عرضية، وإمّا طبيعية. فلنتكلم الآن على الحركة التي يقال إنّها من تلقاء المتحرك؛ فقد وقع في أمرها بين أهل النظر تخالفٌ وتشـاجر ما كان من حق هذا المعنى أن يقع من التفتيش عنه والمناقشـة فيه ما وقع بين

that comes down to naming, some having stipulated one meaning [for the term], while others [have chosen] another. Each one of them can stipulate whatever [sense] he uses without any of them bickering among themselves about it. So there are those who stipulate that what is being[10] moved spontaneously is that whose subject naturally undergoes a motion different from that [spontaneous] motion, and yet that [spontaneous] motion is not the result of some external cause. On these thinkers' supposition, plants would be included in the class of what is moved spontaneously, whereas the celestial sphere would be excluded from what undergoes spontaneous motion. Still others deny that the celestial sphere is excluded from that. There are also those who stipulate that [the celestial sphere] cannot even undergo motion. In that case, if this is taken absolutely, then the celestial sphere is also not included among what undergoes spontaneous motion. If there is added to it *it cannot move when it wishes* without additionally stipulating that it is the sort of thing that can wish, then the celestial sphere would be included in it. When it is something that either does not wish at all or cannot wish, then the necessary consequent of that need not occur if (counterfactually) it were to wish. There are also those who make no stipulation except that the motion proceed from an act of volition. You are under no compulsion to choose any of the uses, should you wish [not to], for it is nothing more than arguing semantics.

10. Omitting **Y**'s *al-muḥarrik* (the mover), which appears only in two MSS consulted by **Y** and does not appear in either **Z** or **T**.

طبقات أهل النظر . فإنّ معوّل ذلك على الإسم ، فقد جعله بعضهم لمعنى ، وبعضهم لمعنى آخر ، ولكلٍ منهم أن يجعل ما جعله ، وليس لأحدٍ منهم أنْ يشاجر فيه غيره . فمنهم مَنْ جعل المتحركَ من تلقائه ما لموضوعه أنْ يتحرك بطبعه حركةً غير تلك الحركة ، وتلك الحركة مع ذلك ليس عن سببٍ من خارج . فعلى وضع هؤلاء يدخل النبات في جملة المتحرك من تلقائه ، ويخرج الفلك من أنْ يكون متحركاً من تلقائه . وهم مع ذلك يمنعون أنْ يخرج الفلك من ذلك . ومنهم مَنْ شرط أن يكون له ، مع ذلك ، أنْ لا يتحرك . فإنْ أُخذ هذا مطلقاً ، لم يكن الفلك أيضاً داخلاً في المتحرك من تلقائه . وإنْ زيد عليه ، وله أنْ لا يتحرك إذا شاء ، من غير زيادة شرط أنَّ من شأنه أنْ يشأ دخل فيه الفلك ، وليس إذا كان لا يشاء أمراً البتة ، أو لا يجوز أنْ يشاء ، يلزم من ذلك أنَّ مقتضاه لا يكون لو شاء . ومنهم مَنْ لم يشترط إلاّ أن تكون الحركة صادرة عن الإرادة . وأنت غير مجبرٍ على اختيار أي الاستعمالات شئت ، فإنه ليس إلاّ مشاجرة في التسمية فقط .

Chapter Fifteen

The states of motive causes and the interrelations between the motive and mobile causes

(1) Since we have discussed at great length what we wanted to say about motions and the mobile, it is fitting that we talk about the states of the movers. So we say that, falling under the head of *mover,* there are those that produce motion essentially and others that do so accidentally. We have already differentiated the accidental mover in past remarks,[1] explaining in how many ways they occur; and [how] sometimes it is something that moves itself accidentally, while at other times it moves another accidentally, and sometimes it produces motion naturally, while at other times [it does so] by force. As for what produces motion essentially, some [movers act] through an intermediary—as, for example, the one who works wood by means of an adz—while others are without an intermediary. [In the case of] that which is through an intermediary, the intermediary might be a single thing or many. Some intermediaries are not spontaneous movers, but produce motion only for the sake of that which precedes them that set them in motion. If [the intermediary] is continuous with the mover, like a human's hand, then it is called an *instrument,* while, if it is distinct, it is called a *tool.* In practice, though, there is often no distinction between the two terms. Other

1. See 1.12.2 and 4.9.3.

<الفصل الخامس عشر>

في أحوال العِلل المحرِّكة والمناسبات بين العلل المحرِّكة والمتحركة

(١) وإذ قد استوفينا القول، بحسب غرضنا في الحركات والمتحركات، فحريٌ بنا أنْ نتكلم على أحوال المحركين، فنقول: إنَّ المحرِّك منه ما هو محرِّك بالذات ومنه ما هو محرِّك بالعرض، والمحرِّك بالعرض قد فصلّنا أمره في الأقاويل الماضية، وبيّنا أنّه على كم وجهٍ يكون؛ وأنّه قد يكون الشيء محركاً لذاته بالعرض، وقد يكون محركاً لغيره بالعرض وقد يكون محركاً بالطبع، وقد يكون بالقَسْر. وأمّا المحرِّك بالذات فمنه ما يكون بواسطة؛ مثل النجار بواسطة القَدُّوم، ومنه ما يكون بغير واسطة. والذي بالواسطة، فربما كانت الواسطة واحدة، وربّما كانت كثيرة. وما كان من الوسائط ليس محركاً من تلقائه، بل إنّما تحرّك لأجل أنَّ ما قبله يحرِّكه، فإنْ كان متصلاً بالمحرّك - كاليد للإنسان - سُمي

intermediaries spring into motion of themselves but, nevertheless, have some other principle producing the motion, on account of which they are an intermediary. In this case, [this other] is most appropriately the mover of [the intermediary], although it is a mover as an end (such as, for example, an object of love), or the contrary of the end (such as, for example, an object of fear that is fled). Some of [these ends] are movers that produce motion inasmuch as they are moved, while others produce motion inasmuch as they are not moved. What produces motion inasmuch as it is moved does so by being contiguous with [what is being moved]. Its action is completed by its coming to rest, at which point it is likewise potentially moved.

(2) Because it is impossible for an infinite number of bodies to exist, it is impossible that there be infinitely many movers simultaneously. So it is impossible that every mover be something moved, and so the [whole] thing terminates either at some unmoved mover or some first-moved mover. [That] is because there is no circle in producing motion and being moved, or in causing and being caused. [That follows] since the circle requires that *x* be a starting point of *y*, and *y* is a starting point of *x*, and so one thing would be prior to what is prior to itself. Now, either the principle of the first-moved mover's motion is in it (and so it undergoes motion essentially), or it is distinct from it and not in it. As we have said,[2] however, there is a principle of motion in every body. So if it is distinct and produces motion in a way that conforms with what a given principle of bodily motion requires, then either that motion proceeds from both [the principle and the body] jointly and yet the principle that is in the body can produce the motion alone, or the principle that is in the body cannot produce the motion alone. On the one hand,

2. See 4.12.

أداة، وإن كان مباينـــاً ســـمي آلة، وربّما لم يميّز بين اللفظين في الاســـتعمال. وما كان من الوسائط ينبعث من نفسه إلى الحركة، ومع ذلك فله مبدأ تحريك آخر لأنّه واسطة، فالأولى أنْ يكــون محرّكه مع أنّه محرِّك غاية مثل المحبــوب، أو ضدّ الغاية مثل المخوف والمهروب عنــه. والمحركات منها ما يحرك بأنْ يتحرك، ومنها مــا يحرك لا بأنْ يتحرك، والمحرك بأن يتحرك يحرك بالمماسّــة، ويتم فعله بالسكون منه، ويكون أيضاً من حيث يتحرك بالقوة.

(٢) ولا ســتحالة وجود أجســام بلا نهاية، يســـتحيل أنْ تكون متحركات معاً بلا نهاية. فيســـتحيل أنْ يكون كل محرّكٍ متحركاً، فينتهي الأمر إلى محرّكٍ لا يتحرك، أو إلى أول محــرّكٍ متحرك، إذ لا دور في التحريك والتحرّك والعلّية والمعلولية، إذ الدور يوجب أن يكون الشـــيء مبدأ لأمر، ذلك الأمر مبدأ له، فيكون أســبق من الأسبق بذاته. وأول محــرّكٍ متحرّك، إمّا أن يكون مبدأ حركته فيه؛ فيكون متحركاً بذاته، أو يكون مبايناً له وليس فيه، ولكن في كل جســم مبدأ حركة، كما قلنا. فإنْ كان المباين، يحرّك التحريك الموافــق لما يقتضيه مبدأ حركة الجســم، لم يخل إمّا أنْ تكــون تلك الحركة تصدر عنهما جميعاً بالشــركة – ومع ذلك فإنَّ المبدأ الذي في الجسم له أنْ يحرّك وحده – وإمّا أنْ لا

if that principle cannot produce the motion alone, then it is not a principle of motion in the body. It was said, however, that it was, which is a contradiction. Now, you know that in every body there is a principle of motion, which we have demonstrated;[3] and so, if the principle of motion can produce motion alone, then the distinct thing is not some mover *qua* what the motion strives after, but, rather, a mover in one of the [following] ways. It might be inasmuch as it provides the body with that principle by which it is moved, and so, through that principle, it moves the body. Alternatively, it might provide [the body] with some other power that mutually helps it in producing the motion and is added to [the body]. Another possibility is that it might produce motion because it is an end and paradigm, or [because it is an] object of imitation, or on account of both things. So, if the distinct thing produces motion in generically the same way that the principle of the body's motion does, like its partner, then, if the distinct thing produces motion different from the motion consistently produced, it is something acting by force, whether a body or not a body.

(3) There was a group[4] who said that what moves fire upward is what makes the matter to be fire. So, when it makes [the matter] to be fire, it makes it something completely prepared for that motion, such that it is moved upward after having been in remote potency. One does not do himself any favors by insisting on that, however. That is because the principle that gives the fire the complete preparedness for that motion

---

3. See 4.12.

4. The reference may be to John Philoponus; cf. his *Contra Aristotelem,* frag. 109. Unfortunately, Philoponus's commentary on book 8 of Aristotle's *Physics,* which is the primary source for much of the discussion in this chapter, exists only in an extremely fragmented state and does not include anything relevant to the issue Avicenna is raising here.

يكون للمبدأ الذي في الجسم أنْ يحرّك وحده، فإنْ لم يكن لذلك المبدأ أنْ يحرّك وحده؛ فليس مبدأ حركة في الجسم، وقد قيل ذلك، هذا خُلْف. وأنت تعلم أنَّ كل جسم ففيه مبدأ حركة، قد برهنا على ذلك. فإن كان لمبدأ الحركة أنْ يحرّك وحده، لم يكن المباين محركاً على أنّه مزاول للحركة، بل محرّك على أحد الوجوه. إمّا بأنّه يعطي الجسم ذلك المبدأ الذي به يتحرك، فيحرّك الجسم بذلك المبدأ أو يعطيه قوة أخرى تعاضده على ذلك التحريك وتزيد فيه، أو يكون محركاً لأنّه غاية ومثال أو مؤتّم، وإمّا للأمرين جميعاً. هذا إنْ كان تحريك المباين من نوع تحريك مبدأ حركة الجسم كالمشارك له، فإنْ كان المباين يحرّك خلاف التحريك الموافق؛ فهو قاسرٌ، إمّا جسم وإمّا غير جسم.

(٣) وقد قال قومٌ إنَّ محرّك النار إلى فوق هو جاعل المادة ناراً، فإذا جعله ناراً جعله تام الاستعداد لتلك الحركة، بعد أن كان بقوةٍ بعيدة، فتحرّك إلى فوق. لكن الإصرار على ذلك غير جميل؛ وذلك لأنَّ المبدأ الذي يعطي النار تمام الاستعداد لتلك الحركة، فقد

might then provide the principle by which it is moved—namely, the power by which it is moved (as you have learned).[5] If this complete preparedness, in itself, necessitates the emergence into act, then, in itself, it is a principle of the motion and a mover, for by *mover* we understand nothing more than the thing that is the principle of motion in this way. So it will be the Giver of Forms by which a given body is moved, producing motion by means of a form, and the form producing motion through itself, without an intermediary. From that, it is not necessary that the form be something that produces motion owing to itself, since it produces motion as a whole when there is some matter possessing a form of corporeality. That is because the universe is not one of the parts. So the body that is essentially the universe produces motion essentially while moving itself for the sake of that motion accidentally, since it is not something that undergoes motion essentially.[6] Even if it were something that undergoes motion essentially, the universe's motion—that is, some part of it—would not require that it move from its natural location (namely, [that it] depart from the universe's surrounding vicinity. Instead, it would only undergo accidental motion, as you have learned;[7] and something may be an accidental mover of itself. Now because there is always motion here—and the more so since the Heavens are something that has [always] been manifested—there is some first mover of infinite power; and, as such, [it] is neither a body nor in a body.

5. See 4.14.5.

6. In other words, the cosmos considered as a whole produces motions in all the parts within it, which change their position with respect to the cosmos as a whole; however, this internal change of position is merely an accidental motion of the cosmos considered as a whole (see 4.13.2). The cosmos as a whole does not, in itself, undergo any local or positional motion, since (1) it has no natural position that it could change (see 2.3.14), and (2) there is nothing outside of it that could determine its position, so that, instead, it is what determines the positions of everything else (3.14.9–10).

7. See 4.13.2.

يعطيه المبدأ الذي به يتحرك؛ وهو - كما علمت - القوة التي بها يتحرك، وهذا إنه كان الاستعداد التام يوجب بنفسه الخروج إلى الفعل، فيكون بنفسه مبدأ للحركة ومحرّكاً - فإنّا لسنا نفهم من المحرّك إلاّ الأمر الذي هو مبدأ الحركة على هذا النحو، فيجب أنْ يكون واهب الصور التي بها يتحرك جسمٌ ما، محرّكاً بالصورة، والصورة محركة بذاتها بلا واسطة. ولا يجب من ذلك أنْ تكون الصورة محرّكة لذاتها، لأنّها تحرّك كُلاً ومادة ذات صورة مجسّمة. وذلك لأنَّ الكل ليس هو أحد الأجزاء، فهو يحرّك الجسم الذي هو الكل بالذات، ويحرّك ذاته لأجل تلك الحركة بالعرض، لأنّه ليس ممّا يتحرّك بالذات، ولو كان ممّا يتحرّك بالذات، لما كان انتقال الكل - وهو جزء منه - يوجب انتقاله عن موضعه الطبيعي وهو غير مفارقٍ لما جاوره من الكل، بل كان، كما علمت، متحركاً بالعرض، وقد يكون الشيء محرّكاً لنفسه بالعرض. ولأن ها هنا حركة دائمة، ما دامت السماء قد ظهر أمرها، فهاهنا محرّك أول غير متناهي القوة، فليس بجسمٍ ولا في جسم.

(4) We ought now to mention the interrelations between movers and mobiles. Let us posit a given mover, mobile, distance, and time. Let us also examine the mover *qua* principle of natural motion, principle of pulling, principle of pushing, and something that carries; and [let us] consider the kinds of interrelations that necessarily follow. Let us also posit some mover that has moved some mobile along the [posited] distance for a given period of time and consider whether half of the mover would move the same mobile[8] along the [same] distance for a period of time half of that, or less, or greater.

(5) Now, we say that it does not necessarily follow that [half of the mover] would move [the mobile] at all, for it might be that what alone moves that mobile from its present condition is only the collective power of the mover. So, when it is halved, then it can play a certain contributing role while not necessarily and inevitably producing motion. An example would be the ship that a hundred men can drag two parasangs[9] in one day. In this case, it does not follow that fifty men must be able to move it at all. For this reason, it does not necessarily follow that, when the hum of a swarm of bees produces a sound, each bee will produce an audible sound. Again, when there comes to be a hollow in a rock as a result of a hundred drops [of water], it does not necessarily follow that each drop produces some sensible effect. Instead, each drop could have a certain contributing role in counteracting the hardness, and so, when the contribution is completed, another [drop] acts in the hollowing process, and continuing in that way until a sensible cavity comes to be. There are even movers that when halved, do not retain their power, such as an animal. With respect to motions that involve inclination, this contributing factor gradually counteracts the inclination inherent in them until a foreign inclination enters into them that the power that is in the inclination cannot eliminate.[10]

---

8. Reading simply *al-mutaḥarrik* with **T** for **Z**'s *fī al-mutaḥarrik,* which has basically the same meaning "whether half of the mover would produce motion *in* the mobile..." and Y's *al-muḥarrik al-mutaḥarrik,* which might mean "whether half of the mover would move the mover that moved itself," but may simply be a typographical error.

9. Approximately seven miles (or eleven kilometers).

10. Reading *tamḥīqihi* with **Z** and **T** for **Y**'s *tamḥīqah* (?), which is probably a typographical error.

(٤) فينبغي الآن أنْ نذكر المناسبات التي بين المحركات والمتحركات. لنضع محرّكاً ومتحرّكاً ومسافة وزماناً، ولنمتحن المحرّك على أنّه مبدأ لحركةٍ طبيعيةٍ، وعلى أنّه مبدأ جذب وعلى أنَّه مبدأ دفع، وعلى أنه حاملٌ، ولنتأمل ما يلزم من أصناف المناسبات. ولنضع محرّكاً حرّك متحركاً في المسافة زماناً، ولنتأمل هل نصف المحرّك يحرّك المتحرّك بعينه في المسافة زماناً، نصف ذلك أو أقل أو أكثر

(٥) فنقول: إنه لا يلزم أنْ يحرّكه شيئاً، فإنّه يجوز أنْ يكون المستقل بتحريك ذلك المتحرّك عن حاله إنّما هو مجموع قوة المحرّك، فإذا انتصفت كان لها أنْ تُحْدث إعداداً، ولم يجب أنْ تحرّك لا محالة؛ مثل السفينة التي تجرّها 〈مئة〉 نفسٍ في يوم واحدٍ فرسخين، فلا يلزم أنْ يقدر الخمسون، لا محالة، على نقلها شيئاً. ولهذا ليس إذا حدث صوتٌ عن صَرّة جاورسٍ يلزم أنْ يحدث عن كل جاورسةٍ صوتٌ لا يسمع، أو إذا حدث عن 〈مئة〉 قطرة نَقْرة في الصخر، يلزم أن تكون كل قطرة تفعل شيئاً لا يُحسّ، بل عسى أنْ يكون لكل قطرة إعداد مّا في إبطال صلابة، فإذا تمّ الإعداد فعل الآخر من النقر، وانْ يستمر على ذلك المنهاج، حتى يحدث قعر محسوس، على أنَّ ها هنا من المحركات ما إذا نصّف لم تبق له قوته، كالحيوان. وهذا الإعداد في الحركات الميلية هو إبطال المَيْل المستقر فيها قليلاً قليلاً، حتى يدخل عليها مَيْلٌ غريب تعجز عن تمحيقه القوة المميّلة التي فيه.

(6) When we posit that the mobile is halved, the common opinion is that the mover [either] moves the mobile twice the distance in that time [or] along the [same] distance in half of that time. In what the independent investigator puts forth, however, he does not put much weight in that [common opinion]. In the case of the natural mover, it is not true that the mover remains in itself when the mobile is halved. That is because the natural power is accidentally divided by dividing that in which it is. So, when the mobile[11] is halved, then the whole[12] of the mover is not moving it, but only the half of it that exists in it—that is, unless it is according to conjecture and approximation.

(7) As for what carries, the power of the carrier might not be sufficient to cover twice the distance across which the load was carried, even if there were no load at all. In that case, how could it be necessary when it is accompanied by half of the weight? If what carries does so by means of a natural motion, then, upon finding its natural terminus, it would not pass beyond it with the load. Also, its natural distance between the two natural directions would be double for it only if it began at the [very] middle. Again, in the case when the load has a certain inclination other than the inclination of [what is carrying it], [that contrary inclination] slows [the carrier] down. Even then, it is not able to conserve this ratio, because the motion of natural things is not uniform from beginning to end. The fact is that, whenever it approaches its extreme, it moves faster, and so [the mobile's] state does not remain uniform in the two halves, whether with or without a load.

---

11. Reading *mutaḥarrik* with **Z** and **T** for **Y**'s *muḥarrik* (mover).

12. Rejecting **Y**'s conjectural emendation of *li,* which would have the sense of "it is not possible for the whole mover to move it."

(٦) فـإذا فرضنـا التنصيف في المتحرك، فالمشـهور هو أن المحـرّك يحرّك نصف المتحرّك في ضعف المسـافة في ذلك الزمان، وفي المسـافة في نصف ذلك الزمان، وأمّا المحقّـق فغيره اعتبر ذلك فيما يورده. أما في المحرّك الطبيعي، فإنّه لا يصحّ أنْ يبقى المحرّك بحاله، والمتحرّك به قد تنصّف، وذلك لأنَّ القوة الطبيعية يعرض لها أن تنقسم بانقسام ما هـي فيه، فإذا تنصّـف المتحرّك لم يكن كلية المحرّك أنْ تحرّكه بل النصف الموجود منه فيه إلاّ على سبيل التخمين والتقدير.

(٧) وأمّا الحامل؛ فيجوز أنْ تكون قوة الحامل لا تفي بأنْ تقطع ضعف المسافة التي حمل فيها ما حمل ولو كان فارغاً، فكيف يلزم ومعه نصف الثقل؟ وإنْ كان الحامل يحمل بحركةٍ طبيعية؛ فإنّه عند وجود نهايته الطبيعية، لا يتعداه بالمحمول ولا تتضعّف له مسافته الطبيعية التي بين الجهتين الطبيعيتين، اللّهم إلاّ أنْ يقع الابتداء من الوسـط. فحينئذ إنْ كان المحمول عليه له مَيْلٌ غير مَيْله، أحدث فيه بُطْئاً، إلا أنَّ ذلك لا يحفظ هذه النسـبة، لأنَّ حركة الطبيعيات لا تتفق من الابتداء إلى المنتهى، بل كلّما أمعن ازداد سـرعة، فلا تتفق حاله في النصفين، كان فارغاً أو حاملاً.

(8) The same status of what carries holds for what pushes while remaining attached [to what it pushes], whereas what pushes so as to produce projectile motion might have a greater effect on what is heavier than it does on what is lighter. In that case, it would have a greater effect on what is doubled than on what is halved; and so, again, that ratio is not maintained. Also, given that the speed does not remain constant at its limiting points but is slower at the end of it while, purportedly, the middle is stronger, this ratio is not preserved. The same holds for what pulls, for what pulls sometimes has the form of what carries by dragging [the load]. At other times, it pulls [in the sense of *attracts*] through a power; but the power emanating from the attracting agent has a limit at which its influence on the distant, attracted object terminates and beyond which it can have no influence. So it does not necessarily follow that, whenever we make a mobile smaller, the mover attracts it [either] from a farther place [or] in half the time. It is, indeed,[13] the common opinion that it moves that mobile in the same way through half the distance, but it is not necessary. [That] is because what is traversed during the two halves of the time of the projectile motion (whether forced or natural) need not be equal, since you have learned that the motion varies with respect to speed.

(9) As for the mover in relation to half the distance, the common opinion is analogous to what has been mentioned [namely, that half the mover would move the same mobile half the distance], but the fact of the matter is as we have reported it. As for half the mover with respect to half the mobile, the common opinion is that the ratio is preserved. As it is, though, the mover might not be halved so as to preserve its power. Also, it might be slower for the whole to move the whole, for the

13. Reading *fa-ʾinna* with **Z** and **T**, which is omitted in **Y**, who conjectures, however, that there should be a simple *fa*.

(٨) وأمّــا الدافع اللازم؛ فحكمه حكم الحامــل، وأمّا الدافع الرامي فربما عرض أنّه يفعل في الأثقل أشدّ ممّا يفعله في الأخف، فيفعل في الضعف أشدّ ممّا يفعله في النصف، فلا تبقى تلك النسبة على أنَّ المرمي لا تتشابه السرعة والبُطْء في حدوده، بل المتأخر منه أبطأ، ويقال إنَّ الوســط منه أقوى؛ فلا تكون هذه النسبة محفوظة. كذلك الجاذب؛ فإنَّ الجــاذب قد يكون على صورة الحامل الجار، وقد يكون جاذباً بالقوة، وللقوة الفائضة عن الجــاذب حدّ إليــه ينتهي تأثيره في المنجذب البعيد منه. فما خرج عن ذلك لا يلزم أنْ يؤثّر فيه، فلا يلــزم أن يكون، كلما جعلنا المتحرّك أصغر، جذبه - من مكانٍ أبعد - المحرّك في نصف الزمان فإنّ المشــهور أنّه يحرّك ذلك المتحرّك بعينه في نصف المســافة، وليس يجب. فإنّه ليس يلزم أنْ يتســاوى المقطوع في نصفي زمان المرمي؛ لا في القَسْري ولا في الطبيعي، لما علمتَ من اختلاف الحركة في السرعة والبُطْء.

(٩) وأمّا المحرّك في نصف المسافة، فالمشهور على قياس ما قيل، والحقّ ما أخبرنا عنه. وأمّا اعتبار نصف المحرّك بنصف المتحرّك، فالمشــهور حفظ النسبة، لكن يجوز أنْ لا يتنصّــف المحرّك حافظاً لقوته، ويجوز أنْ يكون أبطأ من تحريك الكل للكل فإنَّ اجتماع

combination and increase of the power might be a consequence of the strength [of the whole] whose relation to the strength of the part is greater than the ratio of size to size.[14] Concerning half the mover in half the time, the common opinion is that the ratio is preserved. According to what you have learned, however, it would be more fitting that it is not preserved. Also analogous to what you have learned is the case of bringing about half the motion in half the distance. From the cases of halving, you also know those of doubling.

(10) There is still one last school of thought that we have related to you on several occasions—namely, that halving by the mover results in its not producing any motion and by the mobile in its not being moved.[15] These interrelations among the mover, motion, mobile, distance, and time might be considered insofar as they are finite as well as infinite,[16] since when any of these is finite, the others are as well, because a given finite part of one parallels a finite [part] of the other. Now, the parts comparable to that part must be exhausted whenever an infinite is taken in parallel with what is being exhausted of the finite, for, if [the finite] continues on, there was not an exact mapping between [the parts of the finite and the infinite]. In that case, why would there not be an infinite motion in a finite period of time or along a finite space, or why wouldn't the time [of the motion] be infinite while a space is finite? The fact is that one finite thing accompanies another, and, from the mapping process, [the finite always] lacks some excess that is in the infinite. If there were no excess, and [if] instead, the infinite is exhausted with the finite in the way that the supposition requires, then the infinite would be finite.

◆

14. The point is that the ratios between strength and volume need not correspond. Thus, halving the volume may not entail halving the strength; and, in fact, it might reduce the strength by more than half.

15. It would seem that Avicenna has in mind Zeno-like paradoxes here; see 3.3.14 and 3.5.3.

16. Reading *wa-ghayr mutanāhīyah* (as well as infinite) with **Z** and **T**, which **Y** (inadvertently) omits.

القوة وتزيّدها قد يستتبع من الحَمْية ما هو أزيد نسبة إلى حمية الجزء، من نسبة العِظم إلى العظم. وأمّا نصف المحرّك في نصف الزمان؛ فالمشهور حفظ النسبة، والأوْلى أنْ لا تُحفظ على ما علمت، وأمّا نصف التحريك في نصف المسافة فذلك أيضاً على قياس ما علمتَ، وأنت تعلم التضعيفات من التنصيفات.

(١٠) على أنَّ ها هنا مذهباً آخر حكيناه لك مرات، وهو أنَّ التنصيف يؤدي بالمحرّك إلى أنْ لا يحرّك، وبالمتحرّك إلى أنْ يتحرك. وقد يقع اعتبار هذه المناسبات بين المحرّك والحركة والمتحرّك والمسافة والزمان من حيث هي متناهية وغير متناهية، إذ أيّ هذه إذا تناهى، تناهى الآخر، لأنَّ جزءًا من المتناهي منه يكون بإزاء متناهٍ من الآخر، وأمثال ذلك الجزء يجب أن تفنى ما أُخذ غير متناهٍ بإزاء فناء المتناهي. فإنّه أنْ بقيَ لم تكن بينهما مطابقة، فلمْ تكن الحركة غير المتناهية في زمانٍ متناهٍ، أو في مسافةٍ متناهيةٍ، أو لمْ يكن زمانٌ غير متناهٍ مع مسافةٍ متناهيةٍ، بل كان متناه مع متناهٍ وخلا فضل ما ليس بمتناهٍ عن المطابقة. وإذا لم يفضل بل فني غير المتناهي مع المتناهي، على ما أوجبه المفروض، كان غير المتناهي متناهياً.

Glossary of Arabic-English Terms

1.14.4	eternal	ابداً
1.14.4; **4**.15.1	for the sake of	لأجل
1.7.1	effect	اثر
2.8.14; **3**.10.passim; **4**.12.3	influence, to	تأثير
3.10.passim	influenced, to be	تأثّر
1.6.2; **3**.8.3	influence (active)	مؤثّر
1.6.2	influence (passive)	متأثّر
1.3.9	posteriority	تأخير
1.10.6; **2**.2.6; **4**.6.4, 8.16, 11.6	posteriority/posterior	تأخّر
2.10.10, 11.2–4, 12.7, 13.10	later	متأخر
4.15.1	instrument	اداة
1.8.2–3; **3**.4.10, 5.5, 7.2, 8.8, 10.13, 14.10; **4**.6.4, 9.7	earth	ارض
1.2.3, 2.6, 3.2, 10.7; **3**.4.2, 8.11, 10.2, 12.3	element	اسطقسّ

3.5.8, 10.11; **4**.6.7	foundation	اصل
1.8.2; **3**.13.6, 14.9	horizon	افق
3.3.1, 3.3, 4.3, 5.2	aggregation	تأليف
3.3.1, 12.4	aggregate	مؤلّف
1.13.15	God	الله
1.8.2	fact that	انّ
1.8.4–5	human-ness	انسانّية
3.12.5	tool	الة
3.6.4–6	first	اوّل
2.1.3, 3.19, 8.22, 12.passim, 13.1; **3**.3.9, 5.8, 5.10, 10.1, 10.5, 10.10, 11.2; **4**.4.1	instant/now	آن
1.14.8; **2**.1.2, 2.1, 3.10, 3.12, 12.6; **4**.11.2, 11.4, 12.1, 12.6, 13.4	where	اين
4.9.4, 14.1	critical day	برحان
1.13–14.passim	luck	بخت
1.14.4; **3**.7.3; **4**.10.4	vapor	بخار
1.14.4	evaporation, to cause	تبخير
1.14.14	evaporation	تبخّر
1.3.2–3; **3**.11.1, 11.3	creation, atemporal	ابداع
2.1.5; **3**.10.5, 11.2	starting point	مبدأ

1.passim; **2**.9.12; **3**.8.11, 10.11; **4**.12.1	principle	مبدأ
1.13.5	principle, elemental	– اسطقسّي
1.10.3	principle, completing	– متمّم
1.3.11	principle, guiding	– مشير
2.5.5	principle, formal	– صورية
2.5.5	principle, material	– عنصري
1.3.11	principle, auxiliary	– معين
1.3.11; **2**.5.5	principle, final	– غائي
1.3.11, 10.4, 15.6; **2**.5.5; **3**.11.1	principle, efficient	– فاعلي
4.4.3, 4.5	principle, emanative	– فيض
1.11.4; **3**.10.15	principle, proximate	– قريب
1.12.4, 15.6	principle, material	– مادّي
1.10.3	principle, preparing	– مهيّئ
3.11.3	Creator	مبدع
1.5.8, 10.7; **3**.12.5; **4**.12.2, 13.5	body	بدن
2.5.8, 6.4	light of reason	بديهة
1.5.1; **2**.3.3	seed	بذر
3.11.1; **4**.4.3	Creator	بارئ
4.8.18, 10.4	coolness	برد

1.5.8, 6.2; **4**.10.1	coolness	برودة
2.9.20	cooling	تبريد
2.3.18	cold, to become	تبرّد
1.8.2; **3**.2.10	demonstration	برهان
1.1.15; **2**.5.3; **3**.10.14; **4**.11.7	simple	بسيط
1.5.8, 8.5; **3**.13.2; **4**.11.9	extension	انبساط
1.8.1	vision	بصر
3.13.4	vision, sensory faculty of	ابصار
3.10.5	slow	بطء
2.11.4–5, 12.1	after	بَعد
2.6.2, 6.4, 7.3–9, 13.3; **3**.2.5, 4.3, 8.5–6, 8.9, 13.2, 14.1	interval	بُعد
1.2.11; **2**.12.4; **3**.11.2, 14.1	remoteness	بُعد
2.13.8	suddenly	بغت
1.13.5	survive	بقاء
3.2.11	between	بين
3.4.16	self-evident	بيّن بنفسه
4.8.13	separation	مباينة
3.1.1, 2.6, 3.9, 4.3	succession, (follow in)	تأتي
1.10.7	consequent	تلو

4.2.6	completeness	تمام
2.5.5	perfection	تمام
2.5.5	perfection, proper	– خاصّ
2.5.5	perfection, common	– مشترك
4.2.1	complete	تامّ
1.12.8; **2**.13.7; **3**.9.6	persistence/permanent	ثبات
2.13.7	impermanent	غير ثبات
4.9.5	stable	ثابت
1.6.2, 11.2; **2**.5.5, 9.21; **3**.10.2; **4**.12.2	weight/heaviness	ثقل
3.4.14, 13.7; **4**.12.2	heavy	ثقيل
2.8.16, 11.6, 12.4; **3**.11.2; **4**.9.5	renew(al)	تجدّد
1.12.3, 14.7; **3**.10.12; **4**.10.2, 11.10, 14.2, 15.4	pulling/attraction	جذب
1.8.2–3	abstracted	مجرّد
3.10.13; **4**.11.5	body	جرم
4.11.5	simple body	– بسيط
2.13.3	circular body	– مستدر
1.7.3, 8.2	heavenly body	– سماوي
1.13.4	atom	– صغير لا يتجزأ
4.12.1	corporeity	جرمية

جزء	part	**2**.3.15; **3**.5.2, 6.6, 8.11, 10.1
متشبه جزء	homoeomerous part	**3**.12.3–4; **4**.12.4
– لا يتجزأ	atom	**1**.4.1, 4.5; **2**.13.3; **3**.3–5.passim
جزئي	particular	**1**.7.2
جسم	body	**1**.2.2; **3**.3–5.passim, 12.5, 13.2
– بسيط	simple body	**3**.7.3, 8.3; **4**.9.6
– مركّب	composite body	**1**.6.3; **3**.8.3,10.14
جسمية	corporeality	**1**.2.2; **2**.1.12–13, 6.6, 9.12, 14.9; **4**.11.1, 12.1
اجتماع	combination/combine, to/collection/join together, to	**1**.6.3, 10.7; **3**.4.3, 6.2, 11.6, 12.6
مجموع	collection	**2**.1.14
مجتمع	combination	**2**.9.11; **3**.2.11
جملة	whole/total/set	**1**.1.15; **3**.6.2, 8.1, 11.5–6
جمال	beauty	**1**.12.6
جنين	fetus	**1**.14.7; **4**.9.4
جانب	side	**4**.3.6
جنس	genus	**1**.1.3, 5.7, 12.5; **2**.1.4; **3**.3.3, 5.2; **4**.6.9
– ابعد	remote genus	**4**.2.1
– اسفل	lowest genus	**4**.3.2
– اعلى	highest genus	**1**.10.9; **2**.8.2; **4**.3.2

4.2.1	proximate genus	– اقرب
1.2.2, 4.4, 5.8, 14.8; **2**.1.2, 3.2, 4.2; **3**.5.6, 10.2, 11.1; **4**.9.1	substance	جوهر
2.5.2	eternal substance	– ازلي
2.5.2	sensible substance	– محسوس
2.5.2	intelligible substance	– معقول
1.6.6; **2**.8.2	substantiality	جوهرية
4.10.1	substance, to become	تجوهر
1.1.13	lunar node	جوزهر
1.5.8; **2**.6.6, 9.16; **3**.12.8	volume	حجم
1.1.3, 8.7, 12.5; **2**.4.1, 9.11	definition	حدّ
2.1.6, 1.21, 2.1, 2.4, 10.10, 12.5, 13.1; **3**.6.4, 8.1, 8.3, 8.9, 9.1, 12.6, 14.1; **4**.5.6	limiting point	حدّ
1.10.7	minor term	– اصغر
1.10.7	major term	– اكبر
1.10.7; **3**.2.10, 8.6; **4**.6.4	middle term	– اوسط
1.8.1	delimited	محدود
2.1.4	indefinite	غير محدود
2.9.9	delimiting	محدّد
4.3.6, 6.6	convex	حدب
1.2.12	comes to be	حادث

حادث	coming to be, subject to	**1**.2.12, 2.15
حرارة	heat	**1**.14.12, 14.14; **2**.1.7, 13.4; **4**.6.2
حركة	motion	**1**.10.2; **2**.1–3.passim; **3**.9.4, 11.1
اول حركة	motion, first	**3**.6.3–6
– مستديرة	motion, circular	**2**.8.8, 13.3; **3**.8.4, 14.6, 14.8–10; **4**.6.1–2, 12.7–8
– ارادية	motion, voluntary	**3**.13.8
– طبيعية	motion, natural	**3**.8.2, 8.8–9, 10.2, 12.11, 13.9; **4**.2.6, 3.3, 6.2, 9.5
– قسرية	motion, forced	**2**.8.14, 13.3; **3**.8.2, 10.2, 10.15; **4**.2.6, 3.3, 6.2, 14.passim
– قوسية	motion, curved	**4**.6.10
– مستقيمة	motion, rectilinear	**2**.8.9, 13.3; **3**.8.10, 13.9, 14.passim; **4**.6.1
– مكانية	motion, local	**2**.1.5–6; **3**.6.2, 10.5; **4**.5.4
– كيفية	motion, qualitative	**4**.5.5, 9.1
– كائنة من تلقاء	motion, spontaneous	**4**.9.4, 14.7
تحريك	moving [something]/ motion, to produce	**2**.1.12, 1.23
تحرّك	moved, being/ motion, subject to	**2**.1.23
محرّك	mover	**2**.1.7; **3**.9.6; **4**.2.4, 14.1–3, 14.5
– اول	first mover	**4**.15.2–3
– غير جسماني	incorporeal mover	**3**.10.15
– لا يتحرّكه	unmoved mover	**4**.15.2

2.1.7; **3**.5.1, 6.2; **4**.15.6	mobile/moved, what is	متحرّك
2.1.15	self-moved	– بذاته
1.1.8	sensation	حسّ
1.8.1, 8.8	arithmetic	حساب
2.1.3	determinate	حاصل
1.8.2	conjunction	محاذية
3.12.1	preservation	حفظ
1.7.2	conservation	استحفاظ
3.11.9	true one	الاحد الحقّ
3.9.1	reality	حقيقة
1.5.6	subsistent reality	– قائمة
3.6.5; **4**.13.2	real	حقيقية
4.14.6	friction	محاكّ
3.3.7, 5.6, 10.14, 11.1	substrate	محلّ
1.2.6; **2**.6.5, 9.11; **3**.12.2; **4**.4.3–4, 14.6	analysis/decomposition	تحليل
2.1.14	categorical statement	حملي
1.10.8; **2**.10.5	predicate	حمل
4.14.2	carrying	حمل
4.15.7	carries	حامل

انحناء	curve	**1**.8.2; **4**.3.3, 12.8
حيّز	space	**2**.5.8, 8.10; **4**.10.passim, 11.1, 11.4, 12.1, 13.5, 15.10
احاطة	periphery	**3**.14.2
محيط	surrounds, what	**2**.9.2
محاط	surrounded, what is	**2**.9.2; **3**.8.9
احالة	transition	**1**.10.2, 14.16
استحالة	alteration	**1**.5.1; **2**.1.4, 2.1, 3.2–3, 3.6; **3**.6.9; **4**.2.2, 2.5, 4.4, 5.8, 6.1, 9.1
محيل	transmit	**3**.10.12
محال	absurdity	**3**.3.12
حوى	contain, to	**2**.3.14, 3.17, 6.1, 7.3, 9.1, 9.9, 9.12
حاوي	container	**2**.6.2, 9.9, 9.12
حياة	life	**4**.9.6
حيوان	animal	**1**.5.1; **3**.12.3
خروج	passage/pass, to	**1**.10.2; **2**.1.2, 13.3
خاصّة	property	**1**.12.5
خاصّية	property unique to the species	**2**.1.4
خطّ	line	**2**.5.4, 7.4, 8.10, 9.6, 12.5; **3**.3.7, 6.7–8, 8.1, 13.2; **4**.3.6
– مستدير	circular line	**3**.5.9; **4**.3.6
– غير منقسم	indivisible line	**3**.3.2

3.5.9; **4**.3.6	rectilinear/straight line	– مستقيم
1.10.9	outline	تخطيط
3.10.2, 13.7; **4**.12.2	light(weight)/lightness	خفيف
1.8.2	declination	انخفاض
1.5.8; **2**.3.8, 6.12, 8.20, 9.17, 9.20–22; **3**.8.3, 8.13; **4**.5.6, 6.1, 9.1, 10.1, 14.6	rarefaction/rarefy	تخلخل
1.12.4	mixture	خلط
1.4.5; **3**.7.3	mixture	خليط
3.12.4	mixing	اختلاط
1.10.6; **4**.10.4	mixed (with), to be	مخالطة
1.10.6	mixture	مختلط
3.11.5	creation	خَلق
1.5.7	temperament	خُلق
3.11.9	creator	خالق
1.4.1, 4.5, 13.4; **2**.6.11, 8.passim, 9.16; **3**.2.7, 7.3, 8.3, 8.13, 10.8, 14.1; **4**.8.10, 11.9	void	خلاء
2.8.22	interstitial void	– مخلّل
1.12.6, 13.15, 14.9–10, 15.4	good/well-being	خير
1.10.10	real good	– حقيقي
1.10.10, 12.6	apparent good	– مظنون

1.7.4, 13.14, 14.11, 15.3	choice/chosen	اختيار
1.1.8; **2**.1.5, 1.19	imagery [faculty]	خيال
1.1.18, 5.7; **2**.9.11; **3**.7.3	imagination	تخيّل
1.7.2	management	تدبير
2.6.15; **3**.2.3–5, 4.1; **4**.12.2	interpenetration	مدخلة
2.7.6; **3**.4.3; **4**.11.9	interpenetrating	تداخل
2.5.3, 7.4; **3**.3.9, 4.16	interpenetrate	مداخل
3.6.9	perceiving	ادراك
1.12.3; **3**.10.12; **4**.14.2, 15.4, 15.8	pushing	دَفع
2.1.2–3, 3.2, 3.12, 11.6, 12.3, 13.10; **3**.5.8, 6.9; **4**.6.3	all at once/ instantaneous	دُفع
4.8.16, 10.4	repulsion	اندفاع
2.10.7, 13.7	everlasting	دهر
3.8.4, 10.9, 11.6; **4**.2.6, 6.10, 14.2	rotation	دورة
1.8.4	round	تدوير
1.8.2; **2**.8.8; **3**.13.2; **4**.3.7	circular	مستدير
1.13.6, 14.8	always, [to occur]	دائم
1.14.4; **3**.10.10, 10.15	perpetual	دائماً
1.14.5, 14.12; **2**.2.1, 3.7–8; **4**.9.4, 14.1	diminution/ deterioration	ذبول
1.8.4–5; **2**.1.5, 10.5, 13.9; **4**.2.2	mind	ذهن

1.14.4, 14.11	deliberation	رؤية
2.6.10, 9.16; **3**.7.1, 8.1, 8.11, 10.6, 14.5; **4**.10.3, 11.1, 13.3	ordered position	ترتيب
3.4.7	arrangement	ترتيب
1.13.6; **3**.11.2	selectively determine, to	ترجيح
1.14.7	womb	رهم
4.6.8	vice	رذيلة
2.4.4; **3**.2.10	description (definite)	رسم
1.8.2, 14.3	astronomical observation	رصدية
4.3.6	wet	رطب
2.1.9	wetness	رطوبة
2.6.5, 9.11	removal	رفع
1.8.2	elevation	ارتفاع
1.2.19, 3.10, 10.7; **3**.3.8, 4.1, 5.7, 12.3; **4**.11.10	composition	تركيب
1.1.11, 3.7, 10.5; **2**.5.3; **3**.12.3; **4**.11.7	composite	مركّب
1.12.4; **3**.10.14	underlying-element	ركن
2.8.16; **4**.12.2, 14.2–6, 15.8	projectile motion	رمى
1.5.1, 13.6, 15.5; **3**.11.2; **4**.9.4, 12.6, 14.7	volition	ارادة
2.8.10	volitional	ارادي
1.8.1	interval	زمان

زمان	time	**2**.1.2, 2.12, 3.12, 3.19, 4.5, 10–13.passim; **3**.1.1, 7.3, 8.6, 9.6, 9.7, 10.1, 10.10, 11.1, 11.9
لا في زمان	instantaneously	**4**.12.1
زوال	process	**2**.1.4
زاوية	angle	**3**.3.10, 4.11, 5.9, 8.7, 11.4; **4**.5.3
– حادّة	acute angle	**3**.3.10, 5.9; **4**.5.3
زيادة	increase	**4**.8.2, 9.4
تزيّد عدد	numerical increase	**3**.9.1
تزايد	increase	**3**.10.3
سبب	cause	**1**.5.1, 13.1; **4**.11.2
– من خارج	external cause	**1**.13.2; **3**.11.2
– مسعد	fortunate cause	**1**.13.4
– مشقي	unfortunate cause	**1**.13.4
– طبيعي	natural cause	**1**.13.4
– موجّب	necessitating cause	**1**.13.2; **3**.11.2
تسخين	heating	**2**.9.20, 13.6; **4**.2.4
تسخّن	hot, to become	**2**.3.18
سريع	fast/speed	**2**.10.9; **3**.10.5; **4**.3.8
سرع و بطء	speed	**2**.8.11, 10.5, 11.1, 12.7; **3**.10.3; **4**.3.8, 15.8

2.13.7	eternity	سرمد
1.5.7	permeating	سريان
1.7.2; **3**.10.14	permeating	سارية
1.8.4; **2**.5.4, 8.10, 9.1, 9.6; **3**.2.7, 3.7, 5.10, 6.2, 8.1, 8.9, 13.2; **4**.3.6	surface	سطح
3.14.2	simple/two-dimensional surface	– بسيط
2.9.2	internal surface	– باطن
2.9.1	non-containing surface	– غير حاوي
2.9.2	convex surface	– مقبّب
2.9.2	concave surface	– مقعّر
2.9.1	contacting surface	– ملاق
1.7.4	blessed	سعيد
1.7.4, 12.7	flourishing, [life of]	سعادة
3.8.10, 13.5, 13.7; **4**.3.3	down(ward)	سفل
1.5.8; **2**.5.8; **4**.7.2, 9.8, 12.2	down(ward)	اسفل
1.6.1; **2**.1.1, 1.19, 4.passim, 8.10, 8.16; **3**.6.6, 8.8, 9.9, 10.13; **4**.6.9, 11.6	rest	سكون
2.11.1	intervening rest	– متخلّل
3.8.8, 9.4	natural rest	– طبيعي
2.12.3	negation	سلب
2.13.7; **3**.6.6, 7.2, 11.6	negation, absolute	– مطلق

1.7.3; **2**.1.4, 12.3	procession	سلوك
2.8.8; **3**.4.7, 4.10	projection/ projected path	سمت
3.13.6, 14.10	projecting towards	مسامت
1.8.2, 14.3; **2**.9.2; **3**.7.2, 8.9, 13.5, 13.7; **4**.3.3, 6.4	heaven	سماء
4.3.8	celestial	سماوي
1.7.4	evil	سوء
1.8.1, 8.11–12, 6.4; **2**.10.5, 11.1; **3**.3.4, 3.14, 6.passim, 10.5; [**2**.8.11–12]	distance/ spatial magnitude/ [medium]	مسافه
2.1.4	sameness	مساواة
2.5.3, 9.1	coextensive	مساوي
2.2.1, 12.5–6	flow(ing)	سيّال
1.15.6	sensible image	شبح
1.4.1, 8.2; **2**.8.22; **3**.8.9, 14.2; **4**.6.10, 9.7, 11.8	homogeneous/ homogeneric	مشابه
1.1.4	individual	شخص
1.1.9	vague individual	– منتشر
2.2.4, 3.2; **3**.10.2; **4**.2.6, 9.2	intensification/ increase/accelerate	اشتداد
1.13.15	evil	شرّ
3.11.7	condition	شرط
2.1.15	conditional statement	شرطي
1.10.9	law	شريعة

1.8.2	noble	شرف
1.8.2	rising	شروق
1.13.6; **2**.12.3	co-operative cause/ subclass	شريك
1.10.5; **2**.2.1	equivocal	اشتراك
1.14.2; **2**.9.11; **4**.5.7	equivocation [fallacy of]	– الاسم
1.2.8	common	مشترك
1.3.9	equivocal term	اسم المشترك
3.2.4	occupying	شغل
3.4.3	occupied	مشغول
3.6.9	transparency	شفّاف
4.6.3	transparent	اشفّاف
3.6.9	transparent	مشفّ
3.1.1, 2.6	following immediately	تشافع
2.12.3	intervening	مشافعة
3.**4**.8	following immediately	متشافعة
1.3.9; **2**.2.6	analogy	تشكيك
2.2.1	analogical	مشكّك
3.4.4	figure (mathematical)	شكل
1.10.7	figure two (logic)	– ثاني

1.8.1, 10.9, 13.4, 14.8; **2**.3.7, 7.3, 8.10; **4**.11.1–4	shape	شكل
3.8.5; **4**.11.5	circular shape	– مستدير
1.5.7	form	شكل
4.12.4	impure mixture	شوب
3.5.3; **4**.13.4	pointing	اشارة
3.10.15	inherent desire	شوق غريزي
2.5.5	desire	اشتياق
2.8.1	nothing	لا شيء
1.14.10; **4**.5.7	health	صحّة
3.4.16, 14.3	co-mate	صاحب
1.14.3	collision	تصادم
4.3.2–3, 8.6	ascending	صاعد
1.13.4; **2**.3.7; **3**.3.1	solidity/hardness	صلابة
1.9.1, 14.10–11	craft/art	صناعة
1.10.6	metaphysics	صناعة اولى
1.10.9, 11.4	class	صنف
1.2.passim, 3.10, 6.2, 9.1, 10.2, 10.9, 11.4–5, 12.5, 15.4; **2**.1.12, 6.3, 6.5, 9.1, 9.10; **3**.6.4, 7.3, 9.2; **4**.4.2, 4.5, 8.11, 10.1, 11.3, 13.5	form	صورة
3.12.1	simple form	– بسيطة

1.2.4, 3.2; **3**.9.4, 12.1; **4**.15.3	corporeal form	– جسمية
1.3.10; **2**.3.3	substantial form	– جوهرية
1.11.3, 12.5	specific form	– خاصّة
1.11.3, 12.5	proximate form	– قريبة
3.9.4	quantitative form	– كمّية
1.6.7; **3**.12.5	perfecting form	– كمال(ية)
1.9.4	species form	– نوية
2.10.5, 12.3	conceptualization	تصوّر
1.2.23	mystic	صوفي
1.12.2, 12.4; **2**.1.20, 3.4–5, 4.1–2; **3**.9.7	contrary	ضدّ
1.2.16	contrariness	ضدّية
2.3.3; **4**.6.6	contrariety	تضادّ
1.2.16; **3**.14.10	contrary/contrariety	مضادّ
3.14.3	contrary, to be	متضادّة
4.10.1	compressed	ضغط
2.9.21	compression	انضغاط
3.4.6, 13.2	side	ضلع
4.8.1	conjoined	تضامّ
4.2.2	distributive attribute	متضمّن

3.6.9; **4**.4.6	light	ضوء
3.6.9; **4**.4.6	illumination	اضاءة
2.1.2, 3.11; **3**.11.1	relation, [category of]	اضافة
3.11.1; **4**.6.4	correlative	مضاف
1.5.6	medical knowledge	طبّ
1.5.6, 10.2, 12.2, 12.4	physician	طبيب
1.passim [esp. **1**.5, 6.2, 6.5]; **3**.7.2, 8.1; **4**.9.5	nature	طبيعة
2.8.3	generic nature	– جنسية
2.3.2	substantial nature	– جوهرية
2.8.3; **3**.3.3	privative nature	– عدمية
1.13.8	universal nature	– كلّية
2.8.3; **3**.3.3	specific nature	– نوعية
1.7.1; **4**.9.7–8	natural	طبيعي
4.13.5	impressed	انطبع
3.4.6	superimpose, to	تطبيق
2.1.7, 13.1; **3**.5.6; **4**.6.8, 15.10	map onto, to/ coincide with, to	مطابقة
3.3.9, 4.8	superimpose, to	اطباق
4.5.2	coincide, made to	انطباق
3.8.1	correspond (exactly) with	مطابق

3.3.9, 4.8	coincide	منطبق
1.13.6	regularity	اتطراد
2.1.3, 1.5, 1.20, 2.4, 6.5, 10.3, 12.2–3, 13.6; **3**.2.2, 2.5, 2.7, 4.9, 6.4–6, 11.1; **4**.6.3	limit	طرف
1.5.2	course	طريقة
3.3.14–16, 4.1	leap	طفرة
1.2.2, 8.1; **2**.1.12; **3**.4.4, 13.4, 14.9	length	طول
1.2.6	stuff	طينة
4.4.6	darkness	ظلمة
4.4.6	darkening	اظلام
1.8.3; **2**.7.4, 13.6; **3**.1.1, 5.4, 7.3, 8.1, 8.12	number	عدد
3.11.9	ordered numbering	– مرتّب
1.8.3	numerable number	– معدود
4.15.5	contributing role	اعداد
1.2.17, 13.8, 14.13, 15.6; **3**.5.2, 9.3–4, 9.6, 12.5; **4**.7.4, 9.7, 15.3	preparedness	استعداد
2.5.8	preparatory [cause]	معدّ
1.2.12–18, 3.4, 3.10, 12.5, 14.12; **2**.4.1, 9.6, 11.5; **3**.9.8, 10.13; **4**.2.5, 6.4, 7.4	privation/ nonexistence	عدم
2.13.1	absolute nonexistence	– مطلق
3.12.5	source of origin	معدن

1.2.2, 8.1; **2**.1.12; **3**.4.4, 13.4, 14.9	breadth	عرْض
1.2.17, 6.1, 6.4, 10.6; **2**.2.1, 4.2, 8.2, 9.4, 9.12, 10.7; **3**.5.6, 10.2; **4**.3.1, 3.7	accident	عرَض
4.3.7	primary accident	– اوّل
4.3.7	consequential accident	– تابع
3.3.1	relational accident	– مضاف
3.3.1	non-relational accident	– غير مضاف
1.1.1; **4**.3.1	necessary accident	– لازم
4.11.3	non-necessary accident	– ليس لازم
1.1.4–5, 1.13	better known	اعرف
1.13.6	resolve	عزيمة
1.12.3; **4**.13.5	limb/organ	عضو
2.8.14; **3**.4.1, 10.2, 10.4; **4**.15.9	bulk/size/massive	عظم
2.7.4, 8.11	greater/bulk	عظيم
2.5.8, 6.3, 9.9; **4**.4.3, 11.3	replacement	تعاقب
1.14.5, 14.16	congealing	اعقاد
1.1.19; **2**.8.3, 9.11, 9.15, 10.5; **4**.11.2	intellect	عقل
1.10.9; **2**.9.4, 10.5; **3**.13.7	intelligible	معقول
2.8.3	vain intelligible	– مفروغ
1.1.11, 3.12, 8.2, 10.passim, 12.passim; **2**.9.7	cause	علّة

1.12.1	simple cause	– بسيط
1.12.1	remote cause	– بعيدة
1.12.1	particular cause	– جوزئي
3.11.2	temporally originated cause	– حادثة
3.9.7, 11.2	motive cause	– محرّكة
1.12.1	specific cause	– خاصّ
1.12.1	essential cause	– بذّات
1.12.1	compound cause	– مركّب
1.12.1	accidental cause	– بالعرض
1.9.7	material cause	– عنصورية
1.12.1	general cause	– عامّ
4.14.6	auxiliary cause	– معنّية
1.12.1	actual cause	– بالفعل
1.12.2	efficient cause	– فاعلة
1.12.1; **2**.13.4	proximate cause	– قريبة
1.12.1	potential cause	– بالقوة
1.12.1	universal cause	– كلّي
1.1.11; **2**.9.7	effect	معلول
1.5.6, 10.2, 12.4	cure	علاج

2.3.6	embryo	علقة
1.8.3; **2**.13.2	dependence relation/ depend, to	تعلّق
1.8.passim	science	علم
1.8.1	weights	– الأثقال
1.8.1, 8.8	spherics	– الأكر المتحرّكة
1.8.1, 8.8	music	– الموسيقى
1.8.1–2, 8.8	optics	– المناظر
1.8.1–2, 8.8	astronomy	– الهيئة
1.3.6, 13.4; **2**.6.15, 13.7	world/universe	عالم
1.8.2; **2**.5.8	mathematical	تعليمي
3.8.10, 13.5, 13.7; **4**.3.3	upward	علو
1.2.8; **2**.10.4	generality	عموم
3.4.14; **4**.14.4	tendency	عتماد
1.2.2, 8.1; **2**.1.12; **3**.4.4, 13.4, 14.9	depth	عمق
4.6.2	antipathetic	متعاند
1.2.3, 2.6; **3**.12.3; **4**.11.6	component, [elemental]	عنصر
3.2.7	force	عنف
1.15.4; **2**.10.5; **3**.9.6	account/connotational attribute/formal aspect	معنى
1.1.7	common notion	– عامّ

1.5.8, 6.2, 13.6; **4**.12.1	impediment	عائق
1.10.4; **2**.8.16	auxiliary	معين
1.1.4; **2**.1.5, 8.4, 10.1	concrete particular	عين
2.9.19; **4**.9.7	nutrition/nutriment	غذاء
1.8.2	setting	غروب
4.12.6	innate	غريز
1.8.4	inherent disposition	غريزة
1.1.13, 2.7, 10.2, 10.4, 10.10, 11.1, 11.5; 12.6–7, 13.2, 14.2, 14.8; **4**.15.2	end	غاية
3.14.1; **4**.6.1, 6.4	maximal degree	غاية
1.13.15	blameworthy end	– مذمومة
1.13.12, 14.2, 14.16	essential end	– ذاتية
1.14.3	disadvantageous end	– ضارّة
4.9.5, 10.3	natural end	– طبيعية
1.14.3, 14.16	accidental end	– عرضية
1.15.4	ultimate end	– الغايات
2.1.3	otherness	غيرية
1.1.5; **2**.3.15, 11.4; **3**.11.9	change, to	تغيّر
1.2.12	changeable	متغيّر

4.2.6	decelerate	فتور
1.14.6	individual instance	فرد
3.2.11	separate, being	فرادى
2.8.10; **3**.2.8, 6.4, 9.3; **4**.14.6	posit(ing)/supposition	فرض
3.4.3	unoccupied	فارغ
3.3.2, 5.2, 6.6	separation/separate, to	تفريق
3.9.3	separate, becoming	افتراق
1.3.2; **2**.2.1, 12.2; **3**.7.3, 10.13, 12.8; **4**.6.7, 9.4	corruption/cessation	فساد
2.11.2; **4**.4.4	corruption, subject to	فاسد
1.12.5; **2**.1.10, 5.8, 8.4, 11.1–2; **3**.7.2; **4**.1.1, 3.1, 8.13	difference (species/specific)/division	فصل
1.4.5; **2**.2.2	species making difference	– منوّع
2.1.10; **3**.12.5, 12.8	discontinuity	انفصال
3.2.8	discontinuous	منفصل
2.12.3	disjunctive [proposition]	منفصل
1.8.2	remnant	فضل
4.6.8	virtue	فضيلة
1.3.8, 10.2, 13.2; **2**.1.2–3, 2.6, 3.18; **4**.6.5	actuality/act/action/actual	فعل
1.14.14	divine action	– اللهي
1.2.8	first actuality	– اوّل

2.1.22	remote in actuality	بعيد من فعل
2.3.2	absolute actuality	– صرف
2.1.22	proximate in actuality	قريب من فعل
1.5.1; **2**.1.2, 2.1, 3.18; **4**.3.7	passivity/passion	انفعال
1.1.13, 2.7, 3.11, 10.2, 11.1, 11.5, 15.3; **2**.5.5; **3**.5.2, 10.1, 10.4	agent	فاعل
3.10.1	patient	منفعل
3.3.16, 4.11–12	fragment, to	تفكّك
3.12.6	fragmentation	انفكاك
3.9.3	fragmented	مفكّك
2.1.20, 3.4, 3.13–15, 10.8, 10.12; **3**.10.10, 11.6, 13.5; **4**.7.4, 11.9, 13.3, 14.7	celestial sphere	فلك
2.3.14–15, 6.17; **4**.13.2	outermost celestial sphere	– اعلى
2.2.6	concept	مفهوم
1.5.8; **2**.5.8; **4**.7.2, 12.2	up(ward)	فوق
4.4.3	emanation	فيض
3.11.4; **4**.10.4, 10.6	emanation	فيضان
1.13.8	emanating	فائض
3.2.7; **4**.3.8	convex/convexity	تقبيب
1.5.8	contraction	انقباض
2.11.4–5, 12.1	before	قبل

3.14.10	opposition	مقابلة
1.8.2	opposition	مقابل
3.4.1, 5.4	quantity (determinate)	قدر
3.11.1	power	قدرة
1.4.4, 8.1, 8.4; **2**.7.4, 11.2; **3**.1.1, 4.1, 5.4, 6.6, 7.3, 8.1, 12.4; **4**.3.8	magnitude	مقدار
2.9.11, 11.4	measurement/ measure, to	تقدير
3.10.8	supposition	تقدير
3.10.8	supposition	اقتدار
3.11.4	eternal	قدم
1.1.4	prior (by nature)	اقدم
1.3.9	priority	تقديم
1.10.6–7; **2**.2.6; **4**.6.4, 8.16, 11.6	prior/priority/ antecedent	تقدّم
2.9.7	causal priority	– علّية
1.4.3, 10.7; **2**.10.5; **3**.11.7	premise	مقدّمة
1.4.3, 10.7	conjunctive premise	– مقترنة
2.10.10, 11.2–5, 12.7, 1310	earlier	متقدّم
2.1.1	stable	قارّ
1.2.11; **2**.12.4; **3**.11.2, 14.1	proximity	قرب
1.5.6	proximate	قريب

1.5.6	non-proximate	غير قريب
1.1.11	conjunction	مقرن
1.14.3; **4**.11.2	force	قسر
1.14.3	forced	قسري
1.5.6; **2**.8.15	force/agent acting by force	قاسر
2.2.1; **3**.5.7, 9.1, 12.6	divisibility	انقسام
2.10.2, 11.2; **3**.8.11	divisible	منقسم
2.10.2	indivisible	غير منقسم
1.13.10; **4**.9.4, 12.5–6	intention	قصد
3.9.7	secondary intention	– ثاني
1.14.8	natural intention	– طبيعي
1.7.4, 13.6, 13.10; **2**.1.3; **4**.10.3	intended	مقصود
2.4.2	derive	اقتضاب
1.5.8	proper	مقتضي
2.6.2, 6.6, 8.2; **3**.1.1, 4.6, 8.6	dimension/diagonal/diameter	قطر
2.1.7, 12.5, 13.1; **3**.6.1, 9.6; **4**.2.2, 4.1–2, 5.8, 9.5	traversal	قطع
3.6.6	discontinuity	تقطيع
2.8.10, 9.3	discontinuity	انقطاع
2.11.6; **3**.6.6	interrupted/discontinuous part	منقطع

3.2.7; **4**.3.6, 3.8, 6.6	concave/concavity	تقعير
2.12.3, 13.10	gradually	قليلاً
1.13.6	seldom, [to occur]	في اقلّ الامر
4.3.8	general rule	قانون
2.4.1	possession	قنية
4.11.2	coercion	قهر
3.8.4; **4**.5.3, 6.7, 12.8	arc	قوس
1.4.3, 10.7; **2**.9.9; **3**.11.7	syllogism	قياس
2.1.2, 2.passim, 3.1	category	مقولة
1.3.8, 7.1; **2**.3.8	subsistence/ underlying thing	قوام
1.6.2; **3**.5.6; **4**.6.4	subsist, to make/sub- sistence (to bring about)	تقويّم
1.6.2	subsist, to	تقوّم
1.8.2; **3**.13.2; **4**.2.5, 3.3	rectilinear/straight	استقامة
2.11.2	subsisting	قائم
4.3.7	rectilinear	مستقيم
2.12.1	linear extension	– امتداد
1.1.6, 3.8, 5.3, 10.2; **2**.1.2, 3.7, 4.4, 8.6, 9.11; **3**.10.2, 11.1, 12.5; **4**.7.4, 11.10, 12.1, 14.3	potential/potentiality / potency/power / faculty	قوة
2.3.7	impotence	لا قوة
1.5.7	divine power	– الهية

– بعيدة	remote potential	**1**.12.3; **4**.15.3
– جاذبة	attractive power	**2**.8.20–21
– محرّكة	motive power	**2**.8.14, 8.20; **3**.11.2; **4**.8.11
– ارادية	volitional power	**4**.9.6
– صرف	absolute potentiality	**2**.3.2
– غاذية	nutritive faculty	**4**.9.7
– فاعلية	active power	**1**.5.7
– مستفادة	acquired power	**4**.12.2, 14.2, 14.5
– قريبة	proximate potential	**1**.12.3; **2**.12.1; **4**.6.5
– ميلية	inclinatory power	**1**.5.7
كثرة	multiplicity	**2**.12.7; **3**.10.5
كثير	many / multiple	**2**.7.4; **3**.4.1, 5.2
في اكثر	for the most part, [to occur]	**1**.13.6, 14.8; **3**.12.3
تكاثف	condensation/ condense	**1**.5.8; **2**.3.8, 6.12, 9.17, 9.20–22; **3**.8.3; **4**.5.6, 6.1, 9.1
كذب	falsity	**3**.3.12
كرة	sphere	**4**.3.8, 8.4, 13.2
كريرة	sphere	**1**.8.2
انكسار	broken up, to be	**2**.8.10
كسوف	eclipse	**1**.1.13, 8.2, 13.9; **3**.11.6

كلّ	whole / universe	**1**.7.2; **2**.3.15; **3**.8.9, 8.11, 9.1, 10.1, 11.5, 13.5; **4**.11.6, 14.1, 15.3
– واحد	each one	**2**.3.15; **3**.11.6, 11.8
واحد كلّ	unified whole	**3**.2.8
كوليّ	universal	**1**.7.2, 12.5; **2**.1.7
كوليّة	universal / collective kind	**1**.1.6, 3.6; **3**.12.8; **4**.10.2, 11.8
كمّ	quantity	**1**.5.8, 8.1; **2**.1.2, 8.2; **3**.1.1, 11.6
– متّصل	continuous quantity	**1**.4.4
كمّية	quantity	**2**.6.6; **3**.7.1, 9.3, 10.3
كمال	perfection	**1**.14.9; **2**.1.3, 2.3, 2.6, 4.4; **4**.7.2
– اول	first perfection	**1**.2.22; **2**.1.3, 1.6, 1.20, 4.4, 9.10; **3**.9.6; **4**.2.3
– ثاني	second perfection	**1**.2.22; **2**.1.5, 1.20, 4.4
– بالقوة	potential perfection	**1**.2.22; **2**.1.5, 1.20
استكمال	perfection	**1**.3.8; **4**.9.1
مستكمل	perfectible	**1**.2.12–13
كوكب	planet	**1**.8.2
كون	generation	**1**.3.2; **2**.2.1; **3**.7.3, 10.13, 12.8
مكان	place	**1**.5.8, 6.2; **2**.3.10, 3.14, 5–9.passim, 11.1; **3**.2.2, 6.2, 8.2, 12.5; **4**.10.2, 11.passim
– طبيعي	natural place	**3**.10.13; **4**.8.11, 10.2, 11.8, 12.1, 13.3

2.13.1; **3**.12.3	generation	تكوّن
1.2.12, 2.15, 13.5; **4**.4.4	generable/subject to generation	كائن
1.5.8, 14.8; **2**.1.2, 3.7	quality	كيف
3.12.5; **4**.8.13	quality	كيفية
1.8.2	reason why	لماية
1.14.14	why question	مسألة عن لما
1.1.1; **4**.3.7	concomitant	لاحق
3.10.11	inter-connected	ملتحم
1.4.2; **2**.9.7	adhere/inseparable	لزوم
2.9.7; **4**.13.3	necessary concomitant/ inseparable	لازم
3.1.7	cohesion	تلاصق
3.2.9	cohere	ملتصق
4.14.6	attenuation	تلطيف
1.13.1, 13.14	spontaneously	من تلقاء نفسه
3.2.3	coincide	لقاء
2.8.15; **3**.2.3–5, 3.9, 4.1	contact/encounter	ملاقاة
2.1.12, 1.20; **3**.14.10; **4**.2.5, 3.2, 5.5, 8.1	terminus ad quem (towards which)	ما اليه
2.1.12; **4**.3.1, 7.2	that with respect to which [aristotelian category]	ما فيه

2.1.12, 1.20; **3**.14.10; **4**.2.5, 3.2, 5.5, 8.1	terminus a quo (from which)	ما منه
1.2.17, 4.2, 6.1, 10.5, 10.9, 11.2, 11.5, 15.4; **2**.5.1, 8.4, 13.3; **4**.3.1–2, 11.4	essence	ماهية
2.1.2, 3.12	when	متى
4.15.2	paradigm	مثال
2.10.10, 13.7; **3**.10.2	duration	مدّة
1.2.2; **3**.7.3, 13.2; **4**.3.6	dimension/ extension	امتداد
3.6.6	un-extended	ليس امتداد
1.2.3, 2.6, 3.2, 6.6, 7.1, 9.1, 10.passim, 12.4, 15.4; **2**.11.6; **3**.8.13, 9.2–3; **4**.3.7	matter	مادّة
1.9.4	prime matter	– اولى
1.9.4	determinate matter	– معيّنة
1.11.3	proximate matter	– قريبة
4.2.4	a lot	مرار
2.3.4, 13.7; **4**.6.10	extension/ uninterrupted	استمرار
2.11.6	uninterrupted	مستمرّ
1.5.6	patient	مريض
4.12.7	mix	مزج
1.2.19, 6.5, 8.4, 10.7; **3**.10.14, 12.2; **4**.4.3, 5.7	mixture, [elemental and humoral]	مزاج
1.2.19, 10.7	humoral temperament	مزاج

1.5.8	humoral balance	– موافق
4.11.10, 12.7	mixing	امتزاج
3.2.2; **4**.1.3, 6.2	contiguity	تماسّ
2.11.1, 12.1; **3**.2.5; **4**.1.3, 5.8	contiguity	مماسّة
2.12.4; **3**.2.2, 3.9	contiguous	مماسّ
2.3.6	fetus	مضغة
2.11.5	simultaneous	مع
3.2.11	together, being	معاً
1.10.6; **2**.10.4, 10.10, 13.7	simultaneity	معاية
3.11.1, 11.5	possibility	امكان
2.12.1; **3**.12.3	absolute possibility	– مطلق
2.1.8	situated	متمكّن
2.8.5–6, 8.23; **3**.8.13, 10.8, 14.2	plenum	ملأ
1.14.11; **2**.3.7; **4**.4.3	habit/disposition	ملكة
4.10.4	momentary disposition	– وقتية
1.5.6, 12.3, 13.6; **3**.4.16; **4**.9.6	hindering/obstacle	مانع
3.3.12	impossible	ممتنع
4.9.4	sperm	منة

4.13.3	rippling effect	تموّز
1.3.3, 8.3–4, 12.4; **3**.6.4, 7.3; **4**.10.1, 10.4	water	ماء
1.5.6, 6.2; **2**.5.5, 8.14–15; **3**.4.14, 10.2, 10.15; **4**.3.9, 8.11, 8.16–18, 10.4, 11.9, 12.passim, 13.3, 14.2, 14.5, 15.5, 15.7	inclination	ميل
1.5.1; **3**.12.3	plant	نبات
1.10.7	conclusion	نتيجة
2.8.12; **3**.4.1, 10.1, 10.4; **4**.12.2, 15.7	proportion/ratio	نسبة
3.10.7; **4**.3.9, 14.passim, 15.4–9	interrelation/ratio	مناسبة
1.13.5, 14.5	evolution/development	نشوء
1.1.9	vague	منتشر
1.2.19, 5.1, 11.5, 12.4, 14.7; **2**.3.3	semen	نطفة
3.13.3	speculative theologian	اهل النظار
4.4.5	organization	نظم
1.8.2, 14.5, 14.9; **3**.10.11, 14.5	order/system	نظام
1.10.9	regulative order	– مستحفظ
1.8.1; **4**.2.4	tone	نغم
1.5.6, 7.4, 11.2, 14.9, 14.11; **2**.3.7, 8.15, 10.5, 12.7, 13.1; **3**.12.5, 12.5; **4**.9.6, 12.5–6, 13.5	soul	نفس
1.5.3; **4**.12.5	animal soul	– حيوانية

– فلكية	celestial soul	**1**.5.3
– نبتية	vegetative soul	**1**.5.3
متنفّس	principle of animation	**4**.9.6
نافع	beneficial	**1**.12.6
نقص	deficiency	**1**.14.12
نقصاً	fall short of	**3**.10.7, 10.9; **4**.5.2
تنقّص	decrease	**2**.3.2
ناقص	deficient	**4**.2.6
نقطة	point	**2**.5.4, 8.10, 9.6; **3**.2.2, 3.7, 4.9, 5.6, 6.7–8, 7.2, 8.1
نقلة	locomotion/transition	**2**.12.5; **3**.11.2; **4**.2.4, 3.4, 6.1, 12.1
انتقال	locomotion/transition	**2**.5.4, 13.3, 13.6; **4**.12.1
نمو	augmentation	**1**.5.8; **2**.2.1, 3.7–8; **3**.6.9, 8.2, 8.13; **4**.6.1, 9.1
نهاية	extremity	**2**.6.2, 8.10, 9.1, 12.3, 12.5, 13.6; **3**.4.11, 5.6
متناه	finite	**1**.4.2; **3**.1.1, 7–9.passim; **4**.15.10
غير متناه	infinite	**1**.4.2; **3**.1.1, 7–9.passim; **4**.15.10
غير متناه بالفعل	actual infinite	**3**.8.passim
غير متناه بالقوة	potential infinite	**2**.3.2; **3**.5.1, 9.2
نار	fire	**1**.8.3, 11.2, 14.16; **3**.6.4, 7.3, 10.2, 10.10

1.1.3; **3**.7.2; **4**.6.9	species	نوع
3.5.2	specificity	نوعية
4.3.2–3, 8.6	descending	هابط
1.8.2	crescent	هلالي
1.8.1; **3**.10.8	geometer	مهندسة
1.8.1–2, 8.8; **3**.5.9	geometry	هندسة
4.2.1, 3.6	being	هوية
1.3.3; **3**.6.4, 7.3, 8.8	air	هواء
4.13.2, 13.3	higher atmosphere	حواء (جوّ) علي
1.2.16–17, 10.9, 13.2, 13.4, 14.3; **2**.8.2; **3**.6.4; **4**.5.6	disposition/ configuration	هيئة
2.11.2	fixed disposition	– قرّة
2.11.2	unfixed disposition	– غير قرّة
3.9.4	disposition, to produce	تهيئة
1.2.17, 15.6	predisposition/ disposition, to have	تهوّء
1.2.3–6, 3.10, 6.6, 11.5; **2**.3.3, 6.3, 6.5, 7.6, 9.1, 9.11; **3**.9.4	material	هيولى
1.3.7, 9.1	prime matter	– اولى
1.12.2; **4**.5.3, 6.7	sinew/chord	وتر
2.9.7	necessary	واجب

2.9.15	properly necessary	صواب واجب
2.9.15	necessarily to exist	– الوجود
2.10.12	affirmative proposition	موجبة
2.1.2, 3.17	possession, [category of]	جدة
1.4.2, 11.2; **2**.2.1, 2.3, 2.6, 5.1, 11.5; **3**.11.4	existence	وجود
2.1.5; **3**.11.6	reality, in(to)	في وجود
3.11.1, 11.8	possible existence	جائز وجود
2.13.1	absolute existence	وجود مطلق
3.11.2; **4**.2.2	fixed existence	قرّ وجود
2.1.5; **3**.11.4	exist, to necessarily	واجب وجود
3.11.1	impossible existence	ممتنع وجود
1.4.2	necessary existent	موجود واجب
2.8.9, 8.16, 13.3; **3**.2.1, 4.8–9, 13–14.passim; **4**.8.11, 10.2, 13.4	direction/directional side/location of contact/side	جهة
2.1.2–3	advance/advancing	توجّه
1.8.3; **2**.2.3; **3**.5.2, 6.5, 9.5; **4**.2.2	unity	وحدة
1.8.3; **2**.13.6; **3**.4.1, 4.4, 5.2, 8.1, 9.5; **4**.2.1	unit/one	واحد
2.11.1, 12.1, 13.2–3; **3**.11.2	juxtaposition	موازة
3.2.11; **4**.9.2, 12.7	intermediate/mean	وسط

واسطة	intermediary	**1**.5.6; **4**.2.5, 15.1
متوسّط	medium	**2**.8.16
صفة	description/attribute	**1**.2.17, 3.8
اتّصال	continuity/continuous/ continuum	**1**.8.1; **2**.1.3, 1.10, 11.6, 12.5; **3**.1.1, 2.10, 3.1, 4.1, 4.3, 6.2, 6.6; **4**.2.4
– مواحّد	unified continuity	**4**.8.14
– مفرّق	discrete continuity	**4**.8.14
متّصل	continuous	**2**.8.10
واصل	connection	**2**.12.1
وضع	position	**1**.8.2; **2**.1.2, 3.13–16, 5.8, 9.15, 11.1; **3**.2.25, 4.1, 7.1; **4**.6.1, 6.11, 11.1, 12.1, 13.4
وضع	subject (logical)	**1**.10.7; **2**.10.5
وضعية	hypothetical	**3**.6.5
موضع	location	**2**.8.10; **3**.8.7; **4**.7.4, 10.2, 12.1
– كلّي	universal location	**3**.8.8
موضوع	subject	**1**.2.3, 2.5, 8.4, 10.7, 12.4; **2**.1.12, 3.4, 3.7, 9.12, 11.2; **3**.11.9
– بعيد	remote subject	**1**.12.4
– جوزئي	particular subject	**1**.12.4
– قريب	proximate subject	**1**.12.4; **4**.6.5, 6.7
– كلّي	universal subject	**1**.12.4

1.3.9; **2**.2.1	univocity	تواطع
2.2.1; **4**.3.8	univocal(ly)	بتواطؤ
1.13–14.passim	chance	اتفاق
3.6.4	congruent	موافقة
1.3.6; **2**.10.1, 10.10; **3**.5.10, 11.7	moment	وقت
3.4.14; **4**.8.13	pause	وقفة
3.6.6, 11.7	rest, coming to/ dependent, to be	توقّف
1.13.4; **4**.14.4	engender	تولّد
2.12.1; **4**.13.4	adjacent (immediately)	ولي
1.10.3; **4**.15.3	giver of forms	واهب الصور
2.1.14	estimative faculty	وهم
1.8.7; **2**.5.8, 9.11	estimative faculty	توهّم
4.11.5–6	dryness	يبس
4.3.6	dry	يابس

Subject Index

References are to book.chapter.verse (or paragraph #)

abstracted (*mujarrad*), **1**.8.2–3, 8.6–7; **2**.2.7
absurdity (*muḥāl*), **2**.13.3; **3**.3.12
acceleration (*ishtidād*), **4**.2.6
accident (*ʿaraḍ*), **1**.1.3, 1.15, 2.2, 2.5, 2.17, 4.4, 5.6, 6.1–2, 6.4, 8.1–3, 8.8, 9.1–2, 10.6, 10.9, 11.4, 12.5; **2**.1.9, 2.1, 3.2–3, 4.2, 5.1–2, 8.2, 9.4, 9.12, 10.7, 13.2; **3**.3.7, 5.2, 5.6, 10.2; **4**.2.2, 3.1, 3.7, 10.4, 11.3, 13.6
  consequential (*ʿa. tābiʿ*), **4**.3.7
  necessary (*ʿa. lāzim*), **1**.1.1, 6.2, 7.1; **3**.2.10; **4**.3.1, 6.5
  non-necessary (*ʿa. laysa lāzim*), **4**.11.3
  non-relational (*ʿa. ghayr muḍāf*), **3**.3.1
  primary (*ʿa. awwal*), **4**.3.7
  relational (*ʿa. muḍāf*), **3**.3.1
accidental efficient cause, builder example, **1**.12.2, 12.8
accidental end. See end, accidental.
accidental motion, boat example, **1**.12.5
action (*fiʿl*), **1**.13.2; **2**.1.2, 2.6, 3.18
  divine (*f. illāhī*), **1**.14.14
actuality (*fiʿl*), **1**.2.8, 3.8, 10.2; **2**.1.2–3, 1.6, 1.22, 3.2; **3**.9.2; **4**.6.5
  absolute (*f. ṣirf*), **2**.3.2
  first (*f. awwal*), **1**.2.8
  proximate in (*qarīb min f.*), **2**.1.22
  remote in (*baʿīd min f.*), **2**.1.22
agent (*fāʿil*), **1**.1.14, 2.7–8, 2.11, 3.11, 10.2, 11.1–3, 11.5, 12.2–3, 15.3–4, 15.6; **2**.5.5; **3**.5.2, 10.1, 10.4
  acting by force (*qāsir*), **1**.5.6; **2**.9.20
aggregation(s) (*taʾlīf*), **3**.3.1, 3.3, 3.12, 4.3, 5.1–2, 12.4
air (*huwāʾ*), **1**.3.3; **3**.6.4, 7.3, 8.8
  ambient, **1**.12.3
all at once (*dufʿah*), **2**.1.2–3, 3.2, 3.12, 11.6, 12.3, 13.10; **3**.5.8, 6.9; **4**.6.3
Almagest, **1**.8.2
alteration (*istiḥālah*), **1**.5.1, 10.7–8; **2**.1.4, 1.6–7, 3.2–4, 3.6–8, 4.6; **3**.6.9, 12.8–9; **4**.2.2, 2.5, 4.4, 4.6, 5.8, 6.1, 7.4, 8.1, 8.4, 9.1, 10.4, 14.1
  corporeal, **4**.9.1
  forced, **4**.10.1
always [to occur] (*dāʾim*), **1**.13.6, 14.8
analogy (*tashkīk*), **1**.3.9; **2**.2.1, 2.6
analysis (*taḥlīl*), **1**.2.6; **2**.6.5, 9.11
Anaxagoras, **1**.4.1n.14, 14.5n.11, **3**.12.4, 12.5
ancients, **3**.7.1
angle (*zāwiyah*)
  acute (*z. ḥāddah*), **3**.3.10n, 5.9; **4**.5.3
  horn, **3**.3.10n.9; **4**.5.3
  intersecting (*z. taqāṭuʿ*), **3**.8.7
  right (*z. qāʾimah*), **3**.4.11, 11.4
animal (*ḥayawān*), **1**.5.1; **3**.12.3
animation, principle of (*mutanaffis*), **4**.9.6
ant and sandal. *See* Zeno's paradoxes
Antiphon, **1**.6.6, 9.1
appendages, additional, **1**.7.1, 7.4, 13.8, 14.12; **4**.9.8
Appendices (*al-lawāḥiq*), **4**.12.5n.9
arc (*qaws*), **3**.8.4; **4**.5.3, 6.7, 12.8

Aristotle, **1**.2.10n.6, 3.9n.9, 5.5n.6, 5.5n.7, 6.6, 7.3n.6; **2**.1.2n.2, 1.2n.3, 1.3, 2.1n.2, 1.2n.3, 1.3, 2.1n.2, 2.5, 3.7n.8, 6.2n.4, 6.4n.6, 6.9n.8, 8.8n.14, 8.20n.30, 10.8; **3**.8.6n.11, 12.8n.9; **4**.12.6n.10
  Posterior Analytics, **1**.1.4n.7, 10.8n.12
  Categories, **1**.1.4n.8, 14.8n.5; **2**.3.17n.18
  Metaphysics, **1**.4.1n.2, 10.2n.2; **2**.3.15n.17
  Generation and Corruption, **1**.4.5n.11; **2**.9.17n.29
  Nichomachean Ethics, **1**.14.11n.8; **4**.6.9n.9
arithmetic (*ḥisāb*), **1**.8.1, 8.8
ascending (*ṣāʿid*), **4**.3.2–3, 8.6
astronomical observation (*raṣdīyah*), **1**.8.2, 14.3
astronomy (*ʿilm al-hayʾah*), **1**.8.1–2.8
atmosphere, higher (*huwāʾ* [*jaww*] *ʿalī*), **4**.13.2, 13.3
atom (*juzʾ lā yatajazaʾu*), **1**.4.1, 4.5, 13.4; **2**.13.3; **3**.3–5 passim
  infinite in number (*j. ṣaghīr lā yatajazaʾu*), **1**.13.4
  sheet of (*j. lā yatajazaʾu*), **3**.4.4
atomism, **3**.3–5 passim
  mustard seed example, **3**.3.5, 5.4–5
attenuation (*talṭīf*), **2**.9.21; **4**.14.6
attraction (*jadhb*), **1**.14.7; **2**.8.20, 8.25, 9.20; **4**.10.2, 11.10
augmentation (*numūw*), **1**.5.8; **2**.2.1, 3.7–8; **3**.6.9, 8.2, 8.13; **4**.6.1, 9.1

bad planning, **1**.13.15
beauty (*jamāl*), **1**.12.6
beginning, temporal, **3**.11 passim
being (*huwīyah*), **4**.2.1, 3.6
beneficial (*nāfiʿ*), **1**.12.6
better known (*aʿraf*), **1**.1.7, 1.15, 4.3; **2**.10.6
  by intellect, **1**.1.1–4, 1.6–7, 1.16
  by nature, **1**.1.1–4, 1.6–7 1.13, 1.14, 1.16
bile, purging, **1**.12.2, 12.7
blessed (*saʿīd*), **1**.7.4
body (*jism*/*jirm*/*badan*), **1**.1.1, 1.7, 2.2–3, 2.5, 2.12, 2.15–16, 5.8, 6.1, 7.4, 10.7, 14.12; 2.1.13–17, 5.3, 6.11, 7.3, 7.6–7, 8.2, 8.10, 9.1, 9.10, 9.12, 9.20; **3**.2–5 passim, 8.1, 8.11–13, 9.4, 10 passim, 12.5–6, 12.8, 13.2; **4**.3.6, 10 passim, 11 passim, 12 passim, 13.4, 13.5, 15.2–3, 15.5
  circular, **1**.13.3; **2**.13.3; **3**.8.5
  composite, **1**.6.3; **3**.8.3, 10.14; **4**.9.6
  growing, **2**.5.7; **4**.9.4
  heavenly, **1**.7.3, 8.2–3
  simple, **1**.8.2, 14.9; **2**.5.3, 6.7; **3**.7.3, 8.3; **4**.9.6, 11.5
  six directions of, **3**.13.4–5
bone, **3**.12.3
Book of Demonstration, **1**.1.13, 8.7
breadth (*ʿarḍ*), **1**.2.2, 8.1; **2**.1.12; **3**.4.4, 13.4, 14.9

camelthorn, **1**.12.7
categorical statement (*ḥamlī*), **2**.1.14
category (*maqūlah*), **1**.3.9, 4.3, **2**.1.2–2, 1.24, 2 passim, 3 passim; **4**.7.4, 13.6
cause (*sabab*/ *ʿillah*), **1**.1.11, 3.1, 3.12, 5.1, 8.2, 10 passim, 12 passim; **2**.9.7; **4**.11.2
  accidental, **1**.12.1, 13.9, 14.9, 14.12
  auxiliary, **1**.8.18, 10.4; **2**.8.16; **4**.14.6
  compound, **1**.12.1, 12.3
  efficient, **1**.5.6, 10.1, 10.4, 10.8, 11.1–2, 12.2, 13.10, 13.12
  essential, **1**.12.1, 13.13, 13.15
  essentially ordered, **3**.11.2
  external, **1**.5.1, 13.2; **2**.8.4; **3**.11.2; **4**.9.6, 11.2, 12.5, 14.2, 14.7
  final, **1**.10.1, 11.1, 15.6
  general, **1**.2.10, 12.1, 12.3
  material, **1**.9.7, 10.8, 11.4, 12.4, 15.6; **2**.9.7
  motive, **2**.8.23; **3**.9.7, 10.10, 11.2
  natural, **1**.13.4
  necessitating, **1**.13.2; **3**.11.2
  potential, **1**.12.1, 12.8
  proximate, **1**.10.6, 11.3, 12.1, 15.4; **2**.13.4
  specific, **1**.12.1
  temporally originated, **3**.11.2
cave, companions of the, **2**.10.8
celestial sphere (*falak*), **2**.1.20–22, 3.4, 3.13–17, 7.2, 8.10, 10.8–9, 10.12; **3**.10.10, 11.6, 13.5, 13.8–9, 14.9–10; **4**.2.4, 6.4, 4.3, 7.4, 11.9, 13.2–3, 14.7
  outermost (*f. aʿlá*), **2**.3.14–15, 6.17; **4**.13.2

chance (*ittifāq*), 1.13–14 passim
going to market example, **1**.13.2
rain example, **1**.14.4, 14.14
threshing-floor example, **1**.14.4
change (*taghayyur*), **1**.1.1, 1.5, 1.6, 2.2, 2.12, 2.14, 2.16, 3.5, 5.3, 6.1; **2**.3.2–6, 3.11, 3.14–16, 11.4–5, 13.6–7; **3**.2.11, 6.2–4, 2.8–9, 9.3, 11.9; **4**.5.5, 5.8, 7.4, 8.3, 13.5
choice (*ikhtiyār*), **1**.7.4, 13.14, 14.11, 15.3
chord (*watar*), **4**.5.3, 6.7
circle, **3**.4.4, 7.2, 13.3; **4**.6.8
great, **3**.14.9
circular (*mustadīr*), **1**.8.2; **2**.8.8; **3**.13.2; **4**.3.7
circumference (*ṭawq/muḥīṭ*), **3**.4.5, 7.2, 13.3
coercion (*qahr*), **4**.11.2
coincide, **3**.2.3, 3.9, 4.8, 5.6; **4**.5.2, 6.8
cold, to become (*tabarrud*), **2**.3.18; *see also* coolness
collective kind (*kullīyah*), **3**.12.8; **4**.10.2, 11.8
collision (*taṣādum*), **1**.14.3
color, **2**.1.8, 13.4; **3**.7.3; **4**.13.6
combination (*ijtimā^c*), **1**.6.3, 10.7; **2**.9.11; **3**.2.11, 4.3
coming to be (*ḥādith*), **1**.2.12, 2.15
completeness (*tamām*), **4**.2.1, 2.6
component, [elemental] (*^cunṣur*), **1**.2.3, 2.6; **3**.12.3; **4**.11.6
composite (*murakkab*), **1**.1.11, 3.7, 6.3, 10.5; **2**.5.3; **3**.12.3; **4**.11.7, 11.10
composition (*tarkīb*), **1**.2.19, 3.10, 10.7; **3**.3.8, 4.1, 5.7, 12.3; **4**.11.10
compression (*inḍighāṭ*), **2**.9.21; **4**.10.1
concavity (*taq^cīr*), **3**.2.7; **4**.3.6, 3.8, 6.6
conceptualization (*taṣawwur*), **1**.1.2, 7.2; **2**.8.4, 10.1, 10.5, 12.3
conclusion of syllogism (*natījah*), **1**.10.7
concomitant (*lāḥiq*), **1**.1.1; **3**.2.10, 6.0, 12.6, 13.1, 14.9; **4**.3.7, 11.1; *see also* necessary concomitant
concrete particular (*^cayn*), **1**.1.4; **2**.1.5, 8.4, 10.1
condensation (*takāthuf*), **1**.5.8; **2**.3.8, 6.12, 9.17, 9.20–22; **3**.8.3; **4**.5.6, 6.1, 9.1
condition (*sharṭ*), **3**.11.7
conditional statement (*sharṭī*), **2**.1.15
cone, **2**.8.11
configuration (*hay^ʾah*), **1**.13.4, 14.3; **2**.8.2; **4**.3.6, 5.6
congealing (*i^cqād*), **1**.14.5, 14.16
egg example, **1**.14.5
salt example, **1**.14.5
conjunction, **1**.1.11, 8.2
consequent, [logical] (*tulūw*), **1**.10.7
conservation (*istiḥfāẓ*), **1**.7.2
constituent, simplest (*basīṭ*), **3**.12.2
contain, to (*ḥawá*), **2**.3.14, 6.1
container (*ḥāwī*), **2**.6.2–5, 6.14, 9.9, 9.12
contiguity (*mumāssah/tamāss*), **2**.11.1, 11.6, 12.1, 12.4, 12.9, 13.3–6; **3**.2.1–5, 2.8–10, 3.2, 4.1, 4.3, 5.1, 5.8, 6.2, 9.3; **4**.1.3, 5.8, 6.2, 8.14
continuity (*ittṣāl*), **1**.8.1; **2**.1.3, 1.10, 8.10, 10.2, 11.6, 12.5; **3**.1.1, 2.10, 3.1, 4.1, 4.3, 6.2, 6.6; **4**.2.4
discrete, **4**.8.14
unified, **4**.8.14
continuous, [three] senses of, **3**.2.8–10; **4**.8.14
actual division of, **3**.13.6
equivocal term, **3**.2.8
contrariety (*taḍādd*), **1**.2.16; **2**.3.3; **3**.14.10; **4**.6.6
contrary (*ḍidd*), **1**.2.16, 12.2, 12.4; **2**.1.20, 3.4–5, 4.1–2; **3**.9.7, 14.3
small and big, **2**.3.9; **4**.6.1
convexity (*taqbīb/ḥadabah*), **3**.2.7; **4**.3.6, 3.8, 6.6
cooling (*tabrīd*), **2**.9.20
coolness (*bard/burūdah*), **1**.5.8, 6.2; **4**.8.18, 10.1.4
coolness, natural (*b. ṭabī^cī*), **4**.8.11
corporeality (*jismīyah*), **1**.1.9, 2.2–3; **2**.1.12–14, 1.16–17, 6.6, 9.12, 14.9; **3**.2.1, 9.4, 12.1, 14.9; **4**.11.1, 12.1, 12.3, 15.3
corporeity (*jirmīyah*), **4**.12.1
correlative (*muḍāf*), **3**.11.1; **4**.6.4
corruption (*fasād*), **1**.1.6, 2.17, 3.2, 3.3, 3.6–7, 14.5; **2**.2.1, 11.2, 12.2, 13.8; **3**.6.9, 7.3, 10.13, 12.8; **4**.6.7, 9.4
subject to (*fāsid*), **2**.11.2; **4**.4.4
creation (*khalq*), **2**.5.8; **3**.11.5
creation, atemporal (*ibdā^c*), **1**.3.2–3, 3.7; **3**.11.1, 11.3–4
Creator (*bāri^ʾ/mubdi^c/khāliq*), **3**.11.3, 11.9; **4**.4.3

critical day (*burḥān*), **4**.9.4, 14.1
cube, **3**.13.3
cure (*ʿilāj*), **1**.5.6, 10.2, 12.4
curve (*inḥināʾ*), **1**.8.2; **4**.3.3, 12.8

darkness (*ẓulmah*), **4**.4.6
day (*yawm*), **2**.10.2, 10.8
daytime, **3**.11.4
death (*mawt*), **1**.14.12
deceleration (*futūr*), **4**.2.6
decomposition (*taḥlīl*), **3**.12.2; **4**.4.3–4, 14.6
decrease (*tanaqquṣ*), **2**.3.2
deficiency (*naqṣ*), **1**.14.12; **4**.2.6
definition (*ḥadd*), **1**.1.3, 8.7, 12.5; **2**.4.1, 9.11
deformity, **1**.14.4, 14.12
degree, maximal (*ghāyah*), **2**.14.4, 14.6–7; **3**.14.1; **4**.6.1, 6.4, 6.7, 6.9
deliberation (*ruʾyah*), **1**.14.4, 14.11
Democritus, **1**.4.1n.3, 4.5n.10, 13.4, 14.3, 14.3n.1; **3**.3.2, 3.2n.3, 3.3, 3.12, 5.11
demonstration (*burhān*), **1**.8.2; **3**.2.10
dependence relation (*taʿalluq/tawaqquf*), **1**.8.3; **2**.13.2; **3**.11.7
depth (*ʿamq*), **1**.2.2, 8.1; **2**.1.12; **3**.4.4, 13.4, 14.9
descending (*hābiṭ*), **4**.3.2–3, 8.6
description (definite) (*rasm*), **2**.4.4; **3**.2.10
desire (*ishtiyāq*), **2**.5.5
  inherent (*shawq garīzī*), **3**.10.15
deterioration (*dhubūl*), **1**.14.5, 14.12; **4**.9.4, 14.1
development (*nushūʾ*), **1**.14.5
diagonal (*quṭr*), **3**.4.6
diameter (*quṭr*), **3**.8.6
difference (species/specific) (*faṣl*), **1**.12.5; **2**.1.10, 5.8, 8.4; **4**.1.1, 3.1
  species making (*f. munawwiʿ*), **1**.4.5, 2.2.2
dimension (*imtidād*), **1**.2.2
dimension (three) (*quṭr*), **2**.6.2, 6.6, 8.2; **3**.1.1, 3.2
diminution (*dhubūl*), **2**.2.1, 3.7–8
direction(s) (*jihah*), **2**.8.9, 8.16, 13.3; **3**.4.9, 13–14 passim; **4**.8.11, 10.2, 13.4
  concomitant of quantity, **3**.13.1
  six, **3**.13.4–5
directional side, **3**.14.2
discontinuity (*infiṣāl/taqṭīʿ/inqiṭāʿ*), **2**.1.10, 8.10, 9.3; **3**.2.8, 6.6, 12.5, 12.8
disjunctive [proposition] (*munfaṣil*), **2**.12.3
disposition (*hayʾah/malakah*), **1**.2.16–17, 10.9, 13.2, 15.6; **3**.6.4, 9.4
  fixed, **2**.11.2
  inherent, **1**.8.4
  momentary, **4**.10.4
  unfixed (*h. ghayr qārrah*), **2**.11.2
distance (*masāfah*), **1**.8.1; **2**.11.1; **3**.3.4, 3.14, 6 passim, 10.5
divisibility (*inqisām*), **2**.2.1, 10.2, 11.2; **3**.5.7, 8.11, 9.1, 9.3, 12.6, 12.8
  said in two ways, **3**.9.3, 12.6, 12.8
division (*faṣl*), **2**.11.1–2; **3**.7.2; **4**.8.13
down(ward) (*sufl*), **1**.5.8; **2**.5.8; **3**.8.10, 13.5, 13.7; **4**.3.3, 7.2, 9.8, 12.2
drug, **1**.10.7
dry (*yābis*), **4**.3.6
dryness (*yabs*), **4**.11.5–6
duration (*muddah*), **2**.10.10, 13.7; **3**.10.2

earth (element) (*arḍ*), **1**.8.2–3; **3**.8.8, 10.13
Earth (planet) (*arḍ*), **3**.4.10, 5.5, 7.2, 14.10; **4**.6.4, 9.7
eclipse (*kusūf*), **1**.1.13, 8.2, 13.9; **3**.11.6
effect (*athar/maʿlūl*), **1**.1.11, 7.1; **2**.9.7
electuary, **1**.10.7, 12.4; **3**.12.3
element(s) (*usṭuqiss*), **1**.2.3, 2.6, 3.2, 10.7; **3**.4.2, 8.11, 10.2, 12.3
  four, **2**.3.4
  underlying (*rukn*), **1**.12.4; **3**.10.14
elephant, **3**.12.3
elevations (*irtifāʿ*), **1**.8.2
emanation (*fayḍān/fayḍ/fāʾiḍ*), **1**.7.3, 13.8; **3**.11.4; **4**.4.3, 10.4, 10.6
embryo (*ʿalaqah*), **2**.3.6
embryonic palm, **1**.13.6
Empedocles, **1**.13.5, 14.4, 14.5n.4, 14.6
end (*ghāyah*), **1**.1.14, 2.7–8, 10.2, 10.4, 10.10, 11.1–3, 11.5, 12.6–7, 13.2, 14.2, 14.8, 14.12, 14.15, 15.7; **4**.15.2
  accidental, **1**.12.2, 12.6, 13.13, 14.3, 14.16
  blameworthy, **1**.13.15
  disadvantageous, **1**.14.3
  essential, **1**.12.6, 13.13, 14.2, 14.16
  natural, **1**.13.5, 13.13, 14.4; **4**.9.5, 10.3
    incisor example, **1**.13.5, 14.4
  object of love or fear, **4**.15.1
  seeking natural, **1**.2.8; **4**.10.3

ultimate, **1**.15.4
universal, **1**.1.4; **4**.9.8
engender (*tawallud*), **1**.13.4; **4**.14.4
Epicurus, **3**.3.2, 3.2n.3, 3.14–15
equivocation [fallacy of] (*ishtirāk al-ism*), **1**.3.9, 10.5, 14.2; **2**.2.1, 9.11; **4**.5.7
essence (*māhīyah*), **1**.1.2–3, 2.17, 3.6, 4.2, 5.4, 6.1, 6.3, 8.3, 10.5, 10.9, 11.2, 11.5, 15.4; **2**.5.1, 8.4, 13.3; **4**.3.1–2, 11.4
estimative faculty (*wahm*), **1**.8.7; **2**.1.14, 5.8, 9.11
eternal (*abadan/qidam*), **1**.14.4; **3**.11.4
eternity (*sarmad*), **2**.13.7
ethics, **4**.6.8
Euclid, **1**.8.1n.1; **2**.7.3n.4, **3**.3.10, 3.10n.9, 5.9; **4**.5.3ff
evaporation (*tabakhkhur*), **1**.14.4, 14.14
everlasting (*dahr*), **2**.10.7, 13.7
evil (*sū'/shirr*), **1**.7.4, 13.15
evolution (*nushū'*), **1**.13.5
existence (*wujūd*), **1**.1.4, 2.17, 3.7, 4.2, 4.3, 6.6, 7.1–2, 8.7, 9.5, 10.4, 11.2; **2**.2.1, 2.3, 2.6, 5.1, 8.3, 9.7, 10.5, 11.5; **3**.9.2, 11.4
absolute (*w. muṭlaq*), **2**.13.1
fixed (*qārr w.*), **3**.11.2; **4**.2.1
impossible (*mumtaniʿ w.*), **3**.11.1
necessary (*wājib w.*), 1.4.2; **2**.1.5, 9.15, 10.7, 12.1; **3**.11.4
possible (*jā'iz w.*), 2.12.1; **3**.11.1, 11.2, 11.8
existent (*mawjūd*), **1**.4.4
necessary (*m. wājib*), **1**.4.2
extension (*inbisāṭ/imtidād/istimrār*), **1**.5.8, 8.5; **2**.3.4; **3**.7.3, 13.2; **4**.3.6, 6.10, 11.9
extremity (*nihāyah*), **2**.6.2, 8.10, 9.1, 12.3, 12.5, 13.6; **3**.4.11, 5.6

faculty (*qūwah*), **1**.1.6; **3**.12.5
nutritive (*q. ghādhīyah*), **4**.9.7
falsity (*kidhb*), **3**.3.12
fast (*sarīʿ*), **2**.10.9; **3**.10.5; **4**.3.8
faster, definition of (*asraʿ*), **4**.5.1
fetus (*janīn/muḍghah*), **1**.14.7; **2**.3.6; **4**.9.4
fig-olive progeny, **1**.14.9
figure (mathematical) (*shakl*), **3**.4.4
figure two (logic) (*sh. thānī*), **1**.10.7
finger, additional. *See* appendages, additional
finite (*mutanāhī*), **1**.4.2; **3**.1.1, 7–9 passim; **4**.15.10
fire (*nār*), **1**.8.3, 11.2, 14.16; **3**.6.4, 7.3, 10.2, 10.10
first philosophy, **1**.2.11, 2.15, 6.3, 11.2, 12.8, 13.8, 13.13, 14.1; **2**.3.5, 8.2
First Teacher, **1**.6.6; **2**.10.8
flesh, **3**.9.4, 12.3
flood, **2**.10.11; **3**.11.5–6
flourishing, [life of] (*saʿādah*), **1**.7.4, 12.7
flow(ing) (*sayyāl*), **2**.2.1, 12.5–6
for the most part, [to occur] (*fī akthar*), **1**.13.6, 14.8; **3**.12.3
for the sake of (*li-ajli*), **1**.14.4; **4**.15.1
force (*qasr/qāsir/ʿunf*), **1**.14.3; **2**.8.15; **3**.2.7; **4**.11.2
form (*ṣūrah*), **1**.2 passim, 3.10, 6.2, 9.1, 10.2, 10.9, 11.4–5, 12.5, 15.4; **2**.1.12, 6.3, 6.5, 9.1, 9.10; **3**.6.4, 7.3, 9.2; **4**.4.2, 4.5, 8.11, 10.1, 11.3; **4**.13.5
corporeal, **1**.2.4, 3.2; **3**.9.4, 12.1; **4**.15.3
giver of, **1**.10.3; **4**.15.3
perfecting, **1**.6.7, 10.7; **3**.12.5
preservation of, **3**.12.1
proximate, **1**.10.6, 11.3, 12.5
quantitative, **3**.9.4
simple, **3**.12.1
species, **1**.2.5, 9.4, 10.7; **3**.12.5
specific, **1**.11.3, 12.5
spread through out body, **3**.12.6
strengthening and weakening of, **2**.3.5; **3**.10
substantial, **1**.3.10; **2**.3.2, 3.3, 3.5, 3.6–7, 3.10
foundation (*aṣl*), **3**.5.8, 10.11; **4**.6.7
fragmentation (*infikāk*), **3**.3.16, 4.11–12, 9.3, 12.6
millstone example, **3**.3.15, 4.12; **4**.8.2, 8.10
friction (*muḥākkah*), **4**.13.3ff, 14.6
future, **2**.10.2, 10.8, 10.11, 12.5, 13.10; **3**.7.3, 10.9, 11.6; **4**.4.1

gall, **1**.2.19
gastric juice, **1**.10.7
generality (*ʿumūm*), **1**.2.8; **2**.10.4
generation (*kawn/kā'in/takawwun*), **1**.2.12, 2.15–17, 3.2–3, 3.7, 10.9 13.5, 14.5, 14.7; **2**.2.1–6, 13.1; **3**.6.9, 7.3, 10.13, 12.3, 12.8; **4**.4.4, 9.4

genus (*jins*), **1**.1.3, 5.7, 12.5; **2**.1.4; **3**.3.3, 5.2; **4**.6.9
highest (*j. aᶜlá*), **1**.10.9; **2**.8.2; **4**.3.2
lowest (*j. asfal*), **4**.3.2
proximate (*j. aqrab*), **4**.2.1
remote (*j. abᶜad*), **4**.2.1
geometer (*muhandisah*), **1**.8.1; **3**.10.8
geometry (*handasah*), **1**.8.1–2, 8.8; **3**.5.9
Giver of Forms (*wāhib al-ṣuwar*), **1**.10.3; **4**.15.3
gnat, **3**.12.3
goat-stag progeny, **1**.14.9
God (*Allāh*), **1**.13.3, 13.15, 14.7; **2**.5.8, 13.10n.22, **3**.11.1n.5, 11.6
power of, **3**.5.4, 11.5
good (*khayr*), **1**.12.6, 13.15, 14.9, 15.4
apparent (*kh. maẓnūn*), **1**.10.10, 12.6
real (*kh. ḥaqīqī*), **1**.10.10
grain, **1**.14.7

habit (*malakah*), **1**.14.11; **2**.3.7; **4**.4.3
half, [division by], **3**.3.4, 3.14, 5.3, 9.5
hand, human, **4**.15.1
hardness (*ṣalābah*), **2**.3.7; **3**.3.1
health (*ṣiḥḥah*), **1**.14.10; **4**.5.7
heat (*ḥarārah/tashkīn*), **1**.14.12, 14.14; **2**.1.7, 9.20, 13.4.6; **4**.2.4, 6.2
heaven (*samāʾ*), **1**.8.2, 14.3; **2**.9.2; **3**.7.2, 8.9, 13.5, 13.7; **4**.3.3, 6.4
heaviness (*thiql*), **1**.14.4; **2**.4.4, 5.5, 8.11, 8.14, 9.21; **3**.4.14, 10.2, 13.7; **4**.12.2, 12.5, 15.8
Hesiod, **2**.5.8, 5.8n.7
hindering (*māniᶜ*), **1**.5.6, 12.2
homogeneous (*mushābih*), **1**.4.1, 8.2; **2**.8.22; **3**.8.9, 14.2; **4**.6.10, 9.7, 11.8
horizon (*ufq*), **1**.8.2; **3**.13.6, 14.9
hot, to become (*tasakhkhun*), **2**.3.18; *see also* heat
human-ness (*insānīya*), **1**.8.4–5
humoral balance/mixture/ temperament, (*mizāj muwāfiq*), **1**.2.19, 5.8, 6.5, 7.10, 10.7, 12.4, 15.6; **4**.4.3, 5.7
hypotenuse, **3**.4.4, 4.11, 8.7

illumination (*iḍāʾah*), **3**.6.9; **4**.4.6
image, sensible (*shabaḥ*), **1**.15.6
imagery [faculty] (*khayāl*), **1**.1.8; **2**.1.5, 1.19
imagination (*takhayyul*), **1**.1.8, 5.7; **2**.9.11; **3**.7.3
impediment (*ᶜāʾiq*), **1**.5.8, 6.2, 13.6; **4**.12.1
impossibility (*istiḥālah/mumtaniᶜ*), **2**.13.3; **3**.3.12
inclination (*mayl*), **1**.5.6, 6.2; **2**.5.5, 8.14–15; **3**.4.14, 10.2, 10.15; **4**.3.9, 8.11, 8.16–18, 10.4, 11.9; **4**.12 passim, 13.3, 14.2, 14.5, 15.5, 15.7
definition of, **4**.8.16, 14.5
incline of the sun (*māʾilah*), **1**.14.14
increase, numerical, **3**.9.1
individual, vague (*shakhṣ muntashir*), **1**.1.9
indivisible, **2**.10.2
infinite (*ghayr mutanāhī*), **1**.4.2; **3**.1.1, 1.7–9 passim; **4**.15.10
actual (*gh. m. bi-l-fiᶜl*), **3**.8 passim
definition of, **3**.7.2
in magnitude, **3**.9.5
in number, **3**.9.5
potential (*gh. m. bi-l-qūwah*), **2**.3.2; **3**.5.1, 9.2
traversing, **3**.4.1, 8.6
infinitive, **2**.9.5
influence, **3**.10 passim
active (*muʾaththir*), **1**.6.2; **3**.8.3
passive (*mutaʾaththir*), **1**.6.2
influence, to (*taʾthīr*), **2**.8.14; **3**.10 passim; **4**.12.3
innate (*gharīz*), **4**.12.6
inseparable (*luzūm/lāzim*), **1**.4.2; **4**.13.3
instant (*ān*), **2**.1.3, 3.19, 8.22, 12 passim, 13.1; **3**.3.9, 5.8, 5.10, 10.1, 10.5, 11.2; **4**.4.1
first (*ā. awwal*), **3**.11.3; **4**.8.16
indivisible phase, **3**.5.10
instantaneous. *See* all at once.
instantaneously (*lā fī zamān*), **4**.12.1
instrument (*adāh*), definition of, **4**.15.1
intellect (*ᶜaql*), **1**.1.19; **2**.8.3, 9.11, 9.15, 10.5; **3**.2.3, 4.10; **4**.11.2
intelligible (*maᶜqūl*), **1**.10.9; **2**.9.4, 10.5; **3**.13.7
vain (*maᶜqūl mafrūgh*), **2**.8.3
intensification (*ishtidād*), **2**.2.4; **3**.10.2; **4**.9.2
intention (*qaṣd*), **1**.13.10; **4**.9.4, 12.5–6
natural (*q. ṭabīᶜī*), **1**.14.8

object of (what is intended) (*maqṣūd*), **1**.7.4, 13.6, 13.10; **2**.1.3; **4**.10.3
secondary (*q. thānī*), **3**.9.7
intermediary (*wāsitah*), **1**.5.6; **4**.2.5, 15.1
intermediate (*wasaṭ*), **3**.2.11
interpenetration (*mudākhalah*), **2**.5.3, 6.15, 7.4, 7.6, 7.7, 8.6; **3**.2.3–5, 3.9, 4.1, 4.3
interrelation (*munāsabah*), **4**.3.9, 15.4–9
interrupted (*munqaṭ*ᶜ), **2**.11.6
interval (*buᶜd*), **2**.6.2, 6.4, 7.3–9, 13.3; **3**.2.5, 4.3, 8.5–6.9, 13.2, 14.1
interval (musical) (*zamān*), **1**.8.1

law (*sharīᶜah*), **1**.10.9
leap (*ṭafrah*), **3**.3.14–16, 4.1
left (side), **3**.13.4–5; **4**.6.3
length (*ṭūl*), **1**.2.2, 8.1; **2**.1.12; **3**.4.4, 13.4, 14.9
Leucippus, **1**.4.1n.3; **3**.3.2ff
lice, **1**.12.7
life (*ḥayāh*), **4**.9.6
light (*ḍawʾ*), **3**.6.9; **4**.4.6
light of reason (*badīhah*), **2**.5.8, 6.4
lightness (*khafīf*), **2**.5.5, 8.11, 8.20; **3**.10.2, 13.7; **4**.12.2, 15.8
limb (*ᶜuḍw*), **1**.12.3
limit (*ṭaraf*), **2**.1.3, 1.5, 1.20, 2.4, 6.5, 10.3; **2**.12.2–3, 13.6; **3**.2.2, 2.5, 2.7, 4.9, 6.4–6, 11.1; **4**.6.3
limiting point (*ḥadd*), **2**.1.6, 1.21, 2.1, 2.4, 10.10, 12.5, 13.1; **3**.6.4, 8.1, 8.3, 8.9, 9.1, 9.12, 9.6, 14.1; **4**.5.6
line (*khaṭṭ*), **2**.5.4, 7.4, 8.10, 9.6, 12.5; **3**.3.7, 6.7–8, 8.1, 13.2; **4**.3.6
circular, **3**.5.9; **4**.3.6
indivisible, **3**.3.2
rectilinear/straight, **3**.5.9; **4**.3.6
location (*mawḍi*ᶜ), **2**.8.10; **3**.8.7; **4**.7.4, 10.2, 12.1
of contact (*jihah*), **3**.4.8–9.
universal (*m. kullī*), **3**.8.8
locomotion (*nuqlah/intiqāl*), **2**.2.3, 2.6–7, 5.4, 5.8, 8.23, 9.6–7, 12.5; **3**.5.10; **4**.2.4, 2.6, 3.4–5, 6.1, 9.1, 12.1, 12.3
luck (*bakht*), **1**.13–14 passim
digging a well example, **1**.13.2
treasure example, **1**.13.2
lunar node (*jauzahr*), **1**.1.13

magnet, **1**.5.2; **2**.8.10; **4**.2.3, 9.8
magnitude (*miqdār*), **1**.4.4, 8.1, 8.4, 8.6–8; **2**.1.10, 3.8, 7.4, 8.1, 8.3, 11.2–3, 11.6, 12.5–6; **3**.1.1, 4.1, 4.3, 5.4, 6.6, 7.3, 8.1, 8.10, 9.3, 9.5–6, 12.4, 12.6, 12.10; **3**.12.4; **4**.3.8–9, 5.2, 8.14
concomitants of, **3**.12.6
spatial (*masāfah*), **2**.10.5; **3**.6.4, 1.8.1, 6.4
management (*tadbīr*), **1**.7.2
map onto, to (*muṭābaqah*), **2**.1.7, 13.1; **4**.15.10
material (*hayūlá*), **1**.2.3–6, 3.10, 6.6, 11.5; **2**.3.3, 6.3, 6.5, 7.6, 9.1, 9.11; **3**.9.4
imitating female, **1**.2.21–23
mathematical (*taᶜlīmī*), **1**.8.2; **2**.5.8
mathematics, object of (*amr taᶜlīmī*), **1**.15.2; **3**.5.10
matter (*māddah*), **1**.2.3, 2.6, 2.17, 2.23, 3.2, 6.4, 6.6, 7.1, 8.4–6, 8.8, 9 passim, 9.1, 10 passim, 10.6–7, 10.9, 11.4, 12.4, 14.5, 14.12, 14.17, 15.4, 15.6–7; 12.4, 15.4; **2**.1.23, 7.8, 8.2, 8.4, 9.17, 11.6; **3**.8.13, 9.2–3, 13.5, 15.3; **4**.3.7
determinate (*m. muᶜayyanah*), **1**.8.3, 9.4
infinite (*m. ghayr mutanāhīyah*), **3**.7.3
prime (*hayūlá*), **1**.2.8, 3.7, 9.1, 9.4
proximate (*m. qarībah*), **1**.11.3
mean (*wasaṭ*), **4**.9.2, 12.7
ethical, **4**.6.8
measurement (*taqdīr*), **2**.9.11, 11.4
medial rest (L. *quies media*), **4**.8 passim
pebble example, **4**.8.4, 8.10
wheel, rotating, **4**.8.4, 8.11
medium (*mutawassiṭ/masāfah*), **2**.8.11–12, 16
Melissus, **1**.4.1, 4.1n.1, 4.2
meridian, **3**.14.9
metaphysics, science of, **1**.10.6, 14.1
midheaven, **3**.14.9
millstone, **3**.3.15, 4.12; **4**.8.2, 8.10
mind (*dhihn*), **1**.8.4–5; **2**.1.5, 10.5, 13.9; **4**.2.2
mirror, **1**.15.6
mixture (*mizāj/khilṭ*), **1**.2.19, 4.5, 6.5, 10.6, 12.4; **3**.7.3, 10.14, 12.2, 12.4; **4**.11.10, 12.7
elemental, **1**.8.4
humoral, **1**.2.19, 10.7; **4**.4.3, 5.7
impure (*shawb*), **4**.12.4

mobile (what is moved) (*mutaḥarrik*), **2**.1.7, 1.12, 1.14, 1.24, 8.14, 8.16, 11.1; **3**.6.2; **4**.2.3, 13.3, 14.2, 14.7, 15.4–9
moment(s) (*waqt*), **1**.3.6; **2**.10.1, 10.10; **3**.5.10, 11.7
 time as collection of, **2**.10.6
Moon, **1**.1.6, 1.13; **3**.11.6; **4**.13.3
motion (*ḥarakah*), **1**.10.2; **2**.1–3 passim; **3**.9.4, 11.1
 accidental, **4**.13 passim, 15.3
 circular, **2**.8.8, 13.3; **3**.8.4, 14.6, 14.8–10; **4**.6.1–2, 12.7–8
 circular, non-natural, **4**.9.6
 circular, not the contrary of rectilinear, **4**.6.7–11
 circular, of two types, **3**.14.8, 14.9
 comparing, **4**.5 passim
 continuous in two ways, **2**.13.3
 contrariety and opposition of, **4**.6 passim
 curved, **4**.6.10
 definition of, **2**.1.3–6
 directions of natural, **3**.14.1
 first, **3**.6.3–6
 of the first celestial sphere, **2**.10.9; **4**.9.8
 forced, **2**.8.14, 13.3; **3**.8.2, 10.2, 10.15; **4**.2.6, 3.3, 6.2, 14 passim
 form of, **2**.1.6
 genera (categories) of, **2**.3 passim; **4**.6.1
 local, **2**.1.5–6; **3**.6.2, 10.5; **4**.5.4
 multiplicity of, **4**.2.3
 natural, **3**.8.2, 8.8–9, 10.2, 12.11, 13.9; **4**.2.6, 3.3, 6.2, 9.5
 naturally prior, **4**.9.1–2
 numerical unity of, **4**.2 passim, 4.1
 production of (*taḥrīk*), **2**.1.12
 projectile, **2**.8.16; **4**.12.2, 14.2–6, 15.8
 qualitative, **4**.5.5, 9.1
 rectilinear, **2**.8.9, 13.3; **3**.8.10, 13.9, 14 passim; **4**.6.1
 smallest, **3**.6.4
 specific and generic unity of, **4**.3 passim
 specific difference of, **4**.9.3–8
 spontaneous, **4**.9.4, 14.7
 substantiality of, **4**.6.5
 three kinds of, **4**.12.8
 two senses of, **2**.1.5–6
 voluntary, **3**.13.8
mountain, **3**.3.6, 5.4; **4**.13.3
mover (*muḥarrik*), 1.5.4, 5.6; **2**.1.12, 1.14, 1.23, 8.15–16, 8.22; **3**.9.6–7, 10.4, 10.15, 11.2; **4**.2.4, 8.11, 9.3, 14.1–3, 14.5, 14.6, 15.1–9
 first (*m. awwal*), **4**.15.2–3
 incorporeal (*m. ghayr jismānī*), **3**.10.15
 unmoved (*m. lā yataḥarraku*), **4**.15.2
multiplicity (*kathrah*), **2**.12.7; **3**.4.1, 10.5
muscle, **1**.14.11
music (*mūsīqā*), **1**.8.1, 8.8
mystic (*ṣūfī*), **1**.2.23

natural (*ṭabīʿī*), **1**.7.1; **4**.9.7–8
nature (*ṭabīʿah*), **1**.1–7 passim; **3**.7.2, 8.1; **4**.9.5
 definition of, **1**.5 passim, 6.1, 6.5
 as efficient principle, **1**.3.12
 as form, **1**.6.3
 generic, **1**.1.4–5; **2**.8.3
 as matter, **1**.3.10, 6.6–7
 predication of, **1**.7.2
 specific, **1**.1.4–5, 1.9; **2**.8.3; **3**.3.3
 substantial corruption of, **2**.3.2–5
 universal, **1**.1.6, 7.3–4, 13.8
Naẓẓām, Ibrahīm, **3**.2.4n.7, 3.13n.11, 3.13n.12, 3.14n.12
necessary existence. *See* existence
necessary concomitant (*lāzim*), **2**.9.7
negation (*salb*), **2**.12.3
 absolute (*s. muṭlaq*), **2**.13.7; **3**.6.6, 7.2, 11.6
 way of, **2**.7.6
Nicomachean Ethics, **1**.14.11n.8, **4**.6.9n.9
nonexistence (*ʿadam*), **1**.3.10, 7.4; **2**.8.10, 10.2–3, 11.5, 12.1, 13.1; **3**.10.9, 11.1–2, 11.6–7; *see also* privation
absolute (*ʿa. muṭlaq*), **2**.13.1
north pole, **3**.14.9
now (*ān*), **1**.12 passim; **2**.13.9; **3**.10.10
number (*ʿadad*), **1**.8.3; **2**.7.4, 13.6; **3**.1.1, 5.4, 7.3, 8.1, 8.12
numbering, ordered (*ʿa. murattab*), **3**.11.9
numerable number (*ʿa. maʿdūd*), **1**.8.3
nutrition (*ghidhāʾ*), **2**.9.19; **4**.9.7

obstacle (*māniʿ*), **1**.13.6; **3**.4.16; **4**.9.6
one (*wāḥid*), **1**.8.3; **2**.7.3, 12.7; **4**.2.1
 two senses of, **4**.4.2
opposition (*muqābalah*), **1**.8.2; **3**.14.10
 types of, **4**.6.4

optics (*ᶜilm al-munāẓir*), **1**.8.1–2, 8.8
order (*niẓām*), **1**.14.5, 14.9; **3**.14.5
regulative (*n. mustaḥfiẓ*), **1**.10.9
ordered position (*tartīb*), **2**.6.10, 9.16; **3**.7.1, 8.1, 8.11, 10.6, 14.5; **4**.10.3, 11.1, 13.3
organ (*ᶜuḍw*), **4**.13.5
otherness (*ghayrīyah*), **2**.1.3
oversized head, **1**.7.1

palm, embryonic, **1**.13.6
paradigm (*mithāl*), **4**.15.2
paradox. *See* Zeno's paradoxes
Parmenides, **1**.4.1n.1, 4.2; **4**.2.1
part (*juzʾ*), **2**.3.15; **3**.2 passim, 5.2, 6.2, 6.6, 8.11, 10.1, 11.8
discontinuous (*j. munqatᶜ*), **3**.2.8, 6.6
homoeomerous, **3**.12.3–4; **4**.12.2
simple (*j. basīṭ*), **1**.2.6, 1.11
passage (*khurūj*), **1**.10.2; **2**.1.2, 13.3
passion (*infiᶜāl*), **2**.1.2, 2.1, 2.5, 2.7, 3.18
passivity (*infiᶜāl*), **1**.5.1; **4**.3.7
past, **2**.10.2, 10.8, 10.11, 12.4, 13.10; **3**.7.3, 11.5; **4**.4.1
patient (*munfaᶜil/marīḍ*), **1**.5.6; **3**.10.1
pause (*waqfah*), **3**.4.14; **4**.8.13
perfection (*tamām/kamāl/istikmāl*), **1**.1.4, 2.12–14, 2.22, 3.8, 14.9; **2**.1.3, 2.3, 2.6, 4.4, 5.6; **3**.6.9; **4**.2.2, 7.2–3, 9.2
common, **2**.5.5
first, **1**.2.22; **2**.1.3, 1.6, 1.20, 4.4, 9.10; **3**.9.6; **4**.2.3
potential, **1**.2.22, 2.1.5, 1.20
proper, **2**.5.5
second, **1**.2.22; **2**.1.5, 1.20, 4.4
Peripatetic, **1**.13.7; **3**.6.7, 12.2, 12.4–5; **4**.9.6
Philoponus, John, **1**.4.4n.8, 5.5n.6, 5.7n.10, 13.7n.4; **2**.2.1n.1, 2.13n.12, 5.8n.6, 6.4n.4, 6.11, 6.11n.9, 7.3n.3, 7.6n.7, 8.20n.30, 12.5n.9, 12.6n.11, 12.8n.14; **4**.4.3n.4, 12.6n.10, 13.3n.7, 14.2n.3, 15.3n.4
phoenix, **4**.3.1
physician (*ṭabīb*), **1**.5.6, 10.2, 12.2, 12.4
place (*makān*), **1**.5.8, 6.2; **2**.3.10, 3.14, 5–9 passim, 11.1; **3**.2.2, 6.2, 8.2, 12.5; **4**.10.2, 11 passim
as form, **2**.7.1, 9.9
as interval, **2**.7.3–9, 9.1, 9.10
as material, **2**.7.1
as surface, **2**.6.4, 7.2
jar example, **2**.6.17, 9.3
natural, **3**.10.13; **4**.8.11, 10.2, 11.8, 12.1, 13.3
planet (*kawkab*), **1**.8.2
planning, bad, **1**.13.15
plant (*nabāt*), **1**.5.1; **3**.12.3
Platonist, **4**.2.1
plenum (*malaʾ*), **2**.8.5–6.23; **3**.8.13, 10.8, 14.2
Plotinus, **1**.7.3n.5, 7.3n.6; **2**.2.1n.1, 2.3.7n.8
point (*nuqṭah*), **2**.5.4, 8.10, 9.6; **3**.2.2, 3.7, 4.9, 5.6, 6.7–8, 7.2, 8.1
pointing (*ishārah*), **3**.5.3; **4**.13.4
pole [celestial], **3**.14.9
posit(ing) (*farḍ*), **3**.2.8, 9.3; **4**.14.6
position (*waḍᶜ*), **1**.8.2; **2**.1.2, 3.13–16, 5.8, 9.15, 11.1; **3**.2.25, 4.1, 7.1; **4**.6.1, 6.11, 11.1, 12.1, 13.4
two sorts, **4**.13.2
possession, category of (*jidah*), **2**.1.2, 3.17
possibility (*imkān*), **1**.13.8; **3**.11.1–2, 11.5, 11.5
absolute (*i. muṭlaq*), **2**.12.1; **3**.12.3
posteriority (*taʾakhkhur*), **1**.3.9, 10.6; **2**.2.6; **4**.6.4, 8.16, 11.6
potency (*qūwah*), **1**.10.2; **2**.8.6, 9.11; **4**.7.4
potential (*qūwah*), **1**.2.17, 3.8, 6.6, 9.5, 12.3, 12.5; **2**.1.2–3; **3**.11.2; **4**.5.3
proximate, **1**.12.3; **2**.12.1; **4**.6.5
remote, **1**.12.3; **4**.15.3
potentiality (*qūwah*), **1**.3.8; 2.1.2–3, 1.20, 1.22, 3.2; **2**.4.4
absolute, **2**.3.2; **3**.12.4
two kinds, **2**.1.3, 1.22
poultice, **1**.10.7
power (*qūwah/qudrah*), **1**.5.3, 5.7, 7.3; **2**.2.7, 3.7; **3**.10 passim, 11.1; **4**.9.8, 11.10, 12.1, 14.3, 17.7
acquired, **4**.12.2, 14.2, 14.5–6, 15.4–9
active, **1**.5.7, 12.3, 15.6
attractive, **1**.14.7; **2**.8.20–21
divine, **1**.5.7, 13.8; 3.11.5, 11.9
finitude and infinitude of, **3**.10.3–17
inclinatory, **1**.5.7; **2**.8.11
motive, **2**.8.14, 8.20; **3**.11.2; **4**.8.11
volitional, **4**.9.6
predicate (*ḥaml*), **1**.10.8; **2**.10.5
predication, accidental, **4**.9.3
premise (*muqaddimah*), **1**.4.3, 10.7; **2**.10.5; **3**.11.7
conjunctive (*m. muqtarinah*), **1**.4.3, 10.7

preparedness (*isti^cdād*), **1**.2.17, 13.8, 14.13, 15.6; **3**.5.2, 9.3–4, 9.6, 12.5; **4**.7.4, 9.7, 15.3
preservation (*ḥifẓ*), **3**.12.1
principle (*mabda'*), **1**.1.1–4, 2–3 passim, 4.5, 5.3–4, 10.3; **2**.9.12; **3**.8.11, 10.11; **4**.12.1
  auxiliary, **1**.10.4
  completing, **1**.10.3
  efficient, **1**.2.10–11, 3.11–12, 10.4, 12.2, 15.6; **2**.5.5–6; **3**.11.1
  elemental, **1**.13.5
  emanative, **4**.4.3, 4.5
  final, **1**.2.11, 3.11; **2**.5.5–6
  first, **1**.6.3, 7.3; **2**.9.5
  formal, **1**.10.9; **2**.5.5–6
  guiding, **1**.10.4
  material, **1**.2.4–6, 2.16–18, 10.5–6, 12.4, 15.6; **2**.5.5–6
  non-proximate, **1**.11.3
  preparing, **1**.10.3
  proximate, **1**.11.4; **3**.10.15
priority (*taqaddum*), **1**.1.4, 3.9, 10.6; **2**.2.6; **4**.6.4, 8.16, 11.6
  causal, **2**.9.7
privation (*ᶜadam*), **1**.2.12–18, 2.1–20, 3.4, 3.8, 3.10, 12.5, 14.12; **2**.4.1, 9.6; **3**.9.8, 10.13; **4**.2.5, 6.4, 7.4; *see also* nonexistence
process (*zawāl*), **2**.1.4
procession (*sulūk*), **1**.7.3; **2**.1.4, 12.3
Proclus, **3**.3.2, 3.2n.3
progeny, **1**.14.9
projecting towards (*musāmatah*), **3**.13.6, 14.10
projection (*samt*), **2**.8.8; **3**.4.7, 4.10
property (*khāṣṣah*), **1**.1.16, 12.5; **4**.11.8
property unique to the species (*khāṣṣīyah*), **2**.1.4
proportion (*nisbah*), **2**.8.12–13; **3**.4.1
proposition, affirmative (*mūjabah*), **2**.10.12
pulling (*jadhb*), **1**.12.3; **3**.10.12; **4**.14.2, 15.4
pushing (*dafᶜ*), **1**.12.3; **3**.10.12; **4**.14.2, 15.4, 15.8
Pythagorean theorem, **3**.4.4ff
Pythagoreans, **3**.8.11n.19

quality (*kayf/kayfīyah*), **1**.5.8, 14.8; **2**.1.2, 2.1, 2.3, 2.7, 3.7, 3.10, 3.20; **3**.12.5; **4**.2.6, 5.6, 6.1, 8.13, 10.1
quantity (*kamm/kammīyah*), **1**.4.4, 5.8, 8.1; **2**.1.2, 3.8, 8.1–2, 6.6, 8.2; **3**.1.1, 6.4, 7.1, 9.3–4, 10.3, 11.6, 13.1; **4**.10.1
  continuous, **1**.4.4
  determinate (*qadr*), **3**.4.1, 5.4
  equal in two ways, **4**.5.2
quies media. See medial rest

radius, **3**.8.4
rarefaction (*takhalkhul*), **1**.5.8; **2**.3.8, 6.12, 8.20, 9.17.20–22; **3**.8.3.13; **4**.5.6, 6.1, 9.1, 10.1, 14.6
ratio (*nisbah/munāsabah*), **1**.8.1; **3**.4.13, 10.1, 10.4–5, 10.7–8, 10.17; **4**.12.2, 15.7
Rāzī, Abū Bakr, **2**.10.6, 10.7n.8, 13.7n.14
reason why (*lima-īyah*), **1**.8.2
rectilinear (*istiqāmah/mustaqīm*), **1**.8.2; **3**.13.2; **4**.3.7
regularity (*iṭṭirād*), **1**.13.6
relation, [category of] (*iḍāfah*), **2**.1.2, 3.11; **3**.11.1
renew(al) (*tajaddud*), **2**.8.16, 11.6, 12.4; **3**.11.2; **4**.9.5
replacement (*taᶜāqub*), **2**.5.8, 6.3, 9.9; **4**.4.3, 11.3
repulsion (*indifāᶜ*), **4**.8.16, 10.4
resolve (*ᶜazīmah*), **1**.13.6
rest (*sukūn*), **1**.6.1; **2**.1.1.19, 4 passim, 8.10.16; **3**.6.6, 8.8, 9.9, 10.13; **4**.6.9, 11.6
  coming to (*tawaqquf*), **3**.6.6
  intervening (*s. mutakhallil*), **2**.11.1
  natural, **3**.8.8, 9.4
  opposition to motion, **4**.7 passim
resurrection, **2**.10.11
right(side), **3**.13.4–5; **4**.6.3
rippling effect (*tamawwuz*), **4**.13.3
rotation (*dawrah*), **3**.8.4, 10.9, 11.6; **4**.2.6, 6.10, 14.2
round (*tadwīr*), **1**.8.4

sandal and ant. *See* Zeno's paradoxes
scammony, **1**.12.2
science (*ᶜilm*), **1**.8 passim
seed (*badhr*), **1**.5.1, 2.3.3
seldom, [to occur] (*fī aqall al-amr*), **1**.13.6
selectively determine, to (*tarjīḥ*), **1**.13.6; **3**.11.2

self-evident (*bayyin bi-nafsihī*), **1**.1.4, 4.4, 8.11; **3**.4.16
self-moved (*mutaharrik bi-dhātihī*), **2**.1.15–17; **3**.2.4, 4.3
semen (*nuṭfah*), **1**.2.19, 5.1, 11.5, 12.4, 14.7; **2**.3.3
sensation (*ḥiss*), **1**.1.8
sensible image (*shabaḥ*), **1**.15.6
shadow, **3**.4.10; **4**.4.2
shape (*shakl*), **1**.8.1, 10.9, 13.4, 14.8; **2**.3.7, 7.3, 8.10; **4**.11.1–4
  circular, **3**.8.5; **4**.11.5
side (*jānib/ḍil*ᶜ*/jihah*), **3**.2.1, 4.6, 13.2
silk, **1**.12.7
simple (*basīṭ*), **1**.1.11, 1.15–16, 6.2, 6.5; **2**.6.5; **3**.8.3, 10.14; **4**.9.7, 11.7
simultaneity (*ma*ᶜ*īyah*), **1**.10.6; **2**.10.4, 10.9–10, 11.5, 13.7
simultaneous (*ma*ᶜ*a*), **2**.11.5
sinew (*watar*), **1**.12.2
size (ᶜ*iẓam*), **3**.4.1; **4**.15.9
slow (*buṭ*ʾ), **2**.3.18, 8.14, 10.9, 13.1
softness, **2**.3.7
solidity (*ṣalābah*), **1**.13.4
soul (*nafs*), **1**.5.6, 7.4, 11.2, 14.9.11; **2**.3.7, 8.15, 10.5, 12.7, 13.1; **3**.12.5; **4**.9.6, 12.5–6, 13.5
  animal, **1**.5.3, 14.9; **4**.12.5
  celestial, **1**.5.3
  vegetative, **1**.5.3
sound, **3**.7.2
south pole, **3**.14.9
space (*ḥayyiz*), **2**.5.8, 8.7, 8.10; **4**.8.7, 10 passim, 11.1.4, 12.1, 13.5, 15.10
  empty, **2**.9.11; **3**.8.2–3
species (*naw*ᶜ), **1**.1.3; **3**.7.2; **4**.6.9
speculative theologian (*ahl al-naẓar*), **3**.13.3
speed (*sur*ᶜ*ah* [*wa buṭ*ʾ]), **2**.8.11–12, 10.5, 11.1, 12.7, 12.9, 13.1; **3**.4.15, 10.1, 10.3, 10.5; **4**.3.9, 4.9, 5.1, 5.4–5, 15.8
sperm (*minah*), **4**.9.4
sphere (*kurah/kurayyah*), **1**.8.2; **4**.3.8, 8.4, 13.2
  rolling, **3**.3.11, 5.10
spherics (ᶜ*ilm al-ukar al-mutaḥrrikah*), **1**.8.1.8
spontaneously (*min tilqā*ʾ *nafsihi*), **1**.13.1, 13.14; **4**.14.7
square, **1**.12.5; **3**.4.9, 13.2
square root, **3**.4.4
square-ness, **1**.12.5
starting point (*mabda*ʾ), **2**.1.5; **3**.10.5, 11.2
statement, categorical (*ḥamlī*), **2**.1.14
statement, conditional (*sharṭī*), **2**.1.15
striated, **4**.8.13
subject (*mawḍū*ᶜ), **1**.2.3, 2.6, 2.19–20, 4.4 8.4, 10.7, 12.4; **2**.1.10, 1.12, 2.3, 3.3–4, 3.7, 8.2, 9.12, 11.2; **3**.11.1, 11.9; **4**.2.2
subject (logical) (*waḍ*ᶜ), **1**.10.7; **2**.10.5
  particular, **1**.12.4
  proximate, **1**.12.4; **2**.3.7; **4**.6.6–7
  remote, **1**.12.4
  universal, **1**.12.4
subsistence (*qiwām*), **1**.3.8, 6.2, 7.1, 8.7, 10.6; **2**.3.4, 8.2, 11.2; **3**.5.6; **4**.6.5
substance (*jawhar*), **1**.2.2, 2.4, 2.17, 4.4, 5.8, 6.2, 6.5–7, 14.8; **2**.1.2, 2.3, 2.6, 3.1–6, 4.2, 5.2, 7.8; **3**.5.6, 10.2, 11.1–2; **4**.9.2, 11.2–3
  intelligible, **2**.5.2
  sensible, **2**.5.2
  to become (*tajawhar*), **4**.10.1
substantiality (*jawharīyah*), **1**.6.6; **2**.2.3–3, 2.5, 8.2–3; **4**.6.5, 9.2
substrate (*maḥall*), **2**.3.4–5; **3**.3.7, 5.6, 5.8, 10.14, 11.1
succession, (follow in) (*tatātī*), **3**.1.1, 2.6, 3.7, 3.9, 4.3, 6.7; **4**.8.1
suddenly (*baghatah*), **2**.13. 8
Sun, **1**.1.6, 7.3, 14.14; **3**.4.4, 4.10, 4.13, 11.4
superimpose, to (*taṭbīq/iṭbāq*), **3**.4.6, 4.8, 8.1
supplements (*al-lawāḥiq*), **4**.12.5n.9
supposition (*farḍ/taqdīr/iqtidār*), **2**.8.10; **3**.6.4, 10.8
surface (*saṭḥ*), **1**.8.4; **2**.5.4, 8.10, 9.1, 9.6; **3**.2.7, 3.7, 5.10, 6.2, 8.1, 8.9, 13.2; **4**.3.6
  concave, **2**.9.2
  contacting, **2**.9.1
  convex, **2**.9.2
  internal, **2**.9.2
  non-containing, **2**.9.1
  simple, **3**.14.2
  two-dimensional, **4**.8.4
surrounded, what is (*muḥāṭ*), **2**.6.2, 7.1–2, 9 passim; **3**.8.9, 13.3
surrounds, what (*muḥīṭ*), **2**.6.2, 6.17, 7.1–2, 9 passim; **3**.6.2, 14.1–3; **4**.10.2, 13.2

survive (*baqā'*), **1**.13.5
syllogism (*qiyās*), **1**.4.3, 10.7; **2**.9.9; **3**.11.7
system (*niẓām*), **1**.8.2; **3**.10.11

temperament (*khulq*), **1**.5.5–7, 15.6
humoral (*mizāj*), **1**.2.19, 10.7
tendency (*i'timād*), **3**.4.14; **4**.14.4
tendon, **1**.14.11
term (*ḥadd*)
major (*h. akbar*), **1**.10.7
middle (*ḥ. awsaṭ*), **1**.10.7; **3**.2.10, 8.6; **4**.6.4
minor (*ḥ. aṣghar*), **1**.10.7
synonymous, **2**.1.4
terminus a quo (from which) (*mā minhu*), **2**.1.12, 1.20; **3**.14.10; **4**.2.5, 3.2, 5.5, 8.1
terminus ad quem (towards which) (*mā ilayhī*), **2**.1.12, 1.20; **3**.14.10; **4**.2.5, 3.2, 5.5, 8.1
Themistius, **2**.1.4, 1.4n.12, 6.17n.14; **4**.13.5n.12
time (*zamān*), **2**.1.2, 1.12, 3.12, 3.19, 4.5, 10–13 passim; **3**.1.1, 7.3, 8.6, 9.6, 9.7, 10.1, 10.10, 11.1, 11.9
accident of motion, **4**.6.2
as celestial sphere, **2**.10.8, 10.11
as collection of moments, **2**.10.6
as eternal substance, **2**.10.7
as motion, **2**.10.8–9, 10.11
continuity of, **2**.13.4
definition, **2**.10.8–9, 10.11
numbered by motion, **2**.10.8–9, 10.11, 12.7
unity of, **4**.2.2
tone (*nagham*), **1**.8.1; **4**.2.4
tool (*ālah*), **3**.12.5; **4**.15.1
transition (*nuqlah/intiqāl/iḥālah*), **1**.10.2, 14.16; **2**.13.3, 13.6; **3**.11.2
transparency (*shaffāf*), **3**.6.6, 6.9
traversal (*qaṭ'*), **2**.1.7, 12.5, 13.1; **3**.6.1, 9.6; **4**.2.2, 4.1–2, 5.8, 9.5
triangle, **3**.11.4; **4**.5.2–3, 8.14
right, **3**.4.5–6
True One (*al-aḥad al-ḥaqq*), **3**.11.9

undulation, **2**.6.15
unit (*wāḥid*), **1**.8.3; **2**.6.5, 7.3, 12.7, 13.6; **3**.4.1, 4.3, 4.5, 5.2, 8.1, 9.5
unity (*waḥdah*), **1**.3.9, 8.3; **2**.2.3; **3**.5.2, 6.5, 9.5; **4**.2.2, 2.5–6, 4.1
universal (*kullīy/kullīyah*), **1**.1.6, 3.6, 7.2, 12.5, 2.1.7
universe (*kull/'ālam*), **1**.7.2–3, 13.4, 13.10, 14.12; **2**.6.1, 6.15; **3**.13.5; **4**.2.6, 10.3, 14.1, 15.3
univocity (*tawāṭu'*), **1**.3.9, 2.2, 2.1; **4**.3.8
up(ward) (*'ulū/fawq*), **1**.5.8; **2**.5.8; **3**.8.10, 13.5, 13.7; **4**.3.3, 7.2, 12.2

vague individual (*shakhṣ muntashir*), **1**.1.9
vapor (*bukhār*), **1**.14.4; **3**.7.3; **4**.10.4
vice (*radhīlah*), **4**.6.8
virtue (*faḍīlah*), **4**.6.8
vision (*basar/ibṣār*), **1**.8.1; **3**.13.4
void (*khalā'*), **1**.4.1, 4.5, 13.4; **2**.6.11, 8 passim, 9.16; **3**.2.7, 7.3, 8.3, 8.13, 10.8, 14.1; **4**.8.10, 11.9
ashes example, **2**.6.13
cask example, **2**.6.1, 6.13
clepsydrae example, **2**.8.20–21, 8.25
interstitial, **1**.13.4; **2**.8.22
phial example, **2**.6.11, 9.20
proved by the process of elimination and removal (*raf'*), **2**.6.5, 9.11
siphon example, **2**.8.20, 8.25
wineskin example, **2**.6.11, 9.18
volition (*irādah*), **1**.5.1, 13.6, 15.5; **2**.8.10; **3**.11.2; **4**.9.4, 12.6, 14.7
volume (*ḥajm*), **1**.5.8; **2**.6.6, 9.16; **3**.12.8

water (*mā'*), **1**.3.3, 8.3–4, 12.4; **3**.6.4, 7.3; **4**.10.1.4
wax, **1**.14.5; **4**.6.4
weight (*thiql*), **1**.6.2, 11.2; **4**.12.2
weights, science of, **1**.8.1
well-being (*khayr*), **1**.14.10
wetness (*ruṭūbah*), **2**.1.9; **4**.3.6
when (*matá*), **2**.1.2, 3.12
where (*aina*), **1**.14.8; **2**.1.2, 2.1, 3.10, 3.12, 12.6; **4**.11.2, 11.4, 12.1, 12.6, 13.4
whim, **1**.13.13
whole (*kull/jumlah*), **1**.1.15, 10.5; **2**.1.14, 1.17, 3.15–16, 7.4, 8.23, 9.12; **3**.2.3, 5.11, 8.9, 8.9, 8.11, 9.1, 10.1, 11.5; **4**.11.6, 11.8, 15.6, 15.9
unified, **3**.2.8
whole/each one, distinction of, **2**.3.15; **3**.11.6, 11.8

why question (*masʾalah ᶜan lima*), **1**.14.14
wine, grape, **4**.5.6
  honey, **4**.5.6
womb (*rahim*), **1**.14.7
world (*ᶜālam*), **1**.3.6, 13.4; **2**.13.7
world, infinite number of (*ᶜālam*), **1**.13.4

year, **2**.3.12, 10.2

Zeno, **3**.3.4ff; **4**.15.10n.16
Zeno's paradoxes
  the ant and sandal, **3**.3.4, 3.14
  the tortoise and Achilles, **3**.3.4, 3.14; **4**.5.1

About the Editor/Translator

JON McGINNIS is associate professor of classical and medieval philosophy at the University of Missouri–St. Louis. His general research interest is in the history and philosophy of physics as that science is represented within Aristotle and the Aristotelian commentary tradition extending from its early Greek roots up to the Latin scholastics. His particular focus, however, is on the influence of that tradition within the medieval Arabic-speaking world and especially on the philosophy of Avicenna. McGinnis has authored numerous articles on medieval natural philosophy as well as others topics germane to ancient and medieval thought. He is also the author of a general introduction to the philosophy of Avicenna in Oxford University Press' Great Medieval Thinkers Series (2010) and was co-translator with David C. Reisman of *Classical Arabic Philosophy, An Anthology of Sources* (Hackett Publishing Co., 2007).

A Note on the Types

The English text of this book was set in BASKERVILLE, a typeface originally designed by John Baskerville (1706–1775), a British stonecutter, letter designer, typefounder, and printer. The Baskerville type is considered to be one of the first "transitional" faces—a deliberate move away from the "old style" of the Continental humanist printer. Its rounded letterforms presented a greater differentiation of thick and thin strokes, the serifs on the lowercase letters were more nearly horizontal, and the stress was nearer the vertical—all of which would later influence the "modern" style undertaken by Bodoni and Didot in the 1790s. Because of its high readability, particularly in long texts, the type was subsequently copied by all major typefoundries. (The original punches and matrices still survive today at Cambridge University Press.) This adaptation, designed by the Compugraphic Corporation in the 1960s, is a notable departure from other versions of the Baskerville typeface by its overall typographic evenness and lightness in color. To enhance its range, supplemental diacritics and ligatures were created in 1997 for exclusive use in this series.

The Arabic text was set in DecoType Professional NASKH, designed by Thomas Milo (b. 1950), a pioneer of Arabic script research, typeface design, and smart font technology in the digital era. The Naskh calligraphic style arose in Baghdad during the tenth century and became very widespread and refined during the Ottoman period. It has been favored ever since for its clarity, elegance, and versatility. Milo designed and expanded this typeface during 1992–1995 at the request of Microsoft's Middle East Product Development Department. His designs pushed the existing typographic possibilities to their limits and led to the creation of a new generation of Arabic typefaces that allowed for a more authentic treatment of the script than had been possible since the advent of moveable type for Arabic.

BOOK DESIGN BY JONATHAN SALTZMAN

◆